T0329569

GIGABIT-CAPABLE PASSIVE OPTICAL NETWORKS

GIGABIT-CAPABLE PASSIVE OPTICAL NETWORKS

Dave Hood
Elmar Trojer

A JOHN WILEY & SONS, INC., PUBLICATION

Published by John Wiley & Sons, Inc., Hoboken, New Jersey
Published simultaneously in Canada

For general information on our other products and services or for technical support, please contact our Customer Care Department within the United States at 877-762-2974, outside the United States at 317-572-3993 or fax 317-572-4002.

Wiley also publishes its books in a variety of electronic formats. Some content that appears in print may not be available in electronic formats. For more information about Wiley products, visit our web site at www.wiley.com.

Library of Congress Cataloging-in-Publication Data:

Hood, Dave, 1945-
  Gigabit-capable passive optical networks / Dave Hood, Elmar Trojer.
      p. cm.
  Includes bibliographical references and index.
  ISBN 978-0-470-93687-0 (cloth)
  1. Passive optical networks. 2. Gigabit communications. I. Trojer, Elmar. II. Title.
  TK5103.592.P38H66 2011
  621.38′275–dc23
                                                                    2011028223

10 9 8 7 6 5 4 3 2 1

To Kent McCammon

CONTENTS

ACKNOWLEDGMENTS

We wish to thank the many who encouraged our effort, in particular, Dave Allan, Dave Ayer, Chen Ling, Dave Cleary, Jack Cotton, Lou De Fonzo, Jacky Hood, Einar In de Betou, Denis Khotimsky, Lynn Lu, Kent McCammon, Don McCullough, Derek Nesset, Peter Öhlen, Dave Piehler, Albert Rafel, Björn Skubic, and Mara Williams for their help in reviewing various parts of the manuscript and the premanuscript studies. Tom Anschutz, Paul Feldman, Richard Goodson, David Sinicrope, and Zheng Ruobin helped to clarify issues that came up in the course of the work.

We would especially like to recognize Dewi Williams, an ideal match for our target reader profile, who was willing to provide intelligent and thoughtful comment and discussion well beyond the call of duty.

Needless to say, the inevitable remaining errors and inconsistencies are entirely our own responsibility.

Finally, special thanks to Jacky and Antonia for their support and patience through the process.

PUBLISHER'S BRIEF REVIEW

Although thoroughly grounded in the G-PON standards, this book is far more than just a rehash of the standards. Two experts in G-PON technology explain G-PON in a way that is approachable without being superficial. As well as thorough coverage of all aspects of G-PON and its 10 Gb/s evolution into XG-PON, this book describes the alternatives and the reasons for the choices that were made, the history and the tradeoffs.

INTRODUCTION

Fiber optic access networks have been a dream for at least 30 years. As speeds increase, as the disparate networks of the past converge on Ethernet and IP (Internet protocol), as the technology and business case improve, that dream is becoming reality.

The access network is that part of the telecommunications network that connects directly to subscriber endpoints. This book details one of the technologies for fiber in the loop (FITL), namely gigabit-capable passive optical network (G-PON) technology, along with its 10-Gb sibling XG-PON. Figure 1.1 shows how G-PON and XG-PON fit into the telecommunications network hierarchy. This book is about the G-PON family.

For quick reference, Table 1.1 summarizes the common PON technologies, including both the G-PON and EPON families. The G-PON family is standardized by the International Telecommunications Union—Telecommunication Standardization Sector (ITU-T), while EPON comes from the Institute of Electrical and Electronic Engineers (IEEE). Chapter 7 includes a comparison of G-PON and EPON.

Because they share many properties, this book uses the term G-PON generically to refer to either ITU-T G.984 or G.987 systems unless otherwise stated. Where a distinction needs to be made, we make it explicit: G.984 G-PON or G.987 XG-PON.

Figure 1.2 illustrates the fundamental components of a PON. The head end is called the optical line terminal (OLT). It usually resides in a central office and usually

Gigabit-capable Passive Optical Networks, First Edition. Dave Hood and Elmar Trojer.
© 2012 John Wiley & Sons, Inc. Published 2012 by John Wiley & Sons, Inc.

TABLE 1.1 PON Family Values

Technology	Standard		Speed
G-PON	ITU-T	G.984	2.5 Gb/s downstream, 1.25 Gb/s upstream
XG-PON	ITU-T	G.987	10 Gb/s downstream, 2.5 Gb/s upstream
EPON	IEEE	802.3	1 Gb/s symmetric
10G-EPON	IEEE	802.3	10 Gb/s downstream, 1 or 10 Gb/s upstream

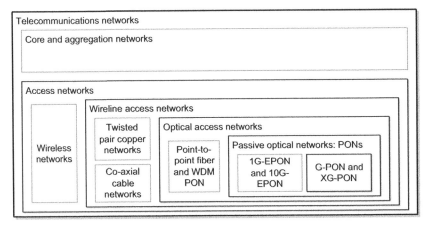

Figure 1.1 G-PON taxonomy.

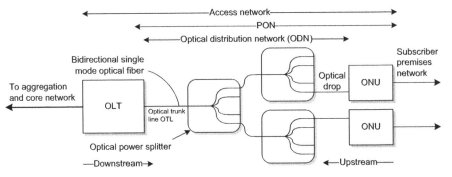

Figure 1.2 G-PON terminology.

serves more than one PON.[*] The PON contains a trunk fiber feeding an optical power splitter, or often a tree of splitters. From the splitter, a separate drop fiber goes to each subscriber, where it terminates on an optical network unit (ONU). ONUs of various kinds offer a full panoply of telecommunications services to the subscriber.

---

[*] The term OLT is therefore ambiguous: it may refer only to the terminating optoelectronics and MAC functionality of a single PON or it may mean the entire access node, terminating a number of PONs and forwarding traffic to and from an aggregation network.

A single-fiber connection is used for both directions, through specification of separate wavelengths for each direction. As we shall see, wavelength separation also allows for coexistence of other technologies on the same optical distribution network (ODN).

After a brief overview and history in this chapter, Chapter 2 outlines the requirements and constraints of a G-PON access network. Chapter 3 explains the optical layer of the network. Moving up the stack, Chapter 4 covers the transmission convergence layer, the home of most of the features that uniquely distinguish G-PON from other access technologies. Chapter 5 introduces the management model in the context of equipment and software management, while Chapter 6 shows how the management model is used to construct telecommunications services. Finally, Chapter 7 describes current and future alternatives, competitors, and partners of G-PON.

Two appendices, a list of references, a guide to acronyms and abbreviations, and an index appear at the end of the book.

1.1 TARGET AUDIENCE

This book is written for the experienced telecommunications or data communications professional whose knowledge base does not yet extend into the domain of PON or, in particular, G-PON and XG-PON. We also address this book to the advanced student, who cannot be expected to have a grounding in the ancient and forgotten lore of telecoms. Hoping to strike a balance against excessive redundancy, we nevertheless include a certain amount of background material, for example, on DS1 and E1 TDM services, and we always try to indicate where to find additional information.

We argue that a simple restatement of the standards adds no value. Accordingly, we structure this book with a view toward explaining and comparing the standards, rather than simply paraphrasing them. This is most evident in the frequent side-by-side comparisons of G-PON and XG-PON. This book also addresses many important aspects of real-world access networks that lie beyond the scope of the standards.

Disclaimers   The complete and authoritative specifications are in the standards themselves. While we make every attempt to be accurate, we have necessarily elided any number of secondary details, especially in the peripheral standards. We trust that the reader who ventures into the formal standards will find few surprises. Although both authors are employed by Ericsson, we should also state that this book is not sponsored by Ericsson, and the views expressed do not necessarily represent Ericsson positions.

1.2 EVOLUTION OF G-PON TECHNOLOGY AND STANDARDS

PON technology began in the 1980s with the idea of a fiber ring dropping service to each subscriber. Ring topology was abandoned early for reasons that will be apparent upon reflection, and subsequent PONs have been based on optical trees. In the early days, several companies[*] developed products around the integrated services digital network (ISDN) standards. The systems delivered plain old telephone service (POTS), but offered few advantages over copper-fed POTS. Although there were some deployments, the technology (cost) and the market need (revenue) were too far apart to justify a realistic business case.

By the 1990s, optical communications were starting to mature in the long-haul network, speeds were increasing, and the industry was starting down the cost curve on the technology side. At the same time, a market for Internet access was developing, and subscribers were increasingly frustrated with modems running at 9.6 or even at the once-impressive speed of 56 kb/s. The PON industry tried again.

The first generation of what might be termed modern PON was based on asynchronous transfer mode (ATM), originally designated A-PON. Significant commercial deployments of ATM PON occurred under the moniker B-PON (broadband PON). B-PON was standardized and rolled out in the last years of the twentieth century.

Like G-PON, B-PON is defined by the ITU-T, in the G.983 series of recommendations. Several data rates are standardized. Early deployments delivered data services only, at aggregate bit rates of 155 Mb/s, both upstream and down. This is one of the bit rates used by the synchronous digital hierarchy (SDH), an optical transport technology developed during the 1980s and 1990s and widely deployed today, albeit having evolved over the years to the higher rates of 2.5 and 10 Gb/s and beyond. B-PON specified the same bit rates as SDH, with the intention to reuse component technology and also parts of the SDH standards for specifications such as jitter.

In the early years of the millennium, B-PON matured in several ways:

- Service definitions expanded beyond best-efforts data, most notably to include POTS and voice over Internet protocol (VoIP).
- The original downstream wavelength spectrum was redefined into two bands, a basic band for use by the B-PON protocols and an enhancement band intended for radio frequency (RF) content such as broadcast video. Spectrum was also identified for use by independent dense wavelength division multiplex (DWDM) access, coexisting on the same fiber plant.
- To improve the utilization of upstream capacity, dynamic bandwidth allocation (DBA) was defined and standardized.
- Although the common upstream rate remained at 155 Mb/s, the technology, cost, and market requirements had evolved to the point that 622 Mb/s—another SDH speed—became the expected downstream rate.

---

[*]One of the original companies—DSC-Optilink—can be tied to today's Alcatel-Lucent; another—Raynet—can be linked to Ericsson.

Reasonable B-PON deployment volume was achieved at these levels. At the time of writing, in 2011, B-PON was still being installed to fill out empty slots in existing chassis.

Although the B-PON standards are ITU-T recommendations, a group called FSAN (Full Service Access Network) guided the requirements and recommendations and continues to guide its successors to this day. FSAN is an informal organization of telecommunications operators founded in 1995. Unlike formal standards development organizations (SDOs) such as ITU-T and IEEE, and non-SDOs such as Broadband Forum (BBF), FSAN has no membership fees and no staff. Equipment and component vendors are members by invitation only. While FSAN emphasizes that it is not an SDO, the same companies and the same people carry their discussions from FSAN into ITU-T for the formal standardization work, often on successive days of the same meeting. In the early days, the text of the recommendations was actually developed under the FSAN umbrella, then passed to ITU-T for formal review and consent.

Around the turn of the millennium, digital video began to come out of the lab. The bandwidth limitations of B-PON became a concern—622 Mb/s distributed across 32 subscribers is only (!) 20 Mb/s average rate per subscriber, much of which would be consumed by a single contemporary high-definition digital video stream—while the cost of technology had continued to improve. Standardization discussions began on a new generation of PON, this one known as gigabit-capable PON, or G-PON. The first G-PON standards, the ITU-T G.984 series, were published in 2003. As with B-PON, the G-PON standards recognize several data rates, but the only rate of practical interest runs at 2.488 Gb/s downstream, with 1.244 Gb/s in the upstream direction, capacity shared among the ONUs. These are also SDH bit rates.

The initial versions of the G.984 series recognized the ATM of B-PON, but ATM was subsequently deprecated as a fading legacy technology. Another capability that was initially standardized and later deprecated was provision for G-PON to directly carry TDM (time division multiplex) traffic. This might have been useful for services such as DS1 (digital signal level 1) or SDH, but detailed mappings were never defined, and it was overtaken by the development of standards for pseudowires, about which we shall learn in Chapter 6.

The only form of payload transport that remains in G.984 today is the G-PON encapsulation method (GEM), usually encapsulating Ethernet frames.[*] It is perfectly accurate to think of G-PON as an Ethernet transport network, notwithstanding marketing claims from the EPON competition that G-PON is not real Ethernet.

Another important evolutionary step from B-PON to G-PON is accommodation of the operators' requirement for incremental upgrade of already deployed installations. The B-PON wavelength plan did not provide for the coexistence of B-PON and G-PON on the same optical network. The eventual need to upgrade access network technology thus presented a dilemma:

---

[*] Other mappings are also defined in G.984 G-PON; some were recognized to be of no market interest and were not carried forward into G.987 XG-PON. The only additional mapping in G.987 XG-PON is multiprotocal label switching (MPLS) over GEM. Time will tell whether it is useful.

- It was usually not economical to install a new optical distribution network, particularly the distribution and drop segments, in parallel with an existing one.
- It was not feasible to replace all ONUs on a PON at the same instant. Imagine an army of 32 service technicians calling on 32 subscribers at precisely 10 AM next Tuesday morning—or at any other time for that matter!
- It was unacceptable to shut down telecommunications service to a group of subscribers for several hours or days to allow for a realistic number of service technicians to schedule realistic service appointments with subscribers.
- And maybe only 1 of those 32 subscribers was willing to pay for upgraded service anyway.

The upshot of this consideration was a requirement for G-PON networks to reserve wavelengths to allow incremental upgrade to the next generation of PON technology, whatever that might be, and to include the necessary wavelength blocking filters in ONUs. Chapters 2 and 3 discuss this in further detail.

G-PON began to be deployed in substantial volume in 2008 and 2009.

Once a standard is implemented and deployed, it is natural to want to confirm that everyone has the same interpretation and that the various implementations will interwork. This led to a series of interoperability test events, beginning with the basic ability of an OLT to discover and activate an ONU on the PON. The first G-PON plugfest occurred in January 2006, and there have been two to four events per year since then. Today's G-PON equipment is largely interoperable, although the final proof remains to be seen: there have not yet been widespread live deployments of multivendor access networks.

As testing moved further up the stack, it became apparent that the flexibility of the ONU management and control interface (OMCI) was not an unmixed blessing. Different vendors supported given features in different ways. If the OLT tried to provision the feature in one way and the ONU supported only some different way, the pair would not interoperate.

Interoperability was a primary motivation for standardization. FSAN therefore created the OMCI implementation study group (OISG) with the charter to develop best practices, recommendations for the preferred ways to implement various features. OISG was and is a vendors-only association, theoretically free to discuss implementation considerations under mutual nondisclosure agreements.

In 2009, OISG released an implementers' guide of OMCI best practices, originally published as a supplement to the OMCI specification ITU-T G.984.4. As G.984.4 was migrated into G.988, the implementers' guide material was incorporated into G.988, where it resides today.

OMCI best practices continue to evolve as minor questions arise, but the issues that spawned OISG have largely been resolved. OISG's charter also evolved and it became an early preview forum for OMCI maintenance, an opportunity for sanity checks and consensus building before new OMCI proposals were formally submitted to the ITU-T process.

Although OISG has not been formally disbanded, it is now dormant, both because of its success at resolving interoperability issues and because of the shift of responsibility from FSAN to BBF. Test plans are published as BBF technical reports, specifically TR-247 and TR-255. Responsibility for plugfests also shifted from FSAN to.

Broadband Forum entered the G-PON scene only recently, but in a major way. Previously known as DSL Forum, BBF changed its name and expanded its scope to include, among other things, the entire access network and everything attached to it. In 2006, BBF published TR-101, which defined requirements for migration of the access network from ATM to Ethernet, but still with a DSL mind-set. The ink had scarcely dried on TR-101 when BBF began a project to define its applicability to the special aspects of G-PON, resulting in TR-156 (2008).

Although there are some rough edges at the organizational boundaries, the scope addressed by BBF is theoretically disjoint from the scope of the ITU-T recommendations. ITU-T specifies an interface between OLT and ONU and the fundamentals of its operation. ITU-T includes tools for maintenance and tools from which applications can be constructed and extends the tool set as necessary when new applications arise.

In contrast, BBF views the overall network architecture as its scope. BBF takes the ITU-T tools as a given, identifies preferred configuration options, and writes network- and service-level requirements to serve these options. Vendors and operators are, of course, free to develop other applications on the same base, but if BBF does its job well, its model architectures prove to be satisfactory for most real-world needs and are suitable as reference models even for applications that lie beyond the strict bounds of the BBF architecture.

If BBF does its job well? In fact, TR-156 has largely been accepted by operators worldwide as a satisfactory model for simple ONUs that deliver Ethernet service to end users. Additional BBF technical reports (TR-142, TR-167) define how the G-PON toolkit can be used to control only the ONU's PON interface, with the remainder of the ONU managed through other means. Chapter 2 expands this topic.

Once the G-PON standards began to mature and vendors busied themselves bringing product to market, FSAN turned its attention to the question of the next logical step after G-PON. Starting in about 2007, as the outline became clearer, the operators launched a white paper project to define the requirements for the next generation. FSAN completed the next-gen white paper in mid-2009. Parallel work had begun in late 2008 to develop the details of the necessary recommendations.

FSAN structured its view of the future into two domains: Next-gen 1 (NG-1) was the set of PON architectures that was required to coexist on the same fiber distribution network with G-PON, while next-gen 2 designated the realm of possibilities freed from that constraint, an invitation to take a long view of technology to see what might make sense at some unspecified time in the future.

As it turned out, NG-1 PON bifurcated further, into versions known as XG-PON1 and XG-PON2, or often just XG-1 and XG-2, where X is the Roman numeral 10, designating the nominal 10 Gb/s downstream rate. Both versions run downstream data at 9.953 Gb/s, another SDH bit rate. The upstream capacity of XG-PON1 is

2.488 Gb/s, while XG-PON2 runs at 9.953 Gb/s upstream. The reason for the distinction was the substantial difference in technological challenge, coupled with the perception that market need was insufficient to justify the significantly higher cost of high-speed upstream links.

XG-PON1 is standardized in the ITU-T G.987 series of recommendations. At the time of writing, XG-PON2 is being held in abeyance, pending the convergence of market demand and technological feasibility. Because IEEE has already standardized a symmetric 10G form of EPON (10G-EPON, Chapter 7), it is possible that XG-PON2 will not be pursued further. If that comes to pass, operators who need symmetric 10G will deploy 10G-EPON, and the G-PON community will work toward next-next-generation access, probably WDM PON.

2

SYSTEM REQUIREMENTS

In this chapter:

- *Overview of power-splitting PONs*
- *Optical network considerations*
  - *Split ratio*
  - *Maximum reach, differential reach*
  - *PON protection*
- *Reach extenders*
- *Coexistence with future generations*
- *Types of ONU*
- *ONU powering*
- *Introduction to*
  - *Dynamic bandwidth assignment; details in Chapter 6*
  - *Quality of service; details in Chapter 6*
  - *Security; details in Chapter 4*
  - *Management; details in Chapter 5*

Before diving into the details of how a G-PON works, we need to understand something about the business case. We return repeatedly to business case questions over the course of this book because, ultimately, everything we do must add value to someone for something.

Gigabit-capable Passive Optical Networks, First Edition. Dave Hood and Elmar Trojer.
© 2012 John Wiley & Sons, Inc. Published 2012 by John Wiley & Sons, Inc.

As with most companies, telecommunications operators are driven forward by market opportunity, cost reduction, and competitive pressure, and they are held back by existing investment and existing practices. New technology is comparatively easy to justify in a greenfield development—we have to do *something*, so let us go for the latest and greatest!—but most of the potential market is already served in one way or another, even if it's no more than ADSL (asymmetric digital subscriber line) from a central office. The difficulty arises in making a business case for the deployment of a new technology that may be of immediate interest to only a small number of existing subscribers, in what is called, for contrast, a brownfield.

Civil works—right of way acquisition, permits, trenching for underground cable, poles for aerial cable—are a very large part of the up-front cost of a change in technology, for example, from copper to fiber. Estimates range from 65 to as much as 80% of the total cost. No matter how economical the equipment itself may be, this cost must be paid. Once the business case has been made to install new fiber in the outside plant infrastructure, it makes sense to place large fiber-count cables, or at least a lot of empty ducts, through which fiber can easily be blown at a later date. The cost of additional ducted fibers is comparatively small, even vanishingly small, in fiber trunks. With the optical infrastructure in place and spare fibers available, it becomes much easier to take subsequent evolutionary steps.

It is easier to develop a business case if all telecommunications services can be provided by a single network. This is the idea behind the oft-heard term *convergence*, a concerted effort to eliminate parallel networks, each of which serves only a subset of the service mix. Software-defined features, Ethernet, and IP are major steps along the road to convergence. The contribution of standards and of the network equipment is to ensure that the investment, once made, can be used for a complete range of services for decades to come. In keeping with the full-service focus of its FSAN parent, G-PON is designed to deliver any telecommunications service that may be needed.

2.1 G-PON OPERATION

2.1.1 Physical Layer

To recapitulate the brief overview in Chapter 1, a PON in general, and a G-PON in particular, is built on a single-fiber optical network whose topology is a tree, as shown in Figure 2.1. The OLT is at the root, and some number of ONUs connect at the leaves. Downstream optical power from the OLT is split at the branching points of the tree. Each split allocates an equal fraction of the power to each branch. The achievable reach of a PON is a tradeoff of fiber loss against the division of power at the splitter. Chapter 3 describes splitters and fiber loss in detail.

A power splitter is symmetric: Its loss upstream is the same as down, 3 dB for each power of 2 in the split ratio.

The OLT transmits a continuous downstream signal that conveys timing, control, management, and payload to the ONUs. The OLT is master of the PON. Based on

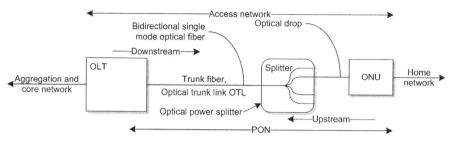

Figure 2.1 Tree structure of a PON.

service-level commitments and traffic offered by the ONUs, the OLT continuously develops an upstream capacity allocation plan for the near future—typically 1 or 2 ms—and transmits this so-called bandwidth map to the ONUs. The ONU is permitted to transmit only when explicitly given permission by a grant contained in a bandwidth map. During its allocated time, the ONU sends a burst of data upstream, data that includes control, management, and payload.

For the bursts to arrive at the OLT at precisely the proper interleaved times, each ONU must offset its notion of a zero reference transmission time by a value determined by its round-trip delay,* the time it takes for the signal from the OLT to reach the ONU, plus ONU processing delay and the time it takes for the signal from the ONU to reach the OLT. The OLT measures the round-trip delay of each ONU during activation and programs the ONU with the compensating equalization delay value.

Another low-level requirement on a PON is the discovery of new ONUs, be they either newly installed devices or existing devices that have been offline for reasons such as fiber failure or absence of power, whether intentional or not. The OLT periodically broadcasts a discovery grant, which authorizes any ONU that is not yet activated on the PON to transmit its identity. Since the round-trip time of a new ONU is unknown, the OLT opens a quiet window, a discovery window, also called a ranging window, a time interval during which only unactivated ONUs are permitted to transmit.

It is possible that more than one unactivated ONU could attempt to activate at the same time; if their transmissions overlapped in time, neither would succeed. Worse, they could deadlock, repeatedly colliding on every discovery grant forever. The ranging protocol therefore specifies that the ONU introduce a random delay in its response to the OLT's invitation. Even though the transmissions from two ONUs may collide during a given discovery cycle, they will sooner or later appear as distinct activation requests in some subsequent interval.

The size of the discovery window depends on the expected fiber distance between the farthest possible ONU and the nearest possible ONU. This is called maximum

---

* True, the correction could instead be applied locally by the OLT as an offset in the bandwidth map, but it is not.

differential reach, standardized with 10-, 20-, and 40-km options in G.984 G-PON. In G.987 XG-PON, the maximum differential reach options are 20 and 40 km.

Chapter 4 goes into detail on all of these aspects of G-PON operation.

2.1.2 Layer 2

In terms of the OSI seven-layer communications model,[*] the access network largely exists at layer 2. Perhaps the single most important concept underlying an Ethernet-based access network—which G-PON is—is that of the virtual local area network (VLAN), specified in IEEE 802.1Q. A very substantial part of the hardware, software, and management of a G-PON is dedicated to classifying traffic into VLANs, then forwarding the traffic according to VLAN to the right place with the right quality of service (QoS). Although an ONU is modeled as an IEEE 802 MAC bridge, MAC addresses are usually less important at the ONU than are VLAN tags.

The access network operates at layer 2, but it judiciously includes some layer 3 functions as well, particularly for multicast management. For practical purposes, multicast means IPTV (Internet protocol television) service; it is expected to represent a large fraction of the traffic and to yield a large part of the revenue derived from a G-PON. The PON architecture is ideally suited for multicast applications because a single copy of a multicast signal on the fiber can be intercepted by as many ONUs as need it. Each ONU extracts only the multicast groups (video channels) that are requested by its subscribers.

To determine which groups are requested at any given time, the ONU includes at least an IGMP/MLD[†] snoop function, about which we shall learn more in Chapter 6. Snooping involves monitoring transmissions from the subscriber's set-top box (STB), based on which the ONU compares the requested channels with a local access control list (ACL). If the requested content is authorized and is already available on the PON, the ONU delivers it immediately without further ado. Also acting as an IGMP/MLD snoop, or more likely as a proxy, the OLT likewise determines whether a given multicast group is already available, or whether it needs to be requested from yet a higher authority. As seen by a multicast router further up the hierarchy, a proxy aggregates a number of physical STBs into a single virtual STB, thereby avoiding unnecessary messages to the router and improving network scalability.

It is an open question what statistical capacity gain should be expected from multicast, now and in the future. Even if 80% of subscribers are watching the same 10 channels, a long statistical tail would require substantial capacity to carry content of interest only to the remaining few. Will a PON with 50 subscribers, each with 2 or more television sets and a recording device, need 50 multicast groups? Thirty? Twenty?

---

[*] See ITU-T X.200.

[†] IGMP: Internet group management protocol (IPv4), MLD: multicast listener discovery (IPv6). We usually spell out acronyms the first time they appear; acronyms are also listed in a separate section at the end of the book.

There is a general expectation that video will move toward unicast, but no one is prepared to say how soon. At the end of the day, it may not matter. The considerations described above suggest that we should expect a busy hour load of two or three multicast groups per subscriber. At bit rates on the order of 5 Mb/s per multicast group, G-PON has enough downstream capacity for that level of loading. If it were to materialize, mass market demand for ultrahigh bandwidth unicast video, up to 65 Mb/s per channel, could motivate further access network upgrade.

2.2 ONU TYPES

A variety of product configurations seeks to fit the range of operators' needs completely and optimally. Here we outline a few of the possibilities.

2.2.1 Single-Family ONU

At least in some markets, the single-family unit (SFU) is the most common form of ONU. Predictably, there are many variations on the SFU theme. The SFU may be located indoors or out. Power is always supplied by the subscriber, but the SFU may or may not include battery backup. The SFU may be regarded as a part of the telecommunications network, owned and managed by the operator, or it may be considered to be customer premises equipment (CPE), owned by the subscriber.

The simplest SFU, such as the one illustrated in Figure 2.2 with its cover off, delivers one Ethernet drop; it is essentially a G-PON-to-Ethernet conversion device. This one is intended to be mounted on a wall, at the demarcation point between the drop fiber and the subscriber's home network. The single Ethernet feed would then

Figure 2.2 Single-family ONU with power brick.

Figure 2.3 SFU with enhanced functionality.

be connected to a residential gateway (RG) at some location convenient to the subscriber's device layout.

Many SFUs, such as the one in Figure 2.3, add value by including several bridged Ethernet drops, suitable for direct connection to several subscriber devices, for example, two or three PCs and a set-top box. Some may also include built-in terminations for one or two POTS lines. Other applications for home use might include low-rate telemetry, for example, to read utility meters or to monitor intrusion detectors. M2M (machine to machine) communications are expected to mushroom over the next few years, and the SFU will surely play a part in backhauling information to centralized servers.

The SFU may also include a full residential gateway, with firewall, NAT (network address translation) router, DHCP (dynamic host configuration protocol) server, 802.11 wireless access, USB ports, storage or print server, and more. This form of SFU is typically managed jointly by the subscriber, by the ONU management and control interface (OMCI, G.988) model of G-PON, and by an access control server (ACS), the latter as defined in various Broadband Forum technical reports and frequently short-handed as TR-69.[*]

2.2.2 Multi-Dwelling Unit ONU

The multiple dwelling unit (MDU) is an ONU that serves a number of residential subscribers. It may be deployed in an apartment building, a condominium complex, or at the curbside. The MDU is always considered to be part of the telecommunications network; that is, its power, management, and maintenance are the responsibility of the operator. Depending on their target markets, MDUs typically serve from 8 to 24 subscribers. Very similar to the MDU, a G-PON-fed digital subscriber line access multiplexer (DSLAM) may serve as many as 48 or even 96 subscribers.

Subscriber drops from an MDU may be Ethernet, but the IEEE 802.3 physical layer is not specified to tolerate the stress of a full outdoor environment, specifically lightning transients. Even if the MDU is housed indoors in the same building as the

---

[*] BBF designates it TR-069. It is always pronounced without the zero, and we like to write it in the same way we say it.

subscriber residences, it may be uneconomical to rewire the building with the cat-5 cable needed for Ethernet.

The alternative subscriber drop technology is DSL. When drops are short, the preferred form of DSL is ITU-T G.993.2 VDSL2; such an MDU may or may not also offer POTS. Existing telephone-grade twisted pair runs from the MDU to the subscriber premises, where there is a DSL modem and a splitter for POTS, if POTS is included in the service. With the short drops implied by fiber to the curb, it is feasible to deliver several tens of megabits per second—even 100 Mb/s and more— effectively overcoming the speed limitations of copper wiring. The rate-reach maximum can be extended through bonding of services across two or more pairs, while G.993.5 vectoring potentially increases attainable speed through crosstalk cancellation.

2.2.3 Small-Business-Unit ONU

As well as the ubiquitous Ethernet service, a small business unit (SBU) is likely to offer several POTS lines to a small-office customer. It may also support a few TDM (time-division multiplex) services such as DS1 or E1 via pseudowire emulation (Chapter 6 explains this). The SBU of Figure 2.4 has eight POTS lines, four Ethernet drops, and four 2.048-Mb/s E1 TDM services.

The cellular backhaul unit (CBU) is a variation of the SBU—perhaps a new category in its own right. In the cellular backhaul application, the ONU carries traffic between the core network and a radio base station. Legacy mobile backhaul requires interfaces such as DS1 or E1. As the cell network migrates from third to fourth generation, Ethernet backhaul is displacing DS1 and E1. As well as the tightly controlled frequency stability required of all TDM services, some wireless protocols require a precise time of day reference, a function described in Chapter 4.

Another variation of the SBU is the multitenant unit (MTU), intended to be shared by several small businesses. The target market is the small islands of commercial activity common along major streets. The important distinction of the MTU from the SBU is its need to isolate services one from another, both in terms of traffic—no bridging between Ethernet ports—and in terms of service-level agreements (SLAs).

Figure 2.4 An SBU.

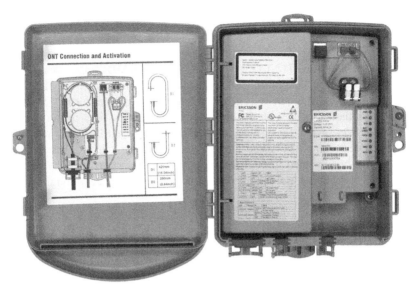

Figure 2.5 Outdoor ONU, outer access cover open.

2.3 NETWORK CONSIDERATIONS

While some operators favor the CPE model, in which the ONU is indoors, located on the subscriber's desktop or perhaps mounted on an indoor wall, other operators wish to deploy ONUs outdoors. To a considerable extent, this reflects a difference in the operator's perspective: ONU as part of the telecommunications network or ONU as subscriber-owned device. Figure 2.5 illustrates such an outdoor ONU, which differs from the device of Figure 2.2 in that it provides two POTS lines, as well as an Ethernet drop, and is accessible to operator personnel without the need to enter the subscriber's home.

ONUs such as MDUs may go into equipment rooms or telecommunications closets in buildings. ONUs may also be designed for curbside pedestals (Fig. 2.6) or other outdoor housings, in which case they need to be fully hardened for outside plant conditions. ONU components must generally be rated for the full industrial temperature range, and ONU enclosures may be required to tolerate extremes of temperature and water exposure, including immersion (Fig. 2.7) and salt fog. Other considerations for outdoor ONUs include lightning protection for all metallic wiring, and insect and fungus resistance. All ONUs must satisfy regulatory requirements for electromagnetic interference (EMI) generation and operator requirements for EMI tolerance.

2.3.1 Power

ONU powering is indisputably a network consideration, but it warrants a separate discussion in its own right. We defer this topic to Section 2.5.

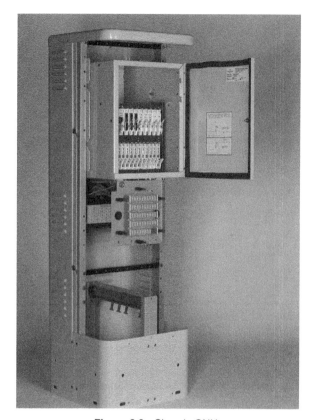

Figure 2.6   Chassis ONU.

2.3.2   Energy Conservation

Reducing the demand for power is an important topic. Power, especially remote power, is difficult and expensive to provide, and heat dissipation is a problem, especially in outdoor deployments subject to high ambient temperatures or direct sunlight.

The natural progress of technology is toward less power consumption for a given function. This is true not only in the silicon of the G-PON ONU itself, but perhaps more importantly, in the efficiency of the AC (alternating current) power converter and the backup battery and its charger.

Not least because it is politically correct, the operator community is interested in saving additional power by shutting down functions when they are not in use. This follows the fine tradition of POTS telephony, in which an on-hook line consumes no power.

Inactive user network interfaces (UNIs) are comparatively easy to power down, but the PON interface presents difficulties: if its PON receiver is powered down, how does the ONU know when a terminating call arrives? And if the ONU's transmitter is shut down, how does the OLT know that the ONU has not failed? The answer is to take only very brief naps, a few tens or perhaps a few hundreds of milliseconds at a time.

Figure 2.7 Underground ONUs.

ITU-T supplement 45 to the G-PON recommendations outlines the energy-saving options, but the topic came to full maturity only in XG-PON. XG-PON defines two energy conservation modes, dozing and sleeping. In doze mode, the ONU keeps its receiver alive at all times. This is especially appropriate for one particular use case, IPTV, in which almost all of the traffic flows downstream. In contrast, sleep mode allows the ONU to shut down both transmitter and receiver. Both modes require the ONU to respond periodically, so that the OLT can confirm that the ONU is still alive and healthy, and to serve whatever traffic that may arrive.

Section 4.5 explores the energy conservation feature in detail.

2.3.3 Plug and Play

MDU ONUs are installed on engineering work orders and are maintained as telecoms equipment. While it is, of course, important that installation and provisioning be no more complex than necessary, it would never be expected that an arbitrary hitherto unknown and unexpected MDU might suddenly appear on the PON, with features and capacities only to be discovered after the fact.

At the other extreme is the desktop ONU. Some operators would like such an ONU to be purchased by the subscriber at the local electronics store and installed by simply plugging it in. The business aspects of installation can be dealt with: the subscriber calls the provider or browses to an introductory web page, signs up for service, provides billing details, receives some kind of license or login credentials, preferably implicit, whereupon everything just comes up and works.

More of an issue for the do-it-yourself subscriber is the physical installation. Although PON optics are rated to be eye safe, it is not really a good idea to leave optical fiber terminations exposed, launching even their small amount of invisible light in whatever random direction they lie. Not only that, but a single speck of dust in an optical connector can render the ONU nonfunctional; cleaning connectors requires tools and training beyond the level of the average subscriber. Because of the optical concerns, it may be that, even when the ONU becomes a commodity item, the operator will roll a truck to install it.

That said, we mention that the ONU of Figure 2.2 is intended to be installed just above floor level, with the optical connector facing down, minimizing concern about dust in the connector. The wall-mount unit and the optoelectronics module, suitably equipped with dust caps, can be uncovered and plugged together within a matter of seconds. So there may indeed be cases in which an ONU can effectively be installed or replaced directly by the subscriber.

2.3.4 How Far?

G-PON parameters specify a maximum reach of 60 km of fiber, with a maximum differential reach that defaults to 20 km.

Figure 2.8 illustrates what we mean by reach and differential reach. The reach is the total fiber distance from the OLT to the farthest ONU, in this case 30 km. Differential reach is the difference in fiber distance between the farthest and the

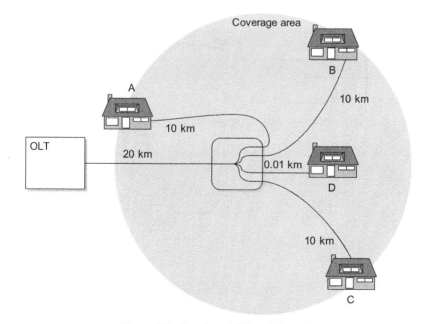

Figure 2.8 Reach and differential reach.

nearest ONU. If our PON included only subscribers A, B, and C, the differential reach would be zero because each is 30 km from the OLT, measured along the fiber run. Add subscriber D to the PON, and the differential reach becomes 10 km.

It would be perverse to run the trunk fiber 20 km to a splitter, and then run a drop fiber back 10 km to subscriber A. In geographically spread-out locations such as imagined here, it often makes sense to deploy a cascade of splitters, as illustrated in Figure 2.9.

The first splitter usually has a lower split ratio, typically 1:4. The shape of the serving area can be tailored by the locations of the splitters.

Reach and differential reach are primarily issues of upstream burst timing, which can be addressed by varying the OLT's delays and quiet intervals. But greater reach, or a larger split ratio, also imply greater optical loss.

The need to go further with more splits made it natural to define G-PON reach extenders (REs). In its simplest form, an RE is simply an optical amplifier or an optical/electrical regenerator in each direction. More sophisticated REs may extend a number of PONs, with the OLT (trunk) side either using a separate fiber for each PON or separate wavelengths on a single fiber. A multi-PON RE is also a likely candidate for trunk-side protection. Sections 2.4, 3.12, and 5.3 discuss reach extenders in more detail.

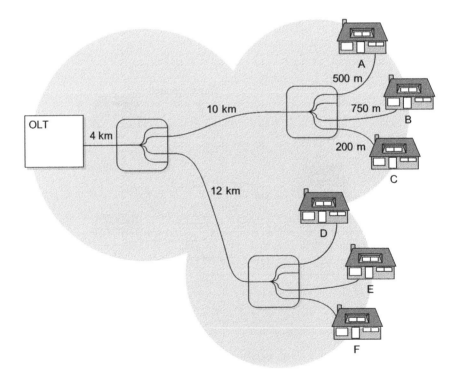

Figure 2.9 Multistage split.

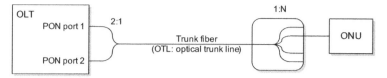

Figure 2.10 OLT port protection.

2.3.5 PON Protection

G-PON protocols do not directly support protection of the nature defined in classical transmission protocols such as SDH—linear or ring protection, for example, with or without bidirectional signaling—but several forms of PON protection are possible. Recommendation G.983.5 describes protection scenarios and works out the details of message exchanges. As far as we know, it was never implemented, and the protection definitions of G.984.3 and G.987.3 omit such details.

Figure 2.10 illustrates the simplest, namely OLT port protection. The trunk fiber is connected to the OLT with a colocated 2:1 splitter, at the cost of an additional 3 dB of loss. Both OLT ports receive the upstream signal, but only one port transmits at any given time. The OLT triggers protection switching if one of its ports fails or is unplugged, or if it declares loss of signal from all ONUs. The ONUs themselves do not know about PON protection. Depending on the OLT's architecture, fast switching is possible, less than the classical target of 50 ms. Depending on the OLT's architecture, it may be necessary to reinitialize or rerange the ONUs after a switch.

OLT port protection covers failures at the OLT itself but does not address issues such as cable cuts in the outside plant. Lack of protection against cable cuts is not necessarily a show stopper because cables in the access network are usually not routed diversely anyway. If a backhoe cuts one cable, it probably also cuts whatever redundant fiber might have been present in an adjacent cable. This is one reason why some operators are considering stationary wireless links for PON protection.

As shown in Figure 2.11, we can readily protect against cable cuts of the trunk fiber by using a 2:N splitter at the remote site. This layout also recovers 3 dB of optical budget that was lost in Figure 2.10. The ONUs cannot tell the difference between the feeders of Figures 2.10 and 2.11. As to the OLT, because the trunk cables are presumably routed diversely (else why protect them?), this protection design requires redetermination of the ONUs' equalization delays after a protection switch,

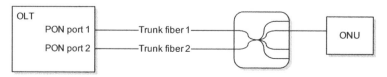

Figure 2.11 Trunk fiber protection.

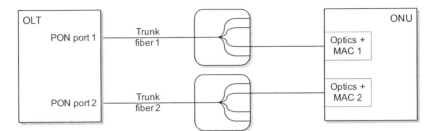

Figure 2.12   Complete redundancy.

although only the trunk delay differs. Both G-PON and XG-PON include the ability for the OLT to minimize recovery time by retiming a single ONU, deriving a correction factor, and broadcasting it to all ONUs.

PON protection may be generalized to use ports on separate OLTs, thereby protecting against complete OLT failure. Further, the separate OLTs may be located in separate central offices, providing at least some degree of protection from large-scale disasters. Dual homing, as this is called, raises additional issues in coordinating PON provisioning, the uplinks from the OLTs, and the real-time switch-over between working and protect PONs.

In Figures 2.10 and 2.11, the ONUs need not know anything about PON protection. Figure 2.12 illustrates an ONU designed for protection, with two optical interfaces. It is possible for such an ONU to have only a single PON MAC (medium access control) device, but if we are going to pay for two optics modules, it could make sense to include two MAC interfaces, with the ability to carry traffic on both PONs at the same time, either duplicate traffic or extra traffic of low priority that could be dropped in the event of a switch.

Another possible merit of Figure 2.12 is that only some, but not all, ONUs need be protected, for example, those serving business customers, large MDUs, DSLAMs, or mobile base stations. Figure 2.12 follows the classical precedent of SDH, a core network technology that generally justifies higher costs. Because of high development cost for a low-volume product, the market for dual MAC ONUs has not yet developed.

There are other ways to do protection, specifically Ethernet link aggregation (originally in IEEE 802.3ad, now in 802.1AX). Figure 2.13 illustrates how individual ONUs may be protected on an end-to-end basis, end-to-end from the layer 2 viewpoint, at least. In this configuration, the ONUs, PONs, and OLTs need know nothing about protection. Standards and equipment already exist, avoiding the need for the PON subnetwork to reinvent the wheel.

2.3.6   How Many?

If we wish to dedicate a 50-Mb/s average downstream data rate to each subscriber, a 2.5-Gb/s G-PON can serve about 50 subscribers; a 10-Gb/s XG-PON about 200.

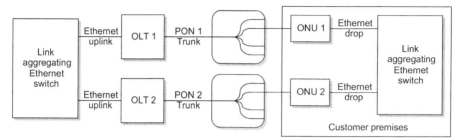

Figure 2.13 Link aggregation protection.

Some operators would like to serve 500 subscribers per PON; others would like to be able to deliver 100 Mb/s to each subscriber. Of course, multicast and bursty traffic patterns mean that these numbers are fairly arbitrary, but they do provide some indication of the capacity available.

When a PON is equipped with MDUs, it may be cost-effective to connect only 16 ONUs, or even fewer. For single-family ONUs, common planning numbers are 32–64 ONUs per PON. Although there is clearly a point of diminishing returns, operators find it economical to pay for higher split ratios, rather than installing additional fibers and OLT blades. Some operators talk about 128-way splits and even more. In the discussions leading up to XG-PON, a PON with 256 ONUs was the largest number anyone could imagine—but understanding how imagination works, the community allocated 10 bits to the ONU-ID, so that in theory, 1023[*] ONUs could be connected to an XG-PON.

The optical loss budget ranges from 28 dB (G-PON class B+) or 29 dB (XG-PON1 class N1), right up to 35 dB (XG-PON1 extended class E2). The standards put the options into the OLT as much as possible. Limiting the number of ONU types recognizes the fact that the ONU is the point at which high-volume components matter, and where the operator's inventory and logistics costs make a big difference. The OLT is also likely to support plug-in optics, while for cost reasons, the ONU is more likely to have integral optics.

Keep in mind that each 1:2 split costs something over 3 dB, so a 10-deep splitter ($2^{10} = 1024$ ONUs) would pretty well use up the most aggressive optical budget, all by itself. Having said that, nothing prevents the development of a reach extender that could indeed support, say eight 128-way splitters from a single PON. Nothing, that is, but the operators' understandable reluctance to deploy powered and managed equipment deep in the field.

2.3.7 Coexistence

Although G-PON will have a long service life, the nature of progress is such that someday, G-PON will be superseded by technologies that better satisfy evolution in demand, in services, in technology, and in revenue. How will we someday replace

---

[*] The 1024th address is used for ONU discovery, described in Chapter 4.

G-PON with the next generation? It is safe to assume that the next generation, whatever it may be, will be based on single-mode optical fiber, to or near the subscribers' premises.

The easiest answer would be to install a new optical distribution network in parallel with the existing one, and when all is said and done, this may well be the least-bad solution at some point. But particularly in residential areas—beyond the first of several possible splitters in tandem—this may be difficult. There is no guarantee that there will be spare fibers or ducts in existing distribution cable, and laying new cable is very expensive. Nor is it feasible to visit 32 or 64 or 128 subscriber premises simultaneously to replace their ONUs.

Indeed, the most complicated factor in the evolution story is that only a few subscribers will need to be upgraded anyway—G-PON ONUs are expected to satisfy the needs of most users for many years to come—and it is hard to justify a large new investment for only that first pioneering upgrade subscriber, especially when the take rate may be quite modest for many years to come.

It is therefore required that G.984 G-PON and next-generation G.987 XG-PON coexist on the same ODN indefinitely, and further, that upgrade not disrupt existing services more than momentarily—zero disruption is the target. Coexistence is achieved through compatible wavelength plans and optical budgets, as discussed further in Chapter 3.

Beyond G.987 XG-PON, the technology options are open. Further migration is sure to be required on existing ODNs, coexisting with at least one of G.984 G-PON or G.987 XG-PON, and possibly both. WDM (wavelength division multiplexing) PON is regarded as a prime candidate, but its parameters remain under discussion. Chapter 7 outlines some of the issues and options of WDM PON.

2.3.8 Unbundling

For business benefit or regulatory compliance, more than one company may be involved in delivering telecommunications services to the subscribers of a PON.

In the context of a G-PON, suppose that company A owns the local network of optical cables or fiber ducts. Physical layer unbundling occurs when company A leases duct space or dark fibers to company B. Generally, this means that the fiber terminates at a fiber distribution frame and is patched to some separate network element that is owned or controlled by company B. Repairs to ducts and cables are the responsibility of company A.

Wavelength unbundling occurs when company A or B[*] leases one or more of the wavelengths on the fiber to company C. Generally, this means that company A is responsible to provide a filter and to break out the contracted wavelengths to a fiber distribution frame for patching to separate network elements. The contract also binds all parties not to cause harmful mutual interference, for example, by transmitting excessive power levels. Physical repairs are the responsibility of company A.

---

[*]Henceforth we omit the subleasing possibilities and just assume that company A is the principal.

In physical and wavelength unbundling, the lessor is free to modulate the fiber with its choice of signal format, subject to contracted channel characteristics and interference constraints.

In layer 2 unbundling, company A lights up the fiber with its own protocol—G-PON, for example—and company D leases capacity within that protocol. Typical lease parameters would include VLAN IDs and service-level commitments. The fiber terminates in an OLT owned by company A, and the unbundled stream is switched at layer 2 into network elements owned by company D. Diagnosis and repair is largely the responsibility of company A.

All of these options are important in terms of the companies' operations and business practices, but duct and fiber unbundling do not affect G-PON. Wavelength unbundling only affects G-PON in the sense of assigning wavelengths. In terms of G-PON requirements, layer 2 unbundling may include requirements to groom traffic into separate bundles, even when the committed QoS of one bundle is identical to that of another, differing only by contractual relationship.

One particular higher layer unbundling feature is wholesale multicast service, in which company A may offer IPTV bundles from companies E, F, G, and so on. This option has implications in the complexity of multicast provisioning, inasmuch as a subscriber may mix and match from a menu of offerings, some of which may overlap. Multicast, and in particular multiprovider multicast, is discussed in Chapter 6.

2.3.9 Synchronization

It is rather taken for granted that a G-PON OLT is timed from a stratum-traceable source, with at least stratum 4[*] and usually stratum 3 or 3E holdover, and a frequency accuracy within four parts per billion. The G-PON itself is synchronous, so a PON-derived frequency reference at the ONU is also stratum traceable. A stratum-traceable frequency reference is important for services such as DS1/E1 circuit emulation. An OLT may derive its timing reference from a building integrated timing supply (BITS), but if it is located in a controlled environment vault or a remote cabinet, the OLT may alternatively be timed via synchronous Ethernet or IEEE 1588.

As well as precise frequency, some radio protocols also require a precise time of day, preferably to be supplied by the mobile backhaul ONU. The underlying reason is that these technologies share spectrum on a time-divided basis among several devices. If separately located transmitters are to know when they are allowed to use the spectrum, they need an accurate time reference. One-microsecond accuracy was provisionally specified for G-PON, in the absence of a better value. As the community works through the standardization issues of next-generation radio systems, it appears that a G-PON system will be asked to reduce its allocation quite considerably, perhaps to as little as 100 ns. This accuracy is a question of hardware design, not a standards issue.

---

[*]See ATIS (Alliance for Telecommunications Industry Solutions) 0900101 [formerly ANSI (American National Standards Institute) T1.101] for definitions of timing strata.

Time of day is not available from a frequency reference. Time of day can be conveyed via IETF (Internet Engineering Task Force) network time protocol [NTP, RFC (request for comments) 5905] or simple NTP (SNTP, RFC 2030). Time of day is also available from GPS (global positioning system) receivers, which are regarded as too expensive to be desirable in every endpoint—nor can every endpoint rely on having a clear view of the sky. The favored candidate for time distribution is IEEE 1588.

The baseline assumption for packet timing is that delay through the network is (a) short, (b) symmetric, and (c) stable. None of these is necessarily true in a G-PON, where the upstream direction is delayed and subject to bandwidth allocation irregularities. Chapter 4 explains how the G-PON protocols include a way to transport time of day over the PON, using the PON ranging parameters for each given ONU. Transparent timing is also possible, in which the equipment merely records and forwards the transit delay of each given timing packet, a delay that can subsequently be used as a correction factor.

2.4 OLT VARIATIONS AND REACH EXTENDERS

The OLT is the interface between the PON and the telecommunications aggregation or core network. Conceptually, it is located in a central office, but in practice it may be located in a controlled environment vault (CEV) or an outdoor cabinet, as a way to extend the reach of the PON. Another way to extend reach is the so-called reach extender (RE; Fig. 2.14). Conceptually, a reach extender is just a repeater, either based on optical amplification or on electrical regeneration. The reach extender is usually located at the same site as the splitter; indeed the splitter may be integrated into the RE equipment itself.

Because a reach extender requires power, management, and possibly facility protection, it makes economic sense to extend several PONs with a single equipment unit. In this case, the reach extender may have one trunk fiber per PON, or may multiplex several PONs onto a single trunk fiber through WDM, either coarse (CWDM) or dense (DWDM).

2.4.1 Why Reach Extenders? The Business Case

In Payne et al. (2006) British Telecom (BT) observes that the demand for bandwidth increases faster than can be supported by the combination of revenue growth—subscribers want more bandwidth but are not willing to pay very much for it—and the

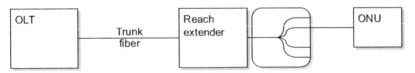

Figure 2.14 Reach extender.

normal year-over-year erosion of equipment cost. This makes it difficult to develop a business case that justifies investment for broadband access, be it G-PON or anything else. Some other economic factor must be folded into the analysis.

In the absence of a clearly visible killer app that will completely redefine the economics of telecommunications, operators look to cost reduction. The BT chapter in Payne et al. (2006) summarizes a study in which a number of best-case assumptions were made as a way to understand the best possible cost savings.

The study concluded that, ignoring the real-world issues, a dual-homed access network with a reach of 100 km could allow as few as 100 well-chosen local exchanges (central offices) to cover the United Kingdom, replacing the 5000 that exist today. Exchange consolidation could represent a major cost savings.

In view of the real world, in particular the capabilities of G-PON:

- Substantial exchange consolidation is possible, even with only 20 km of reach. Twenty kilometers far exceeds the range achievable with the current exchange-fed copper infrastructure.
- G-PON's reach could be extended with C+ optics (explained in Chapter 3). Under this assumption, a very high percentage of the United Kingdom's population could be served with dual-homed G-PON.

It will not come as a surprise to learn that BT is very interested in extending the reach of G-PON in any way possible, or that BT continues to push for dual-homed redundancy.

If the optical network cannot be completely passive, BT would like to see the simplest possible reach extenders, ideally nothing more than optical amplifiers in footway enclosures. BT views this as a better choice than remote OLTs. As much as anything, this preference is a consequence of the increased power demanded by a remote OLT, deployed in an environment where every watt is precious.

In Edmon et al. (2006), SBC (now part of AT&T) considers somewhat the same problem in light of U.S. geography. They conclude that fiber to the home (G-PON) is the right solution for greenfield deployment. There is no question that new cable must be installed to serve new subdivisions, and it might as well be optical fiber. Greenfield developments are likely to be well away from the central city, so reach is an issue. Like most operators, AT&T has consistently pushed for increased optical budgets, just a few decibels more. Each decibel expands the circle that a central-office-based OLT can serve, and like BT, AT&T is keenly aware of the disproportionately higher cost of remote siting. These discussions have led to higher loss budget classes in both G-PON (32 dB C+) and XG-PON (extended classes up to 35 dB).

Edmon et al. (2006) also conclude that the right way to serve brownfield markets is through small remote DSLAMs. These DSLAMs would be sited close enough to the subscriber base that the necessary broadband services could be delivered over DSL copper. Although it could be done with G-PON, the fiber to the DSLAM is proposed to be point-to-point gigabit Ethernet (GbE).

Other operators found that, even without additional loss budget, a differential reach of 40 km, rather than the 20-km default, would assist in picking up isolated

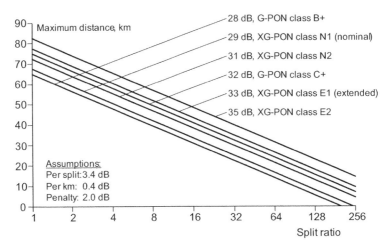

Figure 2.15 Reach–split ratio.

subscribers that would otherwise fall outside the footprint of any of the planned PONs in their territory.

Economic feasibility is a joint effort between subscribers' willingness to pay, between vendors' ability to reduce equipment costs, and between operators' ability to reduce operations costs. Longer reach and central office consolidation are a key aspect of the latter.

2.4.2 Demographics: Population Density

We have seen why operators want high split ratios and long reach, often both at the same time. Figure 2.15 illustrates the tradeoff between reach and split ratio for different ODN loss and power budgets. Chapter 3 goes into further detail.

Figure 2.15 is based on the assumptions shown.

- *Per split 3.4 dB*. Each time we split[*] the flow of light, we send half the power down each side, a loss of 3 dB. That's in the theoretically ideal case. In practice, splitters lose a bit more than 3 dB in each split, nor is the division of power perfectly uniform. Splitter uniformity—specifically the branch that happens to have the greatest attenuation—affects the network design loss.

- *Per kilometer 0.4 dB*. Optical fiber absorbs energy, differing amounts of energy at differing wavelengths, as shown in Figure 2.16. The highest loss is at the 1270-nm upstream wavelength of XG-PON. Newer fiber technology may absorb even less energy than shown, although not by much. Some operators also budget a pro-rated allowance for splices by distance.

---

[*] The device is symmetric, so we also lose 3 dB each time we combine flows from two ONUs into a single upstream output.

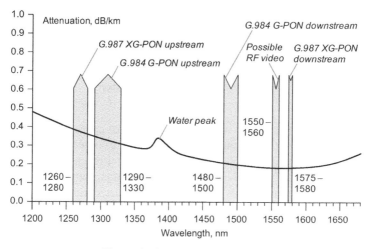

Figure 2.16   Loss vs. wavelength.

If the assumptions change, the results change. In particular, if the installed network is based on original G.652A fiber, the loss may be worse than shown. If an operator has reliable knowledge about the makeup of existing and new fiber plant yet to be installed, the planning margins necessary to determine the reach or split ratio of a proposed deployment can be reduced. Simply reducing the uncertainty could make the difference in determining whether a proposed project is feasible or not.

• *Penalty 2 dB.* Two factors contribute to this value: the recommendations specify an optical path penalty of 1 dB, which is essentially a catch-all for the multitude of little effects that prevent the link from operating at the level we would expect from just adding up the known impairments. A bit of extra margin is also good for unforeseeable contingencies. For example, if a wavelength splitter needs to be added into the PON to support future coexistence, it can use up an extra decibel all by itself.

As we see, the community has responded to the operators' need for decibels by standardizing budgets of as much as 7 dB more in XG-PON than the typical 28-dB budget of G.984 G-PON. Although it will be a challenge, and will certainly carry a price tag, component vendors believe that a 35-dB budget will be possible in XG-PON.

But Figure 2.15 demonstrates that, even with the best optics technology, an operator who wants a 128-way split at 60 km has a problem. Incremental gain from better fiber, increased optical launch power, and improved sensitivity of optical receivers are certainly worth having, but there are limits to this approach. Optical fiber is already very good. High launch power raises issues of eye safety and optical nonlinearity, while receiver sensitivity is ultimately limited by noise. So we need reach extenders.

Without a reach extender, the assumptions of Figure 2.15 tell us that a 32-dB class C+ G-PON loss budget is good for about 24 km with a 64-way split. If we can start

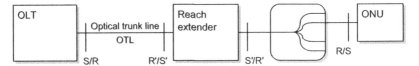

Figure 2.17 Reach extender notation.

the budget from a zero reference at a reach extender that cancels out the loss of the trunk fiber, that 24-km circle could be centered, for example, 36 km away from the OLT (to remain within the 60-km maximum logical reach).

A reach extender may be based on electrical regeneration, with conversion between optical and electrical domains at its interfaces, the so-called optical–electrical–optical (OEO) architecture. It may also be based on optical amplifiers (OA architecture), or it may be a hybrid of both. Chapter 3 describes these options in detail.

A word about notation. For reasons clearly unrelated to human factors, the optical interface at the OLT is designated S/R, and its counterpart optical interface at the ONU as R/S. The proxy interfaces at the RE are designated S'/R' and R'/S', respectively (Fig. 2.17). And yes, we have to refer to the figures too, because we can never remember which interface is where.

Internal Split Multiplier As always, there are variations. In Figure 2.18, we see an RE with an internal split, either electrical or optical, into two separate regenerators or amplifiers. Each serves its own physical splitter, thereby doubling the number of subscribers that can be connected to the PON. Clearly, this approach can be extended to more than two splitters.

In the downstream direction, split multiplication is straightforward. In the upstream direction—at most one splitter can contribute upstream signal at any given time—care must be taken to prevent noise from the silent regenerator from corrupting the merged signal upstream. If the upstream process is optical amplification, this could be done with a squelch circuit, a circuit that enables its upstream output only when it detects the presence of light at its input. An equivalent function would be needed in an electrical regenerator to ensure that upstream receiver noise did not generate random bits toward the OLT.

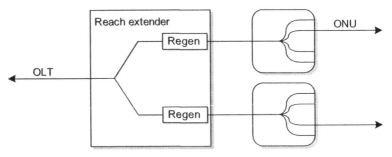

Figure 2.18 Split multiplication.

The split multiplier approach also enables one form of PON capacity upgrade. One or more of the splitters could in theory be relocated onto a new G-PON, effectively reducing the number of subscribers per PON and thereby increasing the capacity available for each. With or without a reach extender, this is known as PON stacking.

It remains to be seen whether PON stacking matches the economic realities of an upgrade, however. In practice, most subscribers will be adequately served by the existing PON, with a few randomly distributed subscribers who are willing to pay for more capacity. Simply halving the split ratio for everyone on the PON may not be the best approach, especially if ONU drops are attached to the splitter through splices rather than connectors, so that they are not easily rearranged.

Extending Multiple PONs   What other variations make sense? At the reach extender site, we have already paid for power, a protected environment, and management access. It is easier to amortize that cost if we can extend several PONs from the same RE equipment. This works especially well if we are serving a village or a small town that might need several PONs to provide complete coverage. So we expect to see composite REs containing 4–12 simple REs in parallel. Another economy, at least in terms of operational complexity, is that a single management agent suffices for a multi-PON RE.

WDM Trunking   But 4–12 parallel REs imply 4–12 trunk fibers, and we are, after all, in the WDM business. Why not a single trunk fiber, with 4–12 wavelengths (Figure 2.19)? Such modest numbers of wavelengths can be inexpensively served with CWDM.

Wavelength conversion is not impossible in the optical domain, but a wavelength-converting reach extender is probably better designed as an OEO equipment. Wavelength conversion is, of course, also necessary at the OLT. We save fibers, but at added equipment cost and complexity.

Protection   Single PONs, especially those serving residential areas, may or may not justify protection–the threshold that justifies protection typically lies in the range of 24–64 subscribers. Aside from the individual ONU, the most likely failures are

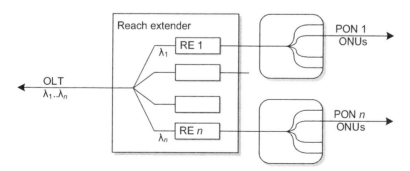

Figure 2.19 Wavelength-converting composite RE.

OLT blade or port failures, power outages, and fiber cuts due to construction, accidents, or severe weather. In many of these cases, the cost of protection would exceed the marginal benefit.

But by the time we have half a dozen PONs served by the same equipment, there is enough traffic, generating enough revenue, that protection may be justified. This would be protection between the OLT and the reach extender, which could utilize most of the same concepts described above in Section 2.3.5.

OLT Considerations   The G-PON recommendations allow for 20 dB of dynamic range between adjacent bursts. That is, if we take a given burst as a reference, the next burst could have a power level as much as 20 dB higher—or lower—subject, of course, to the maximum and minimum levels allowed for received power.

When an RE is based on OEO technology, the upstream signal on the OTL has zero dynamic range. An OA reach extender may also compress or eliminate the OTL upstream dynamic range.

Similar considerations apply to timing. Although an optical amplifier preserves the timing of the upstream signal, an OEO regenerator naturally locks onto the downstream clock—G-PON is synchronous, after all—and retimes upstream signals so that they all have the same bit phase.

Retiming does indeed introduce uncertainty in the ONUs' apparent round trip delay, but this turns out not to be a problem. The retiming uncertainty in an RE is bounded by one bit time, and as described in Chapter 4, delay need only be measured or monitored within eight bit times.[*]

What does this change in dynamic range and signal phase mean to the OLT? The presence of an RE could allow the OLT upstream receiver to be substantially less sophisticated and, therefore, less costly. But unless REs took over a large fraction of the market, developing custom RE optics for OLTs would simply not be worthwhile.

This is not to say that an RE implies no consequences at the OLT. An important difference between a simple PON and a reach-extended PON is that of optical maintenance. For example:

- ONU and OLT optics are often capable of measuring their received power levels; when the ODN includes a reach extender, these measurements represent RE performance, not that of the termination equipment. OLT and ONU values are measurements made by the RE itself. The OLT or element management system (EMS) software must know where to go to find performance information, and how to interpret it.
- The signature of a rogue ONU differs (Section 2.6.3), depending on whether it is behind an RE or connected directly to the OLT.
- Optical time domain reflectometry (OTDR) is a tool to measure faults in optical networks. But an OTDR at the central office cannot see the ODN beyond a reach extender.

---

[*]The RE must *not* perform byte alignment!

Transport Protocol on Trunk Fiber   Our OEO reach extender is now generating a continuous signal upstream on the PON, possibly phase aligned, with interburst gaps possibly filled with padding bits, and with zero dynamic range. We always had a continuous signal downstream on the PON. Why not map the PON signal into the client layer of some suitable transport protocol for backhaul? If we are prepared to do the ranging in the reach extender instead of the ONU, we could use OTN (optical transport network) for backhaul, or even plain old GbE or 10GE. These options could bring carrier-grade protection and operations, administration, and maintenance (OAM) to the optical trunk fiber.

If the subscriber side of an RE is the G-PON part and the network side is a client mapping into a transport protocol, what PON-specific functions remain for the OLT? We could just locate a small OLT at the RE site and be done with it. And, in fact, that is where we started. Do REs make sense? Maybe, but a healthy RE market had not developed at the time of writing. Low demand discourages vendors, and remote powering discourages operators.

2.5 ONU POWERING

In the legacy copper network, the twisted pair delivers power for POTS, and in most venues, it is very reliable. Even when commercial power fails, the telephone works. Indeed, the ability to make emergency calls is likely to be especially important at such times.

With the move toward mobile telephony, backup power is becoming an option, rather than a necessity. But still today, when POTS is provided from a central office, or even from a field site such as a CEV (Fig. 2.20), reliable power is part of the service offering.

There is no twisted copper pair in an SFU deployment, so power for an SFU becomes the subscriber's responsibility. The ONU is furnished with an AC power converter unit, at least a brick (Fig. 2.2), but often including a so-called uninterruptable power supply (UPS; Fig. 2.21), which not only converts AC power to DC (direct current) but also contains a backup battery to keep the ONU alive during power outages. Four to eight hours of battery reserve for lifeline POTS is a typical requirement.

Those who have to start their cars in cold climates know that batteries lose capacity in cold environments. By the same principle, UPS units for ONUs are rated for installation at least in a garage or carport in moderate climates, fully indoors where winters are severe.

How long can the power cable be, from the UPS to the ONU itself? Most residential UPS units are rated at 12 V, and cable length is limited by the wire size. The well-equipped telephone installer will have indoor drop wire and possibly Ethernet cat-5 or cat-6 cable, but will not stock heavy-gauge power wire.

The ordinary twisted pair used for telephone sets—probably what the operator will use for in-house ONU power wiring as well—is typically 24 AWG (American

Figure 2.20 Above-ground part of a controlled environment vault.

wire gauge)[*] (maybe 26 gauge), with a resistance of about 25.7 Ω per thousand feet at room temperature, 20°C. That is the resistance in one direction; in a twisted pair, the total resistance is twice that.

If the power supply delivers 12 V and the ONU consumes 6 W —just to use easy numbers that also happen to be in the right range—the current is 570 mA, and 50 ft of AWG 26 wire drops about 1.5 V.

Now suppose the ONU has an undervoltage shutdown at 10.5 V. If we start at 12.6 V, a typical storage battery output level, and lose 1.5 V in the wiring, we are left with 0.6 V of margin. When commercial power fails, the battery voltage gradually decreases. The lifetime of the ONU after a power failure is the time it takes for the battery voltage to sag by 0.6 V.

The minimum operating voltage of the ONU depends on its design, but we can see that power wiring should be kept as short as possible. If the wiring were to drop only 1 V, for example, instead of 1.5 V, we would have 1.1 V of discharge lifetime, almost twice as much.

The other question about a backup battery in the subscriber's home is that of battery maintenance. Backup batteries have a lifetime of only 2 years or so, 3 at most.

---

[*] The reader outside North America may find the following wire gauge conversion chart convenient:

AWG	mm^2
26	0.13
24	0.20
22	0.33

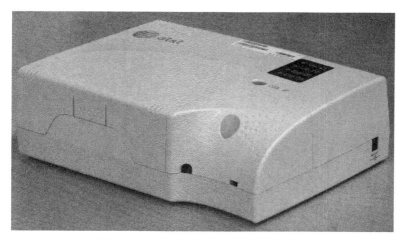

Figure 2.21   Uninterruptable power supply.

The general public has indisputably become more technologically savvy in recent years, but it is unduly optimistic to expect all subscribers to monitor and replace their own backup batteries. If the operator undertakes the task, it represents a continuing operational expense, not to mention periodically annoying subscribers by having to arrange a premises visit to replace the backup battery. This is all workable, but it is not without its problems.

MDU Powering   When the ONU serves multiple subscribers, as in an MDU, the operator must see to the powering arrangements.

- It may be possible for the operator to negotiate reverse powering, back down the subscribers' copper drops. This option is receiving increasing attention, as the simplest technical solution and perhaps not impossibly burdensome from a contractual point of view.

- In an apartment building, the MDU may reside in a utility equipment room, with AC power directly available. Space may even be available for backup batteries.

- AC power is rarely available at a curbside MDU. Even if AC power were available, backup battery technology is not currently economical for small loads—the MDU is limited by the feasible deployment environments to a maximum of perhaps 16 subscribers—in a full outside plant environment.

Therefore, power for curbside ONUs is often delivered over twisted copper pairs from a rectifier plant, which resides in a central office or in a CEV. Even if they are in cabinets (Fig. 2.22), these installations include environmental conditioning, backup battery plant, and often a Diesel emergency generator.

In the real world, it is often the case that MDU power feeds must use existing cables, which may have very few spare pairs. The ultimate irony is having to install new copper cables to deliver power to an optical network unit!

Figure 2.22 Centralized remote power plant.

Power

Most telecommunications equipment is powered from -48-V battery plant. The reason for the negative voltage is to avoid electrolytic corrosion of outside plant conductors whose insulation may have pinhole leaks into damp environments. The reason it is called a battery plant is that there are lead-acid storage batteries in most installations, providing nonstop continuous power. When commercial power is available, the rectifiers normally float the nominal -48 V battery supply at around -53 to -56 V. The display at the far left of Figure 2.23 reads $(-)53.4$ V.

In some places around the world, -60 V battery plant is used. This is becoming less common as time goes on.

International safety standards (60950: the number is the same whether it is UL, IEC, EN, CSA, AS/NZS, ...) define several classes of electrical circuit. Within the constraints of practical telecommunications wiring, the limit of voltage to ground is 200 V DC, and no conductor pair is permitted to deliver more than 100 VA (watts). Unbalanced current in the loop, which indicates a ground fault, is also strictly limited.

For the same reason as illustrated in our 12-V example above, it is desirable to use higher voltages, and therefore lower currents, with lower resistive voltage drops, when equipment is powered remotely. A common voltage for remote power is -130 V. In recent years, -190 V power sources have come into use, as a way to push the voltage as high as permitted, given manufacturing and lifetime tolerances. By going positive with respect to ground, $+190$-V circuits can also be used, creating the potential for 380 V at the source end.

Figure 2.23 Rectifier units.

Figure 2.24 illustrates a simple circuit in which we have a remote power source delivering voltage V_S, connected to a resistive load R_L. The power feed is a twisted pair, each of whose legs has resistance $R_f/2$ (so the total feeder resistance is R_f).

Figure 2.25 illustrates the well-known fact of impedance matching that maximum power is delivered to the load if $R_L = R_f$. At the maximum power point, half the power is dissipated in R_f, and half of the original source voltage V_S is dropped across R_f. It is only 50% efficient, but it is nevertheless the maximum power transfer point. Higher efficiency is possible if we ask for less power.

In a purely resistive circuit such as that of Figure 2.24, we can operate anywhere on the graph of Figure 2.25, extending to 0 and 100%. On either side of center, we get less power to the load, but there are no surprises. So what is this caption in the picture about a Not OK region?

Electronic circuits such as ONUs do not consume power as resistive loads. They are best thought of as constant power devices. If the voltage across their input

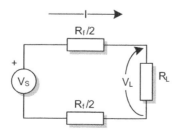

Figure 2.24 Simple Ohm's law circuit.

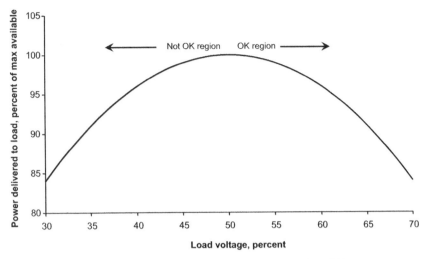

Figure 2.25 Power available at load, as a function of V_L.

terminals decreases, they demand additional current. For practical purposes, the ONU's power demand, $P_L = V_L I$, is constant.

That is fine if the ONU does not ask for too much power, say dropping 60% of the voltage across itself. But if the ONU were to try for 101% of the available power, it would not get it. Not only would it not get the power, but it would crash. That is the Not OK region.

We beg your indulgence for a whimsical analogy (Fig. 2.26). Suppose it is a pleasant winter's day in the Sierra Nevada, and we are hiking to the top of one of its

Figure 2.26 Sierra Nevada, California.

innumerable domes. On the south side, where we are, the sun has evaporated the ice and dried the surface; on the north side, we stipulate that the rock is covered with sheet ice.

Life is good as long as we stay on the south side of the peak. But, well, let us just say it is a bad idea to take even one small step across the crest to the north side. (The first to correctly identify this formation, wins a prize!)

Getting back to serious business: if the ONU ever steps onto the north side of the dome, its voltage will be a bit less than it wants. Being a constant power device, the ONU asks for more current to compensate for its voltage being too low. The drop across the feeder resistance increases, the voltage available to the ONU falls even further, the ONU tries for even more current, and gets even less voltage. Unless the ONU is designed to back off in a hurry, it crashes.

If the ONU's power demand is not a function of its software state, the ONU may just go down and stay down. It may also be that a crashed ONU consumes very little power, in which case the voltage recovers. As the ONU attempts to return to service, its power demand increases and it crashes again.

This is one of two bad things that can happen to a remotely powered ONU. We refer to this as the voltage-limited case because instability occurs when the drop across the feeder equals 50% of the source voltage. It can be shown that, for a given power demand P_L, the maximum survivable feeder resistance is

$$R_{f_{max}} = \frac{V_s^2}{4P_L} \qquad (2.1)$$

The other bad thing that can happen is referred to as power-limited instability. Safety requirements specify that no single source can deliver more than 100 VA (watts) into a load. Because of tolerances, a real-world supply may be limited by the manufacturer to an output of, for example, 95 W, as shown in Figure 2.27.

Suppose we lose 5 W in each leg of the feeder circuit. Then as long as the ONU requires power $P_L < 85$ W, we are okay. Suppose the feeder resistance increases a bit, for example, because it is carried in an aerial cable that gets hot in the summer sun. Suppose the feeder now wants to dissipate 6 W in each leg, while the ONU still demands 85 W. The ONU again tries to increase its feed current to get its full 85 W,

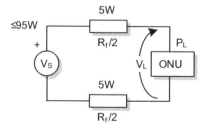

Figure 2.27   Power-limited circuit.

which reduces voltage V_L, and the ONU crashes. This even though the ONU starts off well above the $0.5V_S$ point of the voltage limited case.

In the power-limited case:

$$R_{f_{max}} = \frac{V_S^2}{P_S^2}(P_S - P_L)$$ (2.2)

where P_S is the maximum power available from the source.

As mentioned, the safety standards specify a maximum of 100 VA from any single source, typically margined down to about 95 W. The standards also specify 200 V as the maximum, also typically margined down to 190 V.

In practice, the manufacturer may only guarantee a voltage or a power level that is somewhat lower than our numbers. Because of the exponents in Eq. (2.2), this is expensive in terms of budget. A 1% decrease in the worst-case guaranteed output voltage costs about 2% in maximum resistance and therefore reach. The cost of a 1% reduction in the worst-case guaranteed output power depends on the ONU load P_L and may be either more or less onerous.

Plugging the nominal values ($V_S = 190$ V, $P_S = 95$ W, $P_L = 47.5$ W) into Eq. (2.1), we see that the total feeder resistance in the voltage-limited case must not exceed 190 Ω, or 95 Ω per leg.

For residential wiring, we previously hypothesized 26 AWG wire above, but for remote feeding of a curbside ONU, let us assume we have access to 22 AWG pairs. Their resistance at 20°C is 16.14 Ω/kft. The temperature coefficient of copper is about 0.393% per degree celsius, so at 60°C, our wire increases to 18.7 Ω/kft. If we can tolerate 95 Ω, we can feed an ONU at a distance of 5 kft, as long as it consumes no more than 47.5 W.

That would be 47.5 W absolute max, and with no margin. If a subscriber goes off-hook and increases the ONU's power demand to 47.6 W—well, it is like taking one tiny tiptoe step out onto that icy summit.

Peak and transient demand can be hard to predict. POTS is the worst because ringing a telephone consumes a substantial amount of power. The amount depends on the subscriber's equipage, and subscribers are pretty much free to connect as many devices as they like, each device different from the others. If the subscriber picks up the phone at a peak of ringing voltage, the resulting approximate short circuit causes a spike in power demand until the ring trip circuit operates. Further, a curbside ONU must be able to ring several phones at once, each of them an unpredictable load, with some number[*] of additional lines off-hook.

It is usually not feasible to provide a power circuit that can survive the absolute worst-case transient load conceivable. Short spikes, such as ring trip transients, can be absorbed by a capacitor in the ONU. But capacitors consume space, add cost, and have finite lifetimes.

---

[*] Telcordia GR-909 is a good requirements reference. It contains tables that specify n lines ringing with m lines simultaneously off-hook for ONUs of aggregate traffic capacity t.

On the proactive side, the ONU may be designed to deny power to loads that it cannot support, for example, by reducing the ringing voltage or staggering or abbreviating the ringing phase.

That is the bad news on the ONU side. On the feeder side, transients may be introduced by coupling from parallel power lines or from thunderstorms; if a transient causes the common-mode voltage of a feeder to exceed 200 V for more than a very brief period, the power source shuts down—safety requires it. The power source will restart, but in the meantime, the ONU has crashed. Transients can also generate unbalanced current in the loop, simulating ground faults. The power source shuts down, the ONU crashes.

Well, this is a pretty bleak picture, particularly if our ONU wants more than 47 W. What can we do?

Figure 2.28 shows two approaches, approaches that may be applied jointly or independently.

- If we provide two or more twisted pairs in parallel—such as pairs a and b—we reduce the feeder resistance proportionately. We can either deliver more power to the ONU or place the ONU farther from the power source. It goes without saying that in the real world, these pairs will be in the same cable, so that their resistance is very nearly the same. Downsides: existing cables may not have spare pairs, so it is not necessarily possible to overwhelm the problem by adding pairs. And if one of the pairs were to fail, it could be difficult to diagnose the fault.

- To deliver more than 100 W (in reality 95 W, less wiring loss) to an ONU, it is permissible to provide two or more sources—sources 1 and 2 in Figure 2.28— subject to the strict safety requirement that they be fully isolated from each other. The weak link controls the behavior, so it is important that each circuit deliver 50% of the total power. The downside of this approach is the cost of an additional power source, including its housing, cooling, and such.

- If we provide two pairs (a and b) for source 1, we clearly need to provide two pairs (c and d) for source 2; else we would be creating a weak link as the load attempted to share the power equally. To equalize feeder resistance, it also makes sense that the sources be colocated and that all feeder pairs be in the same cable.

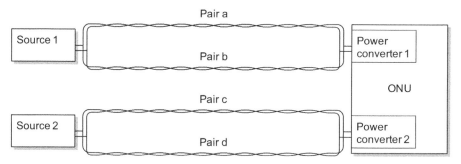

Figure 2.28   Multiple sources, multiple feeders.

In terms of transient immunity, buried cable is better than aerial—the 60°C thermal derating could also be relaxed—and power at 130 V has more margin against the 200 V overvoltage safety limit than does 190 V power. Because of the circumstances of a deployment, neither of these options may be feasible.

If it is at all possible, it is far easier to locate an MDU in the equipment room of a building, an apartment building, for example, a site with limited environmental extremes, with nearby AC and with space for backup batteries. The service drops are then indoor wiring rather than outside plant, so the requirements for lightning protection are also relaxed, saving even more in the total cost.

Even better, of course, is an SFU, located right there next to its power source in the subscriber's home.

Having said all this, it must be stated that powering difficulties are a major operational impediment to the expeditious roll-out of G-PON.

2.6 TECHNOLOGY REQUIREMENTS

2.6.1 VLANs

All Ethernet traffic—all traffic!—through a G-PON access network is carried in VLANs, either a dedicated VLAN per service per subscriber (1:1 model) or a dedicated VLAN per service per group of subscribers, the N:1 model. Layer 2 business customers may be served through so-called transparent LAN service (TLS), in which the operator tunnels the subscriber traffic through a service provider VLAN, with or without recognition of subscriber flow priority.

In terms of traffic flow, the primary function of a G-PON ONU is to classify traffic and forward it to the correct queue for output scheduling. This is true in both directions, although the downstream direction, and the queues serving the downstream subscriber interface, receive less attention in the recommendations.

Somewhere in the access network, Ethernet frame priorities and VLAN tags, and possibly IP differentiated service code point (DSCP) bits need to be added, stripped, translated, or interpreted. The ONU needs to be able to classify traffic based on these fields, as well as others: other fields such as ONU subscriber port or Ethertype.

Chapter 6 goes into considerable detail on the G-PON management model for VLAN classification and tag management.

2.6.2 Quality-of-Service Control

Depending on application, the upstream capacity of a PON is expected to be more than adequate for a few years, until the demand of applications such as peer-to-peer communications overtakes it. Nevertheless, upstream capacity is a finite resource, to be conserved and used wisely. Especially with business services, the day may come when the upstream PON becomes a congestion point.

An ONU is commonly required to offer at least four classes of service, and desired to offer six, with the ability to internally schedule and prioritize traffic among classes. Service priority may be based on VLAN tags, DSCP bits, or other criteria.

Among the several ONUs on a PON, the OLT assigns bandwidth in real time according to the load offered by each class of traffic on each ONU. This is referred to as dynamic bandwidth assignment (DBA). The OLT's assessment of offered traffic may be based on observation of idle upstream frames (called traffic monitoring DBA), upon explicit queue backlog reports provided upon request by each ONU (status reporting DBA), or a combination of both.

We discuss the model and management of quality of service (QoS) in detail in Section 6.3.

2.6.3 Security

There are essentially three areas of concern in G-PON: denial of service attacks, violation of privacy, and theft of service. The latter two are different aspects of the same technical issues.

2.6.3.1 Denial of Service

One of the disadvantages of the PON architecture is that anyone with access to the fiber can transmit an optical signal upstream, a signal that interferes with the legitimate signals and effectively brings down the PON. When this occurs due to an ONU defect, it is known as a rogue ONU. Although fanciful scenarios can be constructed involving malicious reprogramming of otherwise functional ONUs, it is hard to understand why an attacker would go beyond the cost and complexity of a simple laser and optical connector.

That is at the optical level. Various other attacks are also possible, for example, injecting traffic that descrambles to long sequences of zeros or ones, hoping to desynchronize the far end receiver. Higher layer threats such as flooding the network with illegitimate traffic are also recognized. They can be dealt with through management capabilities defined in OMCI, for example, to filter or limit the rate of various kinds of traffic.

2.6.3.2 Privacy and Theft of Service

Downstream traffic on a PON is accessible to anyone with an optical detector, even non-ONU devices of suitable sophistication. With this perspective, the original PON protocol specifications called for encryption of unicast downstream traffic. The characteristics of optical splitters and couplers are such that the upstream direction of a PON was deemed to be intrinsically secure. Splitters with redundant upstream ports would be located in safe places, and the upstream signal available at other subscriber drop fibers would be too weak to detect in practical terms. Upstream encryption was therefore considered unnecessary.

G.983 B-PON specified a weak form of downstream encryption called churning; G.984 G-PON specifies 128-bit AES (advanced encryption standard) encryption. The encryption key is generated by the ONU and communicated in the clear—if the upstream direction is physically secure, then we treat it as such—to the OLT.

Multicast traffic is not required to be encrypted at the PON level, but is expected to be secured at a higher layer, with keys distributed by a middleware server directly to set-top boxes. In this use case, the PON need not add security because the traffic is deemed to be secure already.

When G.987 XG-PON was developed, some operators regarded this security model as inadequate. They feared that physical access to the upstream fiber flow might indeed be possible. There were also concerns about counterfeit ONUs and even counterfeit OLTs (!). Accordingly, the G.987 XG-PON recommendations add upstream encryption to the options, as well as encryption of downstream multicast traffic. Following the principle of layered security, keys themselves are encrypted during key exchange. G.987 also allows for the possibility of mutual authentication of ONU and OLT, using capabilities across a range from a simple password (registration identifier) to IEEE 802.1X, with its open-ended ability to support virtually any authentication protocol.

We discuss G-PON security further in Section 4.6.

2.7 MANAGEMENT REQUIREMENTS

A G-PON ONU can be viewed in two ways, each of which is fully appropriate within its own scope.

ONU as CPE   The ONU may be regarded as customer premises equipment, possibly to be purchased by the subscriber at the local electronics store. Such an ONU is owned, installed, activated, and maintained by the subscriber, who selects service packages from potentially many competing providers. The ONU is integrated with typical RG capabilities: NAT, local DHCP and DNS (domain name service) servers, bridged LAN-side ports, and it may also include any number of additional features such as analog telephony adaptors (ATAs) for VoIP, 802.11 wireless connectivity, USB ports, print and storage servers, ... the list goes on.

ONU as Network Equipment   The ONU may be regarded as telephone network equipment, to be fully managed and controlled as an extension of the OLT, all the way to the UNI. This model is the same as that used for service delivery of POTS and ADSL from the central office. Such a model is also natural for MDUs, especially MDUs that provide POTS service along with DSL. The model also fits ONUs delivering business services such as DS1/E1, and ONUs providing mobile backhaul. For uniformity in management, provisioning, and maintenance, some operators choose to consider all ONUs in this way. This model anticipates that the subscriber will have a separate RG.

Even when the ONU is regarded as CPE, some functions must be controlled and managed from the telephone network equipment perspective, functions such as initialization, at least part of the software upgrade process, PON maintenance and diagnostics, and coordination of traffic mapping between the ONU function and the OLT, especially QoS and VLAN tags.

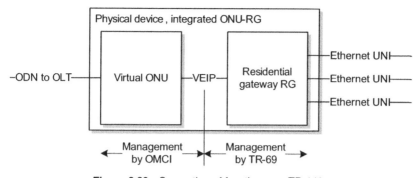

Figure 2.29   Separation of functions per TR-142.

To facilitate this partition, ITU-T and Broadband Forum have cooperatively developed a model that separates the RG (CPE) function from the ONU function, through what is designated a virtual Ethernet interface point (VEIP). Observe that this model best fits an SFU with an integrated RG (Section 2.2.1). The architecture (Fig. 2.29) appears in BBF TR-142, while the details are defined in ITU-T G.988.

This model separates the RG function from the ONU function, with a logical, rather than physical, Ethernet interface between the functions. Its merit is that operators can use much of the same provisioning and management infrastructure for an RG, be it integrated into an ONU or provided as a separate stand-alone device. Separating PON management from CPE management also facilitates service unbundling, in which the subscriber's services may not be delivered by the network operator.

A second application of the VEIP is the case where a G-PON ONU termination may be used as the integrated feeder of a remote DSLAM. Existing DSLAMs are often managed as stand-alone network elements via SNMP (simple network management protocol)[*]; the VEIP allows this management model to be retained, while keeping the PON-specific uplink details within the realm of PON management.

In both cases, a further advantage of the VEIP model is that it helps isolate CPE management or DSLAM management from ONU and PON management. Operators lose a bit of efficiency through coordination that, strictly speaking, would not be necessary, but they gain by having uniform provisioning processes across a range of platforms.

This conclusion is based on the ONU regarded as CPE, deployed in parallel with conventional DSL in the operator's network, and intentionally choosing to minimize the management of the ONU through OMCI. In MDUs, CBUs, SBUs—ONUs that are necessarily network equipment—it is not straightforward to apply the TR-142 model. The operator may yet need two different provisioning models.

OMCI   G-PON is managed via the ONU management and control interface, OMCI. OMCI is defined in the sunsetted ITU-T recommendation G.984.4, and the current

---

[*] IETF RFCs 2578, 3584.

recommendation G.988. Chapter 5 of this book discusses OMCI in some detail. This book has to draw a boundary somewhere, so we choose not to go into the details of TR-69 or SNMP management, which are, in any event, not specific to G-PON.

2.8 MAINTENANCE

2.8.1 Connectivity Fault Management

Although at the time of writing, it was more a wish than a fact, G-PON ONUs are supposed to support Ethernet connectivity fault management (CFM), as specified in IEEE 802.1ag. CFM has three basic aspects:

- Periodic connectivity check messages (CCMs) confirm connectivity of a VLAN (circuit) to the correct provisioned endpoint. Alternatively, if an unexpected CCM appears at a given endpoint, the endpoint can declare a misconfiguration.
- On-demand loopback allows the two-way integrity of a path to be confirmed. Loopback does not disrupt normal traffic, unlike conventional TDM loopback, or for that matter, Ethernet loopback as defined in IEEE 802.3.
- On-demand path trace allows an endpoint to discover its route through the layer 2 network.

Extensions in ITU-T Y.1731 define additional functions such as AIS (alarm incoming signal), a signal that prevents unnecessary alarm propagation. A fault detected at a given network element (NE) causes an alarm at that network element. The NE then generates AIS to the affected downstream clients so they know that the problem has already been detected and reported, and they need not declare their own alarms.

Ethernet CFM is structured in nested layers, where a layer encompasses a pair of endpoints that typically corresponds to a domain of responsibility. Thus, one domain could extend from the service provider's interface (the ONU UNI) into the subscriber's home network. Other domains could extend back into the service provider's network, possibly at several levels in the case of a carrier's carrier. Section 6.1 goes further into Ethernet CFM in G-PON.

2.8.2 Troubleshooting

If an ONU knows that it is about to drop off the PON because of some local action, for example, power failure or simply because the subscriber has switched it off, it can signal dying gasp, DG. This advises the OLT that there is no fault in the fiber plant and assists in fault isolation, should there be a subsequent trouble report. An ONU without a backup battery has limited ability to signal dying gasp after a power failure, so it may not be able to do so. It is also true that power could recover after the ONU signals DG, so the OLT must be prepared for the ONU to stop signaling DG and remain active.

Faults in a direct fiber run are commonly diagnosed by OTDR equipment, which may be used in-service as long as it occupies a separate wavelength, with filters at the necessary points to separate the wavelengths. Wavelengths around 1625 nm are commonly used for this purpose. From the OLT side, it is difficult to see or interpret an OTDR reflection from a fault on the far side of a splitter. Cost-effective in-service centralized and preferably automated fault diagnosis is of considerable interest and remains an opportunity for vendor differentiation.

3

OPTICAL LAYER

In this chapter:

- *Light propagation phenomena*
- *Optical components in PONs*
  - *Fiber*
  - *Splices and connectors*
  - *Passive power splitters*
  - *WDM devices*
- *Optical budgets*
- *Coexistence of G-PON and XG-PON*
- *Transceivers*
  - *Lasers and drivers*
  - *Photodetectors, optical amplifiers*
  - *Receiver sensitivity, threshold determination, clock recovery*
- *Reach extension*
  - *SOAs, EDFAs*
  - *Optical–electrical–optical repeaters*
  - *Optical amplifier-based repeaters*

Gigabit-capable Passive Optical Networks, First Edition. Dave Hood and Elmar Trojer.
© 2012 John Wiley & Sons, Inc. Published 2012 by John Wiley & Sons, Inc.

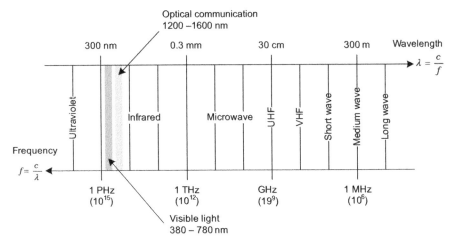

Figure 3.1 Electromagnetic spectrum.

3.1 INTRODUCTION

Before we look into the passive and active components that comprise the physical layer of a PON, we briefly review the fundamental behavior of light: the speed of light in different materials, reflection, refraction, and polarization.

Figure 3.1 illustrates the electromagnetic spectrum, in particular, the range of interest for G-PON and XG-PON systems.

3.1.1 Light Propagation

When the wavelength of light is much smaller than the dimensions of the surrounding geometry, light propagates in straight lines (rays) as it travels through free space or transparent materials. Large-scale effects such as reflection and refraction are easily described in terms of rays.

The speed of light in free space (vacuum), designated as c, is very close to 300,000 km/s (in fact, 299,792,458 m/s) and is a physical constant. The free-space wavelength λ of a wave of frequency f is the distance traveled by the wave during time $T = 1/f$ seconds:

$$\lambda = cT = \frac{c}{f} \tag{3.1}$$

In dielectric or nonconducting material, light travels with reduced speed v. Its relative speed is expressed by the refractive index n of the material, always ≥ 1:

$$n = \frac{c}{v} \tag{3.2}$$

The refractive index of air is essentially 1.0, which means that light travels at the same speed that it would in free space. In water, with an index around 1.33, light travels at about 75% of its velocity in vacuum. The silica glass used in telecommunication fiber has an index around 1.5, so light travels about one third slower in fiber than in free space. When comparing two materials, the material with the larger refractive index, in which light travels slower, is said to be optically more dense.

For a wave of frequency f, the consequence of slower propagation in medium m is that the wavelength inside the medium λ_m is shorter than the wavelength in vacuum:

$$\lambda_m = \frac{\lambda}{n} \tag{3.3}$$

The so-called phase velocity v_p is the rate at which a sine wave (with zero spectral width) travels in a material with refractive index n:

$$v_p = \frac{c}{n} \tag{3.4}$$

For such a wave to travel L meters, the phase propagation delay τ_p is given by

$$\tau_p = \frac{L}{v_p} = L\frac{n}{c} \tag{3.5}$$

Because it repeats identically forever, a pure sine wave conveys no information. In communications systems, carriers are modulated, a process that broadens the bandwidth.[*] The propagation rate of information-carrying composite signals is defined by the group velocity v_g, which is the rate at which changes in the amplitude of the wave—its envelope—are transported in the medium. A group refractive index n_g is defined by analogy to Eq. (3.4).

In physical media, the refractive index varies with wavelength. The variation of index over wavelength, together with the nonzero spectral width of the signal, leads to differential delay between the various spectral components of the signal, called chromatic dispersion, which effectively broadens or smears out the signal as it propagates.

To be precise, the group refractive index n_g is derived from the refractive index by

$$n_g = n - \lambda\frac{dn}{d\lambda} \tag{3.6}$$

The group delay is calculated from the group refractive index by analogy with Eq. (3.5):

$$\tau_g = \frac{L}{v_g} = L\frac{n_g}{c} \tag{3.7}$$

---

[*] The Fourier transform of a modulated carrier signal shows side lobes around the carrier frequency, which can be interpreted as individual spectral components, each of which propagates independently.

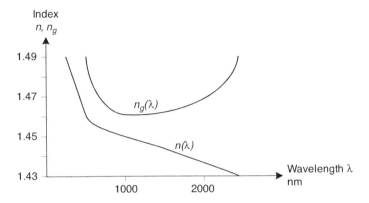

Figure 3.2   Silica glass refractive index and group refractive index.

Figure 3.2 illustrates the experimentally observed refractive index n and group refractive index n_g of the silica glass used in optical fiber.

At 1310 nm, a typical fiber has a group index of 1.4677, resulting in 97.85 μs of delay in a 20-km span. At 1550 nm, the group index increases to 1.4682, and the delay increases to 97.88 μs—a difference of 30 ns. Such propagation delays and delay differences are large enough that the G-PON protocols must take them into account—see Section 4.3.

3.1.2   Reflection, Refraction, and Light Guiding

When light propagating in a material medium encounters a change in refractive index, part of its energy is reflected. The remainder is refracted and continues forward, but in a modified direction. Figure 3.3a shows a light ray with power P_I impinging on an interface between dielectric materials with different refractive indices n_1 and n_2. We see that it is split into two rays, each taking a different direction. A fraction P_R of the light is reflected at angle φ_1, which is equal to the angle of incidence, while the remaining part P_T is refracted, entering the second material at angle φ_2.

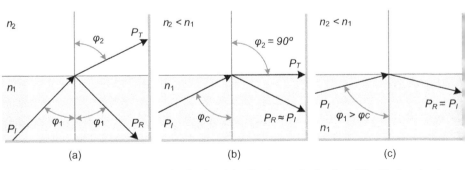

Figure 3.3   Principle of total internal reflection: (a) reflection and refraction, (b) critical angle of reflection, and (c) total reflection.

Snell's law describes the relation between refractive indices and angles:

$$\frac{n_2}{n_1} = \frac{\sin(\varphi_1)}{\sin(\varphi_2)} \tag{3.8}$$

Suppose that the incident ray comes from an optically dense material with higher refractive index n_1 and enters an optically thinner material with lower index $n_2 < n_1$. From Snell's law, $\varphi_2 > \varphi_1$, so the refracted ray enters the second material at an angle greater than that of the incident wave—the refracted ray bends into the thinner material.

As φ_1 increases, a point is reached at which the transmitted ray runs parallel to the interface; that is, $\varphi_2 = \pi/2$ (Fig. 3.3b). The corresponding angle φ_1 is called the critical angle of incidence φ_C. Inserting $\varphi_2 = \pi/2$ into Eq. (3.8) and solving for φ_C results in

$$\varphi_C = \sin^{-1}\left(\frac{n_2}{n_1}\right) \tag{3.9}$$

Figure 3.3c illustrates the fact that an incident ray satisfying $\varphi_1 > \varphi_C$ is totally reflected within the optically dense material, with no refracted light in the thinner material.

The principle of total reflection provides the basis for a light guide. The simplest form of light guide is a glass cylinder in air. Using $n_1 = 1.48$ for glass and $n_2 = 1$ for air, any light ray with incident angle $\varphi_C > 42°$ is totally reflected at the glass–air interface and remains within the glass.

An optical communication fiber is essentially a concentric pair of glass cylinders. The interior cylinder (the core) is encapsulated in an exterior cylinder (the cladding), which has a slightly lower refractive index.[*] The index is controlled by adding impurities such as germanium or boron to the core material during the manufacturing process.

Totally reflected optical power is equal to the incident power, but there is a phase shift at the reflection point, which depends on the value of φ_1. Destructive self-interference limits the number of so-called modes that can propagate in a fiber. If the core diameter of the fiber is sufficiently small, only a single ray parallel to the fiber axis can propagate. Such a so-called single-mode fiber provides the highest transmission capacity available.

3.1.3 Light Coupling

Another important factor is the efficiency of coupling between the light source and the fiber. Clearly, as much light as possible should be coupled into the fiber. Snell's law offers insight into what is possible. Figure 3.4 shows light coupled from a light source via a small air gap (refractive index $n = 1$) into a step-index fiber. Rays with incident angle smaller than α are totally reflected within the fiber; rays entering the fiber at larger angles are dissipated in the cladding. The angle α is called the acceptance angle of the fiber.

---

[*]The index difference ranges from 0.1 to 1% for single-mode fibers. Fiber used in G-PON has a typical index difference of 0.35%.

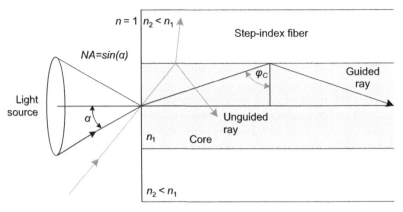

Figure 3.4 Definition of numerical aperture.

The numerical aperture (NA) is a measure of the light-gathering capability of a fiber. It is defined to be the sine of the acceptance angle α. Snell's law yields

$$\text{NA} = \sin(\alpha) = \sqrt{n_1^2 - n_2^2} \tag{3.10}$$

The NA of fiber used in G-PON is typically 0.13 or 0.14, corresponding to an acceptance angle of about 8°.

3.2 OPTICAL FIBER

3.2.1 Attenuation

Together with signal distortion mechanisms, the attenuation of light as it propagates through a fiber is one of the main factors that determines the maximum transmission distance achievable in an optical communications network. The degree of attenuation depends on the fiber material and the wavelength of the light.

The basic attenuation mechanisms in a fiber are energy absorption due to impurities in the quartz silica glass, and scattering due to microscopic variations in material density. These attenuation mechanisms are determined by the manufacturing process.

The basic attenuation is increased by macroscopic bends that have radii larger than the fiber. These are introduced during cable installation and use. As the radius of curvature decreases, the bending loss increases exponentially[*] until a certain critical bend radius threshold is reached, beyond which the loss suddenly becomes very large. Minimum bend radius restrictions are sometimes hard to obey in the real world, and special bend-insensitive fiber has been developed.

---

[*] Linearly in decibels.

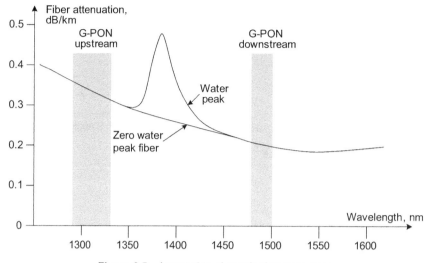

Figure 3.5   Attenuation of standard telecom fiber.

Figure 3.5 illustrates loss as a function of wavelength. The high loss in the 1383-nm water peak area results from OH$^-$ ions introduced during manufacture or that form within the fiber through humidity in the environment. Advances in fiber manufacturing and protection technology have greatly reduced the water peak; some fibers advertise zero water peak. In theory, this frees the full spectrum for use, but because a lot of existing fiber deployment has poor or unknown water peak characteristics, the water peak band is usually avoided.

Maximum fiber losses are specified in various ITU-T standards, as shown in Table 3.1. G.652 is the standard fiber used for G-PON. Commercially available fiber exceeds these standards.

TABLE 3.1   Maximum Fiber Loss in Standard Fiber Cables

Fiber Type	Maximum Loss (dB/km)				
	1270 nm	1310 nm	1490 nm	1550 nm	1577 nm
G.652 A	—	0.5	—	0.4	—
Older fiber					
G.652 B	—	0.4	—	0.35	0.4
Common fiber					
G.652 C, D	0.47	0.4	—	0.3	—
Reduced water peak					
G.657 A	—	0.4	—	0.3	—
Bend-loss insensitive					
G.657 B	—	0.5	—	0.3	0.4
Bend-loss insensitive					
Commercial fiber	0.35	0.33	0.18	0.2	0.23

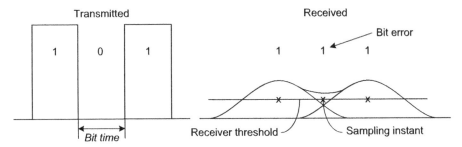

Figure 3.6 Bit errors due to dispersion.

3.2.2 Dispersion

When they traverse a fiber, light pulses are broadened due to dispersion. Figure 3.6 shows the effect of dispersion on received signal quality. Pulses spread out as they travel along the fiber. Pulses that were originally distinct interfere with each other at the receiver. This intersymbol interference (ISI) makes it harder for the receiver to recover the original data. If the spread is an appreciable fraction of the bit duration, error-free detection is impossible. For an ideal receiver that introduces no additional noise, and a nonreturn to zero (NRZ) line code, ISI less than 40% of the bit duration can be overcome. A link in which this condition is not satisfied is called dispersion limited.

Depending on fiber type, we find three different dispersion types contributing to signal distortion, namely modal dispersion, polarization dispersion, and chromatic dispersion.

- Different modes propagating through fiber at different angles arrive at the far end with different delays according to their different path lengths. This is called modal dispersion; it dominates multimode fibers. Figure 3.7 shows that the maximum temporal spread occurs between the ray parallel to the fiber axis

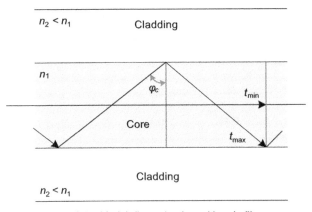

Figure 3.7 Modal dispersion in multimode fibers.

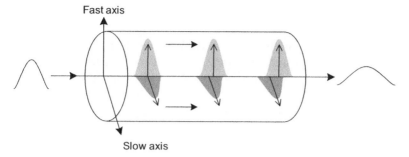

Figure 3.8 Polarization dispersion.

(fastest) and the ray repeatedly reflected at an angle just above the critical angle of incidence (slowest). Modal dispersion is not present in single-mode fibers; G-PON is specified to use single-mode fiber, so we need not consider modal dispersion in further detail.

- Horizontal and vertical polarizations also encounter different delays. The core of a real fiber deviates from perfect cylindrical geometry due to production inaccuracies, bending, mechanical pressure, or thermal deformation. This results in radial nonhomogeneities in the refractive index, and different polarizations of the optical signal travel at different velocities, as illustrated in Figure 3.8. The resulting pulse broadening is described by the term polarization dispersion. We need not go into this effect in detail. Twenty kilometers of today's standard G-PON fiber introduces a polarization dispersion of 0.9 ps, negligible even at 10 Gb/s.

- Because the refractive index of the fiber material varies with wavelength, different spectral components of the modulated light experience unequal transmission delays. At the far end of the fiber, this results in a broadened pulse—a phenomenon called material dispersion. Also, energy in the cladding[*] travels faster than in the core, leading to so-called waveguide dispersion. The two phenomena are combined under the term chromatic dispersion. This one matters; it limits the transmission capacity of single-mode fibers.

3.2.2.1 Chromatic Dispersion

The light produced by a laser has nonzero spectral width. When modulated by a data signal, the spectrum is further broadened, depending on data rate and modulation technique. G-PON and XG-PON employ simple on–off modulation, which produces a comparatively broad spectrum.

Because refractive index depends on wavelength, each component wavelength of this spectrum travels with a different speed through the fiber. This phenomenon is called material dispersion.

---

[*] Yes, energy is carried in the cladding—we are no longer talking about simple ray theory. The cladding must, therefore, also have good geometric and loss characteristics.

A second factor is waveguide dispersion, caused by the higher velocity of energy traveling in the cladding, compared to that traveling in the core. Waveguide dispersion is determined by the fiber material and the fiber geometry.

The total chromatic dispersion is the sum of material dispersion and waveguide dispersion, and is commonly called group velocity dispersion (GVD). Pulse broadening is directly proportional to the length of the fiber, the spectral width of the source and the GVD coefficient. The GVD coefficient is a measure of the time spread per fiber length and bandwidth; its units are picoseconds per nanometer per kilometer. That is, two signals whose wavelengths differ by 1 nm will be skewed by GVD picoseconds after traveling through 1 km of the specified fiber.

The material dispersion coefficient is positive in the range of interest. The waveguide dispersion coefficient is negative. Because it depends partly on fiber geometry, waveguide dispersion can be tuned to create fibers with different overall chromatic dispersion. Figure 3.9 shows chromatic dispersion curves over wavelength for two fiber types. The standard G.652 fiber used in G-PON has minimum dispersion around 1310 nm. Dispersion-shifted fiber shows a zero crossing near 1550 nm.

For a maximum GVD coefficient of 13.7 ps/nm.km in the 1490-nm area where the G.984 G-PON downstream signal is located, an OLT laser linewidth of 0.3 nm, and a transmission distance of 20 km, pulse broadening amounts to 82 ps. This is approximately 20% of the 400-ps pulse width of a 2.5-Gb/s signal, showing that the G.984 G-PON downstream link under such conditions is not limited by dispersion.

At 40 km, the effect is 40%, right at the limit. For a long-reach system covering 60 km, the effect of GVD is 60% and clearly limiting. Dispersion compensation is

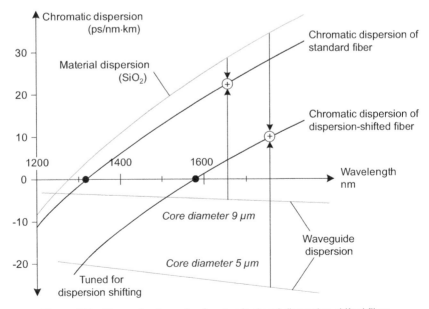

Figure 3.9 Chromatic dispersion for standard and dispersion-shifted fiber.

necessary, through equalization at the receiver, by means of OEO regeneration along the fiber path or strong error-correcting codes at the data layer.

The downstream bit duration of G.987 XG-PON is only 100 ps and the GVD coefficient increases to 18.8 ps/nm.km at 1577 nm. Given a high-grade laser with 0.1 nm spectral width, GVD amounts to 38 ps on a standard G.652 fiber at 20 km. The ratio is 40%, again, right at the limit. For 40 or 60 km of reach, we need an even better laser.

3.2.3   Fiber Nonlinearities

Besides attenuation and dispersion, fiber nonlinearities constrain the transport of modulated light over a fiber. Stimulated Raman scattering (SRS) shifts power between adjacent wavelength channels on the fiber, reducing the power at the higher energy wavelength and adding noise to the lower energy wavelength. Stimulated Brillouin scattering (SBS) reflects a fraction of the power back toward the transmitter.

It should be stated at the outset that these effects are not significant at the power levels of G-PON. Or to consider it in another light, the G-PON specifications limit the power levels to avoid provoking these nonlinearities.

3.2.3.1   *Stimulated Raman Scattering*

Photons propagating through a fiber interact with silica molecules by exchanging energy, as shown in Figure 3.10. A photon can make the molecular structure vibrate, generating an acoustic wave. The impinging photon of wavelength λ_P and energy level $E_P = hc/\lambda_P$ surrenders energy, resulting in a so-called stroke photon of longer wavelength λ_S and lower energy $E_S = hc/\lambda_S$. The newly generated wave at λ_S is effectively pumped by the original wave as both move in the forward direction along the fiber.

If another optical signal is present at the longer wavelength λ_S, the stroke wave interferes with this optical signal. This crosstalk can affect system performance considerably. SRS also attenuates the original signal, but only on the order of parts per million.

The Raman effect is not always bad. A continuous (no modulation to cause crosstalk) pump signal at λ_P can be used to amplify an optical signal at λ_S. This process is called Raman amplification.

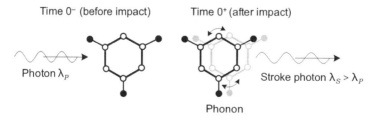

Figure 3.10   SRS process.

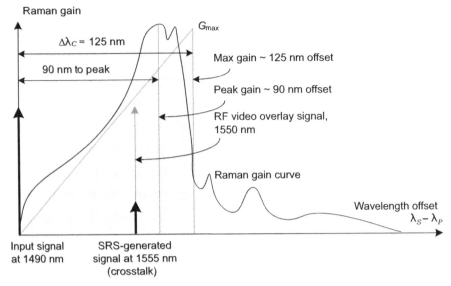

Figure 3.11 SRS gain curve.

SRS occurs over the very wide range of 125 nm (16 THz) across the whole fiber spectrum and in both directions of transmission. Any two channels with wavelength spacing within the Raman cutoff wavelength $\Delta\lambda_C$ are coupled by the Raman effect. As an approximation, the coupled power increases linearly with channel spacing until cutoff. Coupled power also increases linearly with incident power and the length of the fiber.

Figure 3.11 shows an example of a 1490-nm G.984 G-PON downstream signal interfering with a 1555-nm RF video overlay signal. The 1490-nm signal would also pump a G.987 XG-PON downstream signal at 1577 nm. However, the optical powers of G-PON and XG-PON are intentionally specified to be far below the threshold needed to stimulate Raman scattering (+20 dBm), so in practice, the effect is negligible.

3.2.3.2 Stimulated Brillouin Scattering

Photons propagating in a fiber generate phonons (acoustic waves), which cause variations in the density of the material and therefore in the refractive index, scattering light back toward the transmitter. A subsequent photon with wavelength λ_1 passing a refractive index variation strikes a phonon and produces a photon of longer wavelength $\lambda_2 > \lambda_1$ in the backward direction. The backscattered light effectively experiences a Doppler shift to a lower frequency (lower energy). The backscattered wave is amplified by the forward-propagating signal, which leads to the depletion of the forward signal. The interaction—in contrast to SRS—occurs in the forward direction and over a very narrow Brillouin linewidth of approximately 20 MHz, resulting in a very small wavelength shift confined to a single wavelength channel.[*] Figure 3.12 illustrates the SBS process.

---

[*] In a standard single-mode fiber, a signal at 1550 nm is shifted down by 11 GHz, or 0.09 nm.

Time 0⁻ (before impact)   Time 0⁺ (after impact)

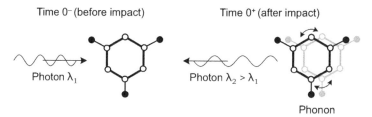

Photon λ_1 Photon $\lambda_2 > \lambda_1$

Phonon

Figure 3.12   SBS process.

The effect of SBS is negligible for low input powers but increases rapidly once a material-specific SBS threshold is crossed. See Figure 3.13. For input powers above the threshold, SBS is excessive and the power received at the far end cannot be increased by increasing the transmitted power.

The SBS threshold for standard fiber is +17 dBm across the telecommunications spectrum. Below the threshold, the backscattered power is down 77 dB from the input power at 1310 nm; at 1550 nm, it is 82 dB down.

3.3   CONNECTORS AND SPLICES

3.3.1   Splices

A splice is a permanent joint between two fibers. A splice may be used to extend optical cable or to join different cable types, for example, from an outside plant cable

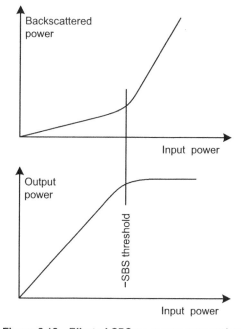

Figure 3.13   Effect of SBS on energy propagation.

to a central office cable or jumper. A splice may be used during the operational life of the network to repair damaged cable. Often, a connector pigtail is spliced onto the main fiber (e.g., see Fig. 3.24). A splice avoids the insertion loss of a connector in situations where disconnection is never needed.

The procedure called fusion splicing is done by melting together the fiber ends with an electric arc or laser pulse. Fusion splicing produces a very low-loss joint, 0.05–0.1 dB, but the joint is mechanically weak. Because of this, the joint must be sealed in a splice closure and fixed in a splice holder.

Mechanical splicing is a less common method in which fiber ends are physically clamped together in a fixture, which is then sealed in a thermoplastic splice closure. Insertion losses of 0.5 dB are achievable by filling the structure with gel that matches the index of refraction of the fiber core. But low loss is one of the primary reasons for splicing, and a connector can achieve loss lower than a mechanical splice, and very nearly as good as a fusion splice.

3.3.2 Optical Connectors

Optical connectors are based on a butt–joint coupling principle in which the mating fiber ends are accurately aligned. A connector comprises a cylinder called a ferrule, with a small hole in the center that precisely matches the fiber cladding diameter. The fiber is fixed in the hole with epoxy and the end of the ferrule is polished nearly flat. Most ferrules are made of ceramic due to the material's good strength, low temperature coefficient of expansion, small elasticity coefficient, wear resistance, and dimensional stability during production to achieve the exact alignment necessary for low-loss connections.

Fibers terminated in ferrules are mated in an alignment adaptor, which provides physical contact between end faces. The ferrules themselves are encapsulated in plastic housings containing snap-in mechanisms and strain relief boots, shielding the junction of the connector to the fiber from bending and pulling.

G-PON commonly uses the standard connectors (SCs) specified in IEC 61754-4 and Telcordia GR-326-CORE. Figures 3.14 and 3.15 illustrate the SC/UPC (ultra-polished connector) and the SC/APC (angled physical contact). The outer plastic housing or the boot is color coded blue for SC/UPC connectors and green for SC/APC connectors.

The difference between UPC and APC connectors is the end face geometry. UPC ferrule and fiber end faces are polished at $0°$, whereas APC end faces are polished at an angle of $8°$, which just happens to coincide with the fiber's acceptance angle (Section 3.1.3). APC connectors have slightly higher insertion loss, but they have far lower return loss—that is, reflected energy—which is a large advantage in an optical tree. ONUs are almost always equipped with APC connectors; OLTs normally use UPC.

Light that may be reflected from a UPC connection is reflected straight back to the source; in an APC connector, reflected light is directed into the cladding, where it is dissipated. When a connector is unmated, for example, at an unused ONU drop, the return loss of the APC is at least -65 dB. An open UPC connector has a return loss in the neighborhood of -55 dB.

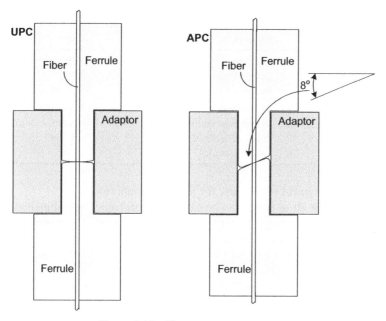

Figure 3.14 Fiber connector structure.

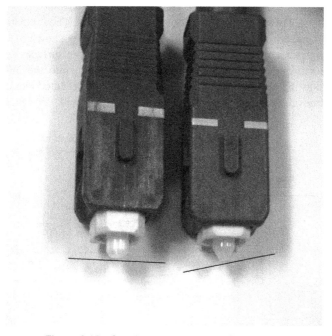

Figure 3.15 Standard connector—UPC and APC.

TABLE 3.2 SC Connector Specifications

Connector Type	Parameter	Value
SC/UPC	Insertion loss (max)	0.2 dB (0.09 dB typical)
	Return loss (min)	−57 dB
	Radius of curvature	10–25 mm
SC/APC	Insertion loss (max)	0.35 dB (0.15 dB typical)
	Return loss (min)	−65 dB
	Polished endface radius	5–15 mm
	Endface angle	$8° \pm 0.5°$

Table 3.2 lists typical parameters for standard connectors.

Loss in a fiber joint or splice can be caused by mechanical misalignment, either offset or angular. With a single-mode fiber whose core diameter is around 9 μm, even the smallest misalignments cause large insertion and return loss variations. However, connector design and manufacture have largely solved this problem. In current practice, loss at a connector is usually caused by foreign material, such as dust or oily films. It is important to protect ferrules when they are not connected and to clean them before mating.

3.4 WDM DEVICES AND OPTICAL FILTERS

3.4.1 Thin-Film Filters

Thin-film filters (TFFs) are used as optical bandpass filters. They transmit a narrow wavelength band, while reflecting all other wavelengths. As shown in Figure 3.16, the filter comprises a cavity formed by two parallel dielectric surfaces with partially transmissive mirrors on the inner surfaces. Light passing into the interferometer structure strikes the opposite surface, where most of it is reflected and only a small fraction is transmitted. Reflected yet again, a phase-shifted version of the light

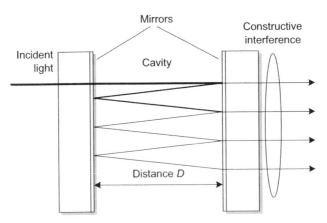

Figure 3.16 Fabry–Pérot interferometer structure.

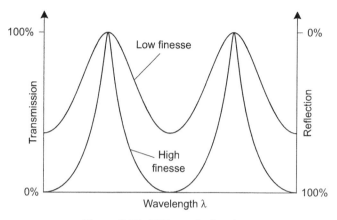

Figure 3.17 TFF transfer function.

appears at the opposite surface. Again, a small fraction is transmitted and most is reflected. The phase shift depends on the distance D between the inner mirror surfaces. The small transmitted fractions interfere constructively for wavelengths (or integral multiples) equal to $2D$, called resonant wavelengths. These wavelengths are passed through the cavity filter; all other wavelengths are reflected.

Figure 3.17 illustrates the so-called finesse of a TFF, which is determined by the reflectivity of the mirrors. Finesse is a measure of the sharpness of the filter transfer function.

The filter characteristics of a TFF can be further controlled by stacking several layers of cavity filters. Figure 3.18 illustrates a multilayer TFF structure. Dielectric thin films

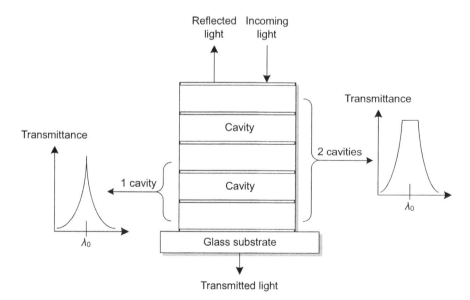

Figure 3.18 Multilayer TFF.

acting as mirrors separated by cavities are deposited on a glass substrate. The substrate can be made as thin as 0.1 mm, with each film layer around 25 μm thick.

A high-finesse TFF with a single layer has a comb of sharply peaked passbands at the different resonant frequencies of the cavity. By adding more layers, the transmission characteristics can be shaped to have a broader flat-top passband and steeper skirts. Each additional layer increases the order of the filter. The multilayer structure, each layer slightly detuned from the others, also rejects harmonics, so that the resulting filter has only a single passband.

TFFs, more precisely dichroic filters, are commonly used in G-PON receivers to pass the wavelength of interest to the photodetector, while blocking foreign wavelengths. The TFF plays a major role in ONU optics, in anticipation that multiple PON flavors may coexist via WDM on the same ODN.

There is a tradeoff between the transfer quality and the cost of an optical filter. A high-order filter provides clear-cut channel filtering, which allows wavelengths to be allocated densely without large guard bands, but, of course, costs more to produce. If a fiber's spectrum is sparsely populated with only a few widely separated wavelengths, inexpensive low-order filters can be used.

Table 3.3 lists typical wavelength-blocking filter (WBF) specs to isolate the downstream wavelength in an ONU.

3.4.2 Diffraction Gratings

A diffraction grating is an optical component used to combine and separate different wavelengths. A periodic reflective structure as depicted in Figure 3.19 splits and diffracts incident multiwavelength light into several monochromatic beams, each wavelength traveling in a different direction.

The direction of the separated wavelengths depends on the spacing of the grating and the angle of incidence. Let Λ be the period of the grating; let φ_1 be the incident angle; and let λ be a wavelength component of the incident wave. The grating

TABLE 3.3 TFF Parameters for WBFs

System	Parameter	Value
G-PON ONU wavelength blocking filter	Passband	1480–1500 nm
	Stopband	1260–1360 nm
		1550–1560 nm
	Transmission loss	<0.04 dB
	Reflection isolation	>20 dB
	Number of layers	4
XG-PON ONU wavelength blocking filter	Passband	1574–1582 nm
	Stopband	1260–1560 nm
		1610–1660 nm
	Transmission loss	<0.05 dB
	Reflection isolation	>35 dB
	Number of layers	6

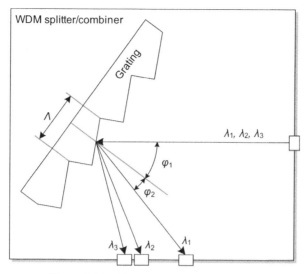

Figure 3.19   WDM device based on grating.

equation defines the necessary conditions for constructive interference, relating the diffraction angle φ_2 to wavelength λ:

$$\Lambda[\sin(\varphi_1) - \sin(\varphi_2)] = \lambda \tag{3.11}$$

As well as reflective gratings, transmissive structures (phase gratings) are possible by periodically changing the refractive index of a transmissive medium. The advantage of a phase grating is mechanical simplification.

A grating is fully reciprocal, so wavelength separation and combination are both possible. As we have described, if the single input contains several wavelength components, they are spread to different outputs. In the reverse direction, individual wavelengths may be injected at the proper points on the device, which then combines them into a single output. Multiplex and demultiplex directions can be mixed on the same device at the same time.

Gratings have important applications in G-PON, where several wavelengths are multiplexed onto a single fiber for transmission but must be separated by direction and wavelength in OLT and ONU transceivers. Figure 3.20 shows how a phase grating device is used in a G.984 G-PON ONU.

Several optical systems can be overlaid on the same fiber infrastructure. Figure 3.21 shows a set of WDM filters that provides for G.984 G-PON, G.987 XG-PON, and an additional RF video overlay wavelength.

3.5   PASSIVE OPTICAL SPLITTERS

One of the fundamental components used in G-PON is the passive power splitter, which distributes input optical power equally to its output ports. A basic 2:2 splitter,

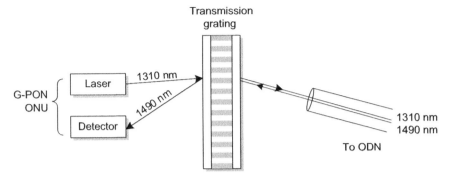

Figure 3.20   G.984 G-PON ONU WDM splitter/combiner.

shown in Figure 3.22, is based on a fused-fiber design. Two single-mode fibers are pulled, twisted, and melted together over a uniform coupling region. The fiber diameter decreases gradually from the input toward the coupling region, where it exchanges power with the second fiber.

By changing the coupling region width W and the tapered region L, the coupling ratio can be designed to be anywhere from 1% or less (a so-called tap coupler) to 50%. The 50% splitters used in G-PON applications attenuate each output by 3 dB from the input power.

Higher order splitters can be constructed as K-stage arrays of such couplers. They have one or two input ports and $N = 2^K$ output ports, as shown in Figure 3.23. The number of output ports is called the split ratio, which corresponds to the maximum number of ONUs that can be connected.

In the downstream direction, the splitter distributes light from its input to all outputs. In the upstream direction, the splitter combines light paths from all ONUs back to the input port(s). The cost of doubling the split ratio is a 3-dB reduction in output power. The upstream signal suffers the same loss as the downstream signal, even though only a single port is connected to the OLT. This is a natural consequence of the reciprocity of passive symmetric devices.

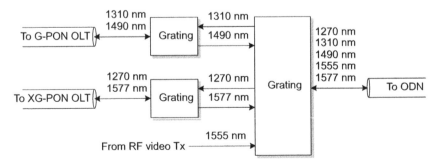

Figure 3.21   G-PON family coexistence filters.

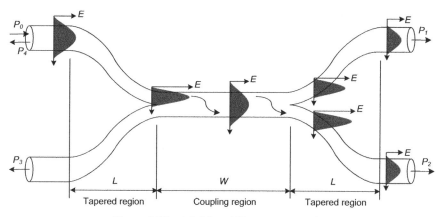

Figure 3.22   A 2:2 fused-fiber power coupler.

So far, we have considered the splitter to be lossless—the 3 dB is not lost energy but merely accounts for the split itself. A practical splitter, of course, introduces additional loss, known as excess loss.

Another significant splitter parameter is its uniformity, which measures how evenly the power is distributed among the output ports. It is the maximum difference in loss between output ports.

Splitters are highly directional. When light is injected at an input (output) port, very little light is scattered back to other input (output) ports. This behavior is captured by the directivity D, also called the near-end crosstalk or near-end isolation:

$$D = 10 \log\left(\frac{P_0}{P_3}\right) \tag{3.12}$$

where P_0 and P_3 are defined in Figure 3.22.

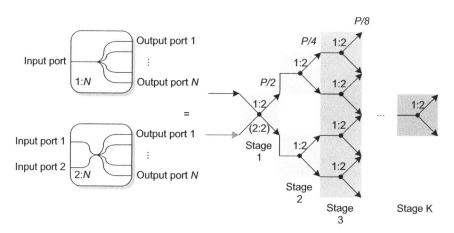

Figure 3.23   1:N and 2:N splitter structures.

The return loss R is the ratio of optical power launched into a port to the optical power returning to the same port. Both directivity D and return loss R are measured with all other ports optically terminated, that is, with zero reflection from the other ports.

$$R = 10 \log\left(\frac{P_0}{P_4}\right) \tag{3.13}$$

with P_0 and P_4 again as defined in Figure 3.22.

Real-world splitters show uniform performance across the whole spectrum of interest, from 1260 to 1600 nm. Figure 3.24 shows a 1:4 splitter for lab use, based on several 1:2 fused-fiber couplers. Observe that the size of the splitter assembly is constrained by the minimum bend radius of the fibers.

For large split ratios such as 32 and more, fused-fiber splitters perform poorly in terms of optical characteristics and especially reliability—the 1:4 fused-fiber splitter of Figure 3.24 contains three 1:2 splitters and seven splices—a lot of components that can fail, and a lot of manufacturing effort.

Planar lightwave circuit (PLC) technology allows splitters to be made with techniques much like those used to manufacture semiconductors. These techniques allow for high split ratios in compact, low-loss, reliable devices. PLC splitters are used in G-PON ODNs where large, concentrated splits are needed—as distinct from trees made of several separately located 1:4 splitters, for example. Table 3.4 lists typical parameters for single-input PLC splitters, including connectors. Table 3.5 lists the same parameters for 2:N splitters.

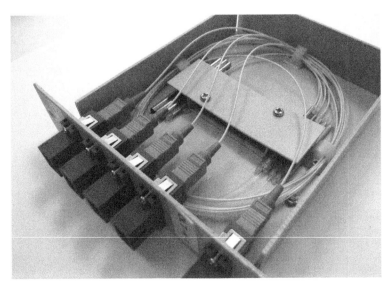

Figure 3.24   A 1:4 passive optical splitter.

TABLE 3.4   Parameters of Standard Connectorized 1:*N* G-PON Splitters

Split Ratio	Insertion Loss, dB max	Uniformity, dB max	Return Loss, dB min	Directivity, dB min
1 : 2	3.4	0.3	50	55
1 : 4	7.5	0.5		
1 : 8	10.7	0.8		
1 : 16	13.7	1.0		
1 : 32	16.9	1.3		
1 : 64	20.4	2.0		
1 : 128	23.6	2.7		

3.6 POWER BUDGET

The optical budget determines the physical performance of a PON in terms of the maximum number of ONUs supported and the reach of the farthest ONU from the OLT. We encounter a wide range of optical power levels in G-PON applications. Launch powers may be +5 dBm or more, while receivers may be expected to perform at input powers below −30 dBm.

3.6.1 Budget Planning

The loss budget available for a link is the difference between the minimum launch power at the transmitter and the required minimum power level at the receiver to guarantee the necessary transmission quality. The budget must be allocated to the lossy components in such a way that it is not exceeded. Spending too little of the budget is also a concern, as insufficient attenuation can cause receiver overload and even damage.

Each direction has its own budget. In G-PON and XG-PON, the loss budgets are specified to be symmetric. Different factors limit the upstream and downstream budgets:

- In the upstream direction:
  o Fiber attenuation is roughly 0.1 dB/km higher in the upstream band, 1260–1360 nm, as compared to the downstream direction.

TABLE 3.5   Parameters of Standard Connectorized 2:*N* G-PON Splitters

Split Ratio	Insertion Loss, dB max	Uniformity, dB max	Return Loss, dB min	Directivity, dB min
2 : 2	4.2	1.1	50	55
2 : 4	7.7	1.2		
2 : 8	11.0	1.6		
2 : 16	14.6	2.4		
2 : 32	17.8	3.0		
2 : 64	21.5	3.7		
2 : 128	25.1	4.0		

o For cost reasons, the ONU laser is specified with a lower minimum launch power than that of the laser at the OLT. The most extreme difference is in XG-PON, where +2 dBm is required from the ONU, with as much as +14.5 dBm required from the OLT.

o Burst-mode upstream transmission introduces a penalty of 2–3 dB.

• In the downstream direction:

o The bit rates are higher. A ratio of 2 in the bit rate is generally regarded as a 3-dB cost (G.984 G-PON). The 4x ratio of G.987 XG-PON is worth 6 dB.

o The downstream wavelength is far from the zero dispersion point of a G.652 fiber. Chromatic dispersion is a problem, especially at 10 Gb/s. It can be reduced to some extent by requiring a spectrally pure OLT transmitter but, of course, at a cost.

The reach–split ratio rule of thumb is simple: doubling the split ratio costs about 3.5 dB, which is worth roughly 10 km of good fiber. A 28-dB loss budget is good for about 10 km at 1:64 split, 20 km at 1:32, and 30 km at 1:16.

Figure 3.25 is a graphic view of a more sophisticated budget plan. In this example, the loss of the fiber distribution network is 26.8 dB. Observe that the connectors at OLT and ONU are not included in the loss. This particular budget has a bit more than 1 dB of margin for maintenance splices, but it must also be said that it does not

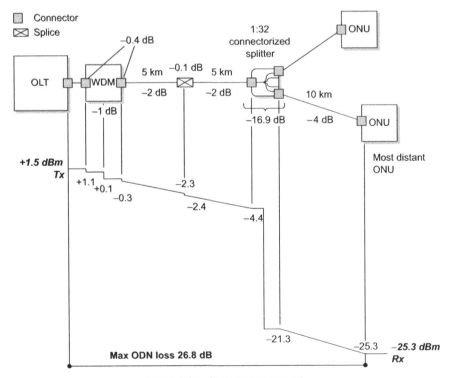

Figure 3.25 Budget plan example.

TABLE 3.6   G.984 G-PON Budget Classes

Link Budget Class	ODN Loss (min–max)	Standard
Class A	5–20 dB	G.984.2 (2003)
Class B	10–25 dB	
Class C	15–30 dB	
Class B+	13–28 dB	G.984.2 Amendment 1 (2006)
Class C+	17–32 dB	G.984.2 Amendment 2 (2008)

include the additional connectors and splices that are surely needed in the real world. Chapter 7 mentions some of them.

3.6.2   G.984 G-PON Budget Classes

The initial release of G-PON standard G.984.2 specified three loss budget classes— A, B, and C. In subsequent years, two additional industry best-practice classes followed—B+ and C+. Table 3.6 lists their parameters.

Each class is defined by a maximum and a minimum value of optical loss. These values bound the total loss of the ODN bidirectionally between the OLT S/R and ONU R/S interfaces. This is shown in Figure 3.26, where NE indicates provision for some form of future network element that might coexist on the same ODN. The reference points represent the optical connectors at the OLT and ONU. Components inside the transceivers, such as filters and WDM devices, are not regarded as part of the ODN.

Class B+   Early G-PON experience revealed that the initial classes A, B, and C covered neither what the market needed nor what optical module vendors were able to deliver. Class B was clearly not enough for 1:32 at 20 km, and class C was too expensive for the technology of the time. As a result, a step between class B and class C was defined—the industry best-practice class B+ budget of 28 dB, providing a reach of 20 km with a 32-way split.

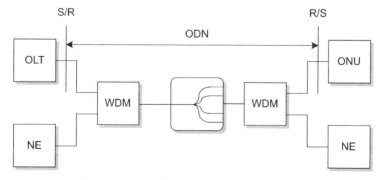

Figure 3.26   ODN budget reference architecture.

Class C+   Class B+ is today's de facto standard for G.984 G-PON. However, there is always a push for more. In 2008, G.984.2 was amended again to introduce the extended 32-dB C+ budget class, with the idea of supporting the following deployment constellations:

- 1:16 split at 40 km
- 1:32 split at 30 km
- 1:64 split at 20 km
- 1:128 split at 10 km

The standard is designed to support a single-ended upgrade from B+ to C+. That is, only the OLT optics must be replaced. This is an important real-world consideration because it means that only one class of ONU need be qualified, purchased, stocked, inventoried, and installed. Figure 3.27 illustrates the upstream optical budget, showing how the increased ODN loss is taken up by the OLT receiver, and the ONU does not change.

A −32 dBm receiver at the OLT is nontrivial. The standard defines the C+ OLT sensitivity under the assumption that forward error correction (FEC) is enabled, compensating in the electrical domain for what is difficult to achieve in the optical domain. Unfortunately, there are problems with FEC, particularly upstream FEC. FEC presupposes byte synchronization and frame alignment, and if the burst header cannot be properly recognized in the presence of a high bit error rate, FEC is of no use. Early G-PON equipment showed almost no upstream FEC gain. This was addressed in part by modifying the receiver's burst header decoder to ignore a few

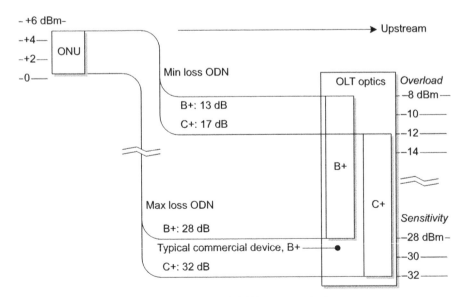

Figure 3.27   G.984 G-PON upstream budget.

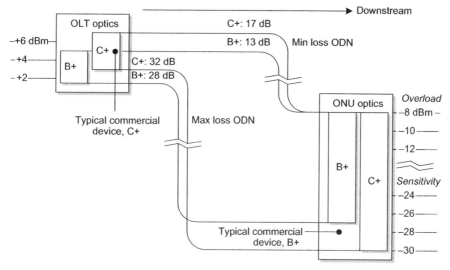

Figure 3.28 G.984 G-PON downstream budget.

bits in error, that is, to recognize an approximate synchronization pattern, rather than requiring an exact match.

Chapter 4 and Appendix I discuss FEC in some detail.

Figure 3.28 shows the G-PON downstream budget. But wait! We said the ONU optics would not change, and here we find a requirement for an additional 3 dB of ONU receiver sensitivity. Although existing B+ optics typically support −28 dBm, we are still short 2 dB. As with upstream, the standard likewise defines C+ ONU sensitivity under the assumption that FEC is enabled, so that lower optical input power will yield satisfactory results from the same physical B+ receiver.

3.6.3 G.987 XG-PON Budget Classes

G.987.2 defines two nominal and two extended loss budgets for XG-PON, listed in Table 3.7. The nominal 1 and extended 1 classes are counterparts of the G.984 G-PON B+ and C+ classes and are intended for coexistence on already deployed G-PON ODNs. The additional 1 dB is for a WDM device that may be needed.

Nominal 2 is an intermediate budget that may be useful if there turns out to be a large cost difference between N1 and E1. As for the extended 2 budget, this was

TABLE 3.7 XG-PON Budget Classes

Link Budget Class	ODN Loss (min to max)	Standard	Maximum Differential Distance
Nominal 1 (N1)	14–29 dB	G.987.2	
Nominal 2 (N2)	16–31 dB		20 km (DD20)
Extended 1 (E1)	18–33 dB		40 km (DD40)
Extended 2 (E2)	20–35 dB		

recognized as a challenge in terms of 2010 technology but was standardized as a target for the components industry. The expectation is that optical amplification will be needed at the OLT.

Regardless of budget class, two maximum differential distances are specified, 20 km (DD20) and 40 km (DD40), allowing different dispersion ranges:

- 0–400 ps/nm for DD20 downstream, 0–140 upstream
- 0–800 ps/nm for DD40 downstream, 0–280 upstream

Basically, what this means is that a 40-km OLT transmitter must be twice as good as a 20-km transmitter in terms of modulated linewidth.

The downstream XG-PON budget is designed against a BER (bit error ratio) of 10^{-3}, which is then improved by FEC to 10^{-12}. FEC is always on in the downstream direction.

As to upstream, the XG-PON budget is designed to a worst-case BER of 10^{-4}, to be improved by FEC to 10^{-10}. The OLT switches upstream FEC on or off depending on the actual measured BER. An ONU connected with only a few decibels less loss than the budget limit may well exceed 10^{-10} without FEC.

To keep ONU optics reasonably inexpensive, the upstream transmit power need be only +2 dBm, which is achievable by common uncooled 2.5-Gb/s distributed feedback (DFB) lasers.

Figures 3.29 and 3.30 illustrate the upstream and downstream budgets for G.987 XG-PON. As with G-PON, the idea is to minimize the number of ONUs that must be

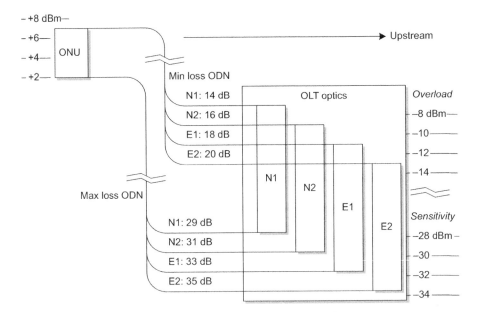

Figure 3.29 G.987 XG-PON upstream budget.

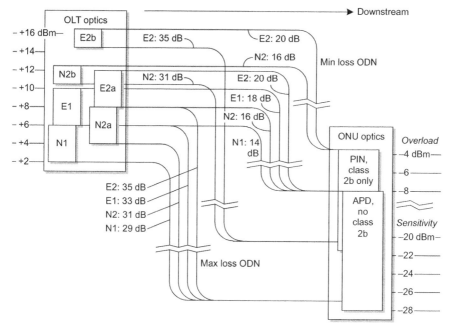

Figure 3.30 G.987 XG-PON downstream budget.

supported. Observe that the downstream budget includes 1 dB of additional margin against otherwise unspecified impairments.

At the time the standards were developed, it was not apparent whether a receiver with −28 dBm sensitivity would be the most economical choice, given that an avalanche photodiode (APD) receiver would be required.[*] The alternative was a −21.5-dBm PIN-based receiver, which had the potential to be much less costly, given the anticipated emergence of super-TIAs (transimpedance amplifiers). Both options were standardized, as a and b power classes, respectively. The evolution of technology will doubtless favor one or the other as the XG-PON market develops.

The a and b power classes are not compatible; that is, they cannot be deployed together on the same PON. To compensate for the reduced ONU receiver sensitivity of the b classes, cost has been shifted into the OLT. The OLT transmitter requires an optical amplifier to boost the power of a standard DFB laser to levels of at least +10.5 and +14.5 dBm, for classes N2b and E2b, respectively. The maximum launch power is bounded by the onset of fiber nonlinearities.

3.7 COEXISTENCE

A single-mode fiber provides a great deal of bandwidth. Even excluding the E band (1360–1460 nm), which contains the water peak, a total of 315 nm is available for

---

[*] Section 3.8 goes into detail on optical receivers.

optical signals. G.984 G-PON allocates 40 nm for the upstream signal, 20 nm for the downstream signal. If an RF video overlay is present, it uses another 10 nm. This adds up to 70 nm. Spectrum is clearly available for next-generation systems such as XG-PON or WDM PON.

When the original G.984 G-PON recommendations were developed, the spectral assignments for future systems such as XG-PON were unknown, and the best that could be done was to specify enhancement bands to allow future systems to be overlaid whenever they might appear. These additional bands allowed for coexistence options between existing and future PONs.

Enhancement bands are important to protect operators' investment in the ODN, but a clear specification of vacant bands to be used by future access technologies is not enough to enable coexistence, that is, the simultaneous presence of two or more PON technologies. Optical detectors respond to light over the entire spectrum. If several wavelengths are present, they interfere with one another, unless each detector is sheltered by a suitable filter from all wavelengths except the one carrying its specific service. We pointed this out in Section 3.4.

In addition, each service must be multiplexed onto the ODN by means of a WDM device, a device that introduces additional insertion loss. If a new service is added to an existing ODN, the new WDM device may jeopardize the optical budget of the existing system.

These factors impose requirements both on the current system and on the new service. A G.984 G-PON ONU deployed today must contain a WBF. The ODN, as laid out initially, must allocate budget to add a wavelength multiplexer at least at the OLT.

ITU-T developed a coexistence standard, G.984.5, to future-proof the G.984 ODN. Its first release was somewhat nonspecific because the spectrum for XG-PON had not yet been defined. Once the spectrum assignments were agreed for G.987 XG-PON, G.984.5 was amended accordingly. The WDM1 filter described in the original G.984.5 was deprecated—beware! Amendment 1 specifies a revised filter called WDM1r.

3.7.1 Enhancement Bands

G.984.5 specifies a total of five bands. The upstream 1260–1360 nm band was originally fully dedicated to G.984 G-PON. G.984.5 defines three different options for this band.

- The *regular band* option (100-nm width), was intended for low-cost ONUs using broad-spectrum Fabry—Pérot lasers with low power budgets such as class A. This option was never adopted by the market.
- The *narrowband* option (20-nm width) was intended for indoor ONUs, that is, without wide temperature fluctuations, and possibly with a requirement for device selection at manufacture. This option was also not adopted by the industry.
- The *reduced band* option (40-nm width) became the industry standard. It is well suited for class B+ ODNs—the ONU uses a 1310-nm DFB laser without special temperature control or selection.

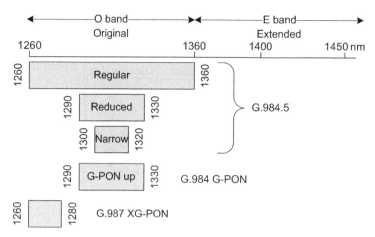

Figure 3.31   G.984.5* spectrum assignment, upstream.

*G.984.5 identifies several additional optional bands for possible future use. We omit them in the interest of clarity. The figures show the important assignments.

With 40 nm consumed de facto by G.984 G-PON, two 30-nm segments became available in the O band. Twenty nanometers of the lower segment was assigned to the upstream signal of G.987 XG-PON. Figure 3.31 illustrates the upstream spectrum assignments.

As to downstream wavelengths, Figure 3.32 shows that G.984 G-PON occupies the basic band from 1480 to 1500 nm. Figure 3.32 also illustrates the shape of the WBF that is found in all G-PON ONUs. As we see, it protects the G-PON downstream wavelength from RF video, downstream XG-PON, and possible OTDR maintenance signals. The filter slope at 1530 nm is mandatory; the slope at 1450 nm

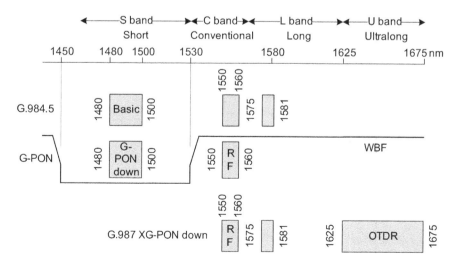

Figure 3.32   G.984.5 spectrum assignment, downstream.

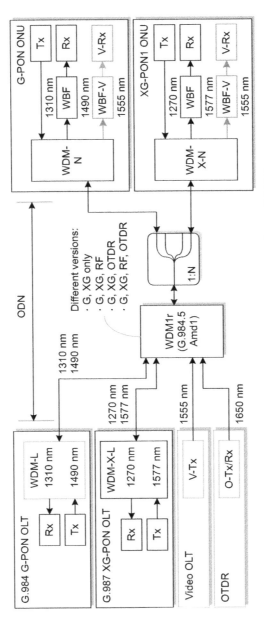

Figure 3.33 Coexistence architecture according to G.984.5.

is optional. Beyond 1539 nm, the WBF is required to provide at least 22 dB of rejection relative to the G.984 downstream signal.

Although it lies beyond the scope of G.984.5, it is clear that XG-PON ONUs also need WBFs to separate the 1577-nm downstream signal from the rest of the spectrum. G.987.2 addresses this filter by specifying the immunity required from foreign signals, rather than the filter stopband itself.

There is a tradeoff between the quality of the WBF filter and its cost. High-order TFFs provide good channel separation (steep filter skirts) and reduce the required guard bands but introduce cost. Devices such as arrayed waveguides (AWGs) have extremely narrow passbands but therefore require very predictable and stable transmitter wavelengths. In general, G-PON spectrum assignments favor lower cost over spectral efficiency. Along with the water peak, this explains the comparatively large gaps in Figures 3.31 and 3.32.

Looking to the future, if WDM PON is to coexist with either or both of G.984 and G.987, and possibly with RF video as well, spectrum becomes considerably more valuable. At the time of this writing, many proposals focused on the L band, which offered the largest unbroken block of spectrum—the industry is still reluctant to use the E band.

3.7.2   Coexistence Architecture in G-PON

The generalized overlay architecture comprises several generations of OLTs and ONUs operating on the same ODN infrastructure, as depicted in Figure 3.33. The WDM1r and WBF optical components are necessary first-day add-ons to enable future coexistence without service disruption.

G-PON OLT transmit and receive wavelengths are multiplexed by WDM-L, which is integrated into the OLT optics. XG-PON OLTs may be expected to have their own WDM devices, according to the wavelength used, designated in Figure 3.33 as WDM-X-L.*

An external WDM1r coexistence device at the central office multiplexes the wavelength pairs for each system—one pair per physical port. The WDM1r device may also be asked to multiplex a downstream RF video overlay or an OTDR maintenance wavelength.

At the ONU, the WDM-N and wavelength blocking filter is integrated with the transmitter and receiver into a device called a diplexer or triplexer (Section 3.10), depending on whether RF video is present.

Individual blocking filters extract the wavelength of interest in the downstream band. In case of a triplexer device with an RF video overlay—see Figure 3.34—the first WDM stage isolates the video wavelength at 1555 nm, followed by a second stage that isolates the G-PON downstream wavelength at 1490 nm. If there is no RF overlay, the first stage is not needed.

For XG-PON, the same design is used but with different wavelengths.

---

*The *L* designates o*L*t. Filters at the o*N*u are identified with the letter *N*.

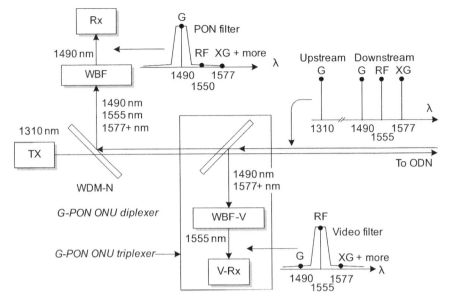

Figure 3.34   G-PON ONU WDM/WBF filter structure.

3.8   OPTICAL TRANSMITTERS

To understand light amplification by stimulated emission of radiation (LASER), we consider the interplay between photons and electrons in a material. There are basically three interactions, namely spontaneous emission, stimulated emission, and absorption.

In Figure 3.35a, we see that an electron in energy level E_2 (excited state) can spontaneously drop to a lower level E_1 (ground state) according to the material properties. In the process, it releases a photon of energy $E_2 - E_1 = hf$, where h is Planck's constant and f is the frequency of the photon. The rate of photon generation is proportional to the density N_2 of electrons in energy state E_2. Spontaneous emission causes narrowband Gaussian noise in lasers and optical amplifiers.

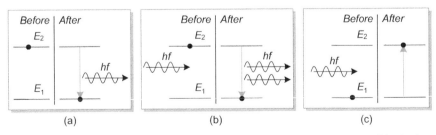

Figure 3.35   Interactions between light and material: (a) spontaneous emission, (b) stimulated emission, (c) absorption.

Figure 3.35b illustrates a photon of energy $E = hf = E_2 - E_1$ interacting with an electron on level E_2, causing it to drop to lower level E_1, and in the process releasing a second in-phase photon with the same energy hf. The rate of interaction is proportional to N_2, as well as to the field strength of the incident light. This effect is called stimulated emission. It constitutes the main principle behind lasers and optical amplifiers.

It is also possible to reverse the process of spontaneous emission by boosting an electron in state E_1 to a higher level E_2, thereby absorbing a photon of energy $E_2 - E_1 = hf$. The rate of absorption is proportional to the electron density N_1 at E_1, as well as the optical field strength. This effect is called absorption, depicted in Figure 3.35c.

When the rate of emission is larger than the rate of absorption, a net photon gain is achieved. This is only possible if $N_2 > N_1$. In thermal equilibrium, $N_2 < N_1$, and the material is absorptive. So-called population inversion is possible by pumping the material with optical or electrical energy. Material in this state is called an active medium.

The energy of a photon need not be exactly the bandgap $E_2 - E_1$ of the material. The structure of the material offers incident waves with optical field intensity I_f a range of interaction around a center frequency f_0, a range described by a probability function $g(f)$. The energy spectrum of light produced by spontaneous emission has a corresponding nonzero linewidth, depicted in Figure 3.36.

If such an active medium is placed between two mirrors, the resulting Fabry–Pérot (FP) cavity oscillates if the optical gain in the round trip exceeds the loss in the structure, as suggested in Figure 3.37.

We are interested in free-space wavelengths λ and frequencies f, that differ from those in the cavity by a factor of the medium's group refractive index n_g. In effect, the length of the cavity is dn_g, rather than d. Constructive interference occurs at any free-space wavelength λ for which an integer multiple q of the half wavelength matches the length dn_g of the resonator:

$$d = \frac{q\lambda}{2n_g}, \text{ for some integer } q. \text{ Equivalently, } \lambda = \frac{2dn_g}{q} \qquad (3.14)$$

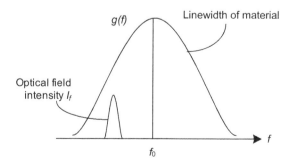

Figure 3.36 Spectral pattern of active material.

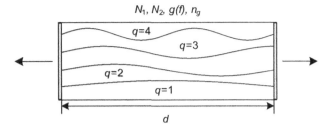

Figure 3.37 Fabry–Pérot resonator.

Corresponding to adjacent integers q, $q-1$, the cavity resonates at free-space frequencies spaced Δf apart:

$$\Delta f = f_q - f_{q-1} = \frac{c}{2dn_g} \tag{3.15}$$

Approximating $\Delta\lambda/\lambda = -\Delta f/f$, we can express the lines in wavelength space by a comb whose spacing is

$$\Delta\lambda = \frac{\lambda^2}{2dn_g} \tag{3.16}$$

Figure 3.38 shows the output spectrum, as shaped by the material gain (line width) curve.

3.8.1 Semiconductor Lasers

In a semiconductor laser, electrical current I in a P–N junction operated in the forward direction (diode mode) causes population inversion and allows lasing, which begins due to noise and continues due to device resonance.

3.8.1.1 Fabry–Pérot Lasers

An FP laser is based on the structure of Figure 3.39, whose active region corresponds to the resonant cavity of Figure 3.37. It comprises an active material sandwiched

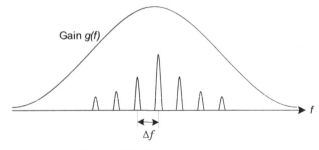

Figure 3.38 Resonator spectrum.

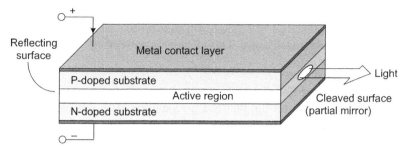

Figure 3.39 FP laser structure.

between doped semiconducting substrates. A dielectric reflecting layer at the back facet acts as a mirror. The cleaved front surface functions as a partially reflecting mirror, which releases some fraction of the laser light to the outside.

The key characteristic of FP laser performance is the relation between laser current I and output power P, as illustrated in Figure 3.40. When the bias current I is below a threshold current I_{th}, population inversion is not achieved, and spontaneous emission dominates—the laser emits broadband noise similar to a LED (light-emitting diode). For I above the threshold current, the device generates coherent and quite monochromatic light, whose power increases linearly with bias current.

Near the threshold, multiple spectral lines exist: the laser operates in multiple modes, with an amplitude envelope determined by the linewidth of the active material. As the bias current increases, the side modes disappear, and a single mode dominates. For a bias current only slightly above the threshold current, and a large

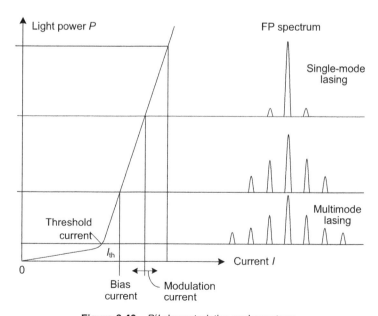

Figure 3.40 P/I characteristics and spectrum.

modulation current to achieve a high difference between laser on/off power levels (extinction ratio), the average spectrum is still quite broad, and fiber dispersion limits the achievable bit rate and distance. This effectively prevents the use of FP lasers in G-PON applications.

3.8.1.2 Distributed Feedback Laser

To achieve the narrow spectrum that is a precondition of acceptable transmission quality for high-speed data in the presence of dispersion, we need a resonator that only supports one mode, regardless of current.

By periodically varying the refractive index structure along the length of the laser (distributed feedback, DFB), we can create a geometry whose harmonics lie far outside the material's response curve. Such a structure, therefore, oscillates at only one single mode. DFB lasers are normally built with Bragg gratings with corrugated index profiles (Fig. 3.41).

A DFB laser provides launch powers on the order of 0–+10 dBm, with a spectral width less than 1 nm (20 dB down) and suppression of side modes by more than 30 dB. This performance makes the DFB laser perfectly suited for transmission up to 10 Gb/s over tens of kilometers of standard fiber. The DFB laser is the only laser type used in G-PON and XG-PON.

3.8.2 Laser Modulation

In the G-PON family, the laser is modulated directly by varying the diode current with the bit stream. To provide stable average and on/off power levels as well as low jitter, the laser driver must take the P/I characteristics of the laser diode into account.

The key parameters of a laser are its threshold current and its slope efficiency. The latter measures the efficiency of current-to-light conversion. Unfortunately, the transfer function of a laser changes with temperature and age, as illustrated in Figure 3.42. As temperature increases, the threshold current increases exponentially and the slope efficiency decreases linearly. Similar changes occur over the lifetime of a laser. At the beginning of life (BOL), a laser operates with a low threshold current and high slope efficiency. Both of these deteriorate as the laser ages toward its end of life (EOL).

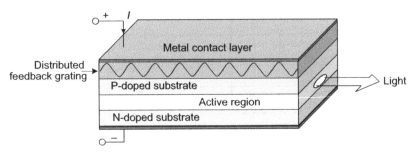

Figure 3.41   DFB laser structure.

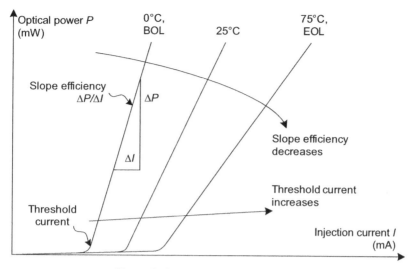

Figure 3.42 Laser diode P/I variation.

To operate a directly modulated laser in its linear region, we bias it slightly above threshold to produce the logical zero power level P_0. It is important to understand that the logical zero level corresponds to a low but nonzero light output. We then modulate the laser by summing dynamic current $I_M > 0$ onto the bias current, generating an optical output that ranges between the zero level P_0 and the one level P_1. Figure 3.43 illustrates the modulation process.

System transmission requirements demand that both P_0 and P_1 values be kept constant over temperature and lifetime. Figure 3.43 shows that both bias current and

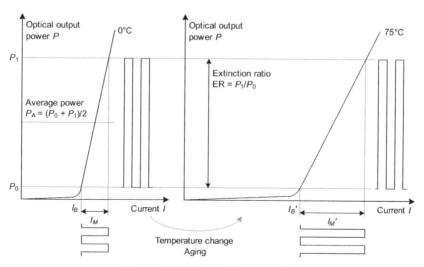

Figure 3.43 Laser driver operation.

modulation current must be adjusted to achieve this. Automatic bias and drive current compensation is the challenge of the laser driver.

Assuming an equal probability of zeros and ones, the average output power is

$$P_A = \frac{P_0 + P_1}{2} \tag{3.17}$$

The extinction ratio ER is the ratio of P_1 and P_0 power levels:

$$ER = \frac{P_1}{P_0} \tag{3.18}$$

ER relates directly to the achievable bit error rate of an optical link because it affects the receiver's ability to distinguish zero from one levels. G-PON and XG-PON specify a minimum extinction ratio of 8.2 dB.

3.8.2.1 Laser Drivers

The need to maintain average power and extinction ratio stable over component tolerances, temperature, and age can be addressed in three ways: open-loop control, single-loop control, and dual-loop control. As might be expected, the performance improves with the degree of complexity, although we will discover one surprise (Section 3.8.3).

Open-Loop Control Open-loop control comprises the measurement of laser characteristics during manufacture and the setting of calibration points in nonvolatile memory. During operation, the driver circuit measures the laser temperature with a sensor. Then it interpolates and applies bias and modulation current values from its calibration table. Open-loop control can compensate for temperature but not for aging, and it knows nothing about the actual optical output power or extinction ratio.

Single-Loop Control In single-loop control, as depicted in Figure 3.44, the laser diode is incorporated into an assembly that contains a monitor photodiode (MPD).

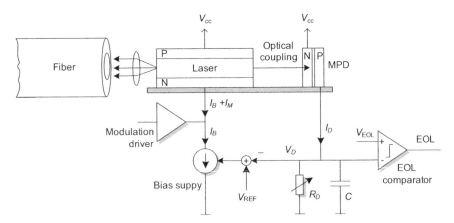

Figure 3.44 Single-loop laser driver.

A small fraction of the light is coupled from the laser's back facet to the MPD, which conducts in proportion to the incident light power. During factory calibration, resistor R_D is adjusted until the desired average power P_A is achieved. The feedback loop holds P_A constant by keeping the MPD current I_D constant. Because the feedback loop stabilizes the average output power, this method is called automatic power control (APC).

The single-loop laser driver compensates for threshold variability by adjusting bias current I_B, but it does not consider variation in slope efficiency. P_A is indeed held constant, but the extinction ratio deteriorates with increasing temperature and age. At the other extreme, low temperatures may improve the slope efficiency so much that the bias current drops below the threshold, introducing pulse width distortion and jitter.

To maintain extinction ratio as well as average power, a single-loop driver may be extended with an open-loop extinction ratio control. A sensor measures the actual temperature and adjusts the modulation current swing, hence the extinction ratio, by approximately the rate at which the slope efficiency drops with temperature. Different lasers exhibit different slope efficiency changes, so transmitters must be calibrated with the correct slope setting—hence the open loop nature of the feedback. Calibration is usually done at three different temperatures.

The photocurrent from the MPD readily provides an EOL indication, as depicted in Figure 3.44. Voltage V_D measures the average launch power; when the laser nears EOL, the feedback loop is no longer able to match V_D and V_{REF}. If V_D drops below a certain EOL reference voltage V_{EOL}, which is typically $V_{REF}/2$, the EOL comparator indicates an alarm. The laser degradation alarm allows time for scheduled replacement well in advance of complete failure.

Dual-Loop Control   A dual-loop laser driver maintains P_A with the same circuit as that of Figure 3.44 but also includes feedback control to hold ER constant. One way to do this is to measure the laser signal swing $P_1 - P_0$ with a peak detector circuit. The result is then fed back to adjust the modulation current I_M. A disadvantage of this approach is that it requires a high-speed MPD.

Another way to do this is to modulate I_M with a low-frequency signal, a tone. The amplitude of the tone, greatly exaggerated in Figure 3.45 for effect, is proportional to the magnitude of I_M. The MPD easily tracks the low-frequency tone in the optical output. The feedback loop holds the MPD tone current swing constant, and thereby compensates for changes in slope efficiency.

Most G-PON transmitters in both OLTs and ONUs are based on the single-loop/open-loop laser driver design, although dual-loop designs are gaining popularity. Dual-loop drivers greatly simplify and rationalize optics design and calibration and reduce production cost, even considering the increased component count.

3.8.3   Burst-Mode Laser Drivers

Thus far, we have implicitly assumed continuous operation of the laser driver. Burst-mode laser drivers, as used in the ONU, face additional challenges, both in terms of interburst emission and in rapid reacquisition of the proper operating point.

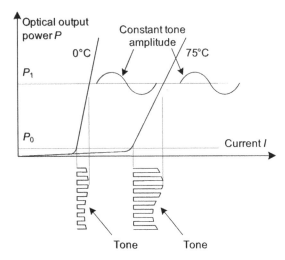

Figure 3.45 Tone control dual-loop laser driver.

For burst-mode transmission, the ONU laser is switched off between bursts. If the laser were kept at the P_0 bias point, the resulting nonzero light output would degrade other ONUs' transmissions. Instead, the ONU's transmitter is biased into its active region only just prior to and during each burst. To allow the laser to stabilize, the laser turn-on interval is part of the guard time between bursts.

The recommendations require the average power emitted by an ONU transmitter outside a burst interval to be at least 10 dB below the minimum downstream burst-mode receiver sensitivity. The numbers vary with optical budget classes but are on the order of −40 dBm or less. To achieve this, the laser bias must be set far below the lasing threshold—better yet, to zero—between bursts.

We increase the bias as we prepare to transmit a burst, but there is a laser turn-on delay on the order of several nanoseconds, accompanied by jitter and pulse width distortion, during which we cannot expect to transmit valid signals. At the end of a burst, when we reduce the bias to zero, there is a turn-off delay before the output is fully extinguished. G-PON specifications reserve part of the guard time between bursts for transmitter start-up and power-down. This time cannot be used productively because the optical levels on the fiber are indeterminate.

The APC loop as depicted in Figure 3.44 must be modified to provide proper burst-mode operation. If the driver were to average MPD current over both burst intervals and interburst gaps, the feedback voltage V_D would be systematically too low, depending on the burst duty cycle. To avoid this, burst-mode transmitters must contain some form of clamp that holds the feedback mechanism in abeyance between bursts.

A related issue affects burst-mode transmitters when they are initialized. The feedback circuit initially contains some arbitrary value of V_D, unrelated to the actual output of the device, and requires a not insignificant transmit opportunity to stabilize. If the OLT's discovery grant (Chapter 4) specifies only a short burst header, the ONU may not be able to successfully respond for quite some number of cycles.

Interestingly enough, this affects only the more sophisticated (automatic power control) drivers; open-loop control needs no feedback convergence time.

3.9   OPTICAL RECEIVERS

An optical receiver comprises a photodiode, a multistage amplifier, and a clock and data recovery circuit, as shown in Figure 3.46. Incident light power P carrying the digital data is converted by the photodiode into a small current I_{PD}. A TIA converts the current into a small voltage V_S, which is further amplified by a postamplifier (PA). We also model two noise sources, and their result V_N, which we discuss below as factors in receiver sensitivity.

Following signal reception, data recovery is a two-step process. In the first step, the front end evaluates the signal amplitude and sets a decision threshold. In the second step, the clock and data recovery (CDR) or clock phase aligner (CPA) determines an optimum sampling instant. It then presents the recovered clock and data to the MAC function for decoding and further processing.

The optimum decision threshold for a binary signal is essentially at the midpoint of the signal swing. The threshold is usually chosen to be the average signal amplitude, under the assumption of an equal density of zeros and ones. Clearly, if this assumption is not valid, the threshold is suboptimal, and the bit error rate can increase.

In the continuous-mode receiver at the ONU, a phase-locked loop (PLL) CDR tracks the frequency of the input signal. A PLL can be designed with an appropriate phase offset to sample the received data, normally at the center of each bit time.

At the OLT, the burst-mode receiver has a different challenge. It has the advantage that the frequency is known precisely, because the ONU uses the downstream signal as its reference for upstream transmission. On the other hand, the OLT's CPA has only the burst preamble time in which to determine the correct sampling instant for each new burst.

3.9.1   Photodetectors

If the essence of a laser is a semiconductor diode operated in the forward direction, the essence of a photodetector is a semiconductor diode operated under reverse bias. A photodiode exploits the photoelectric effect of semiconductor material.

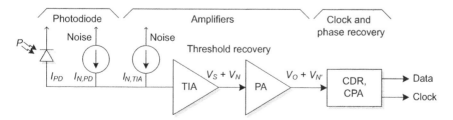

Figure 3.46   Optical receiver structure.

The material absorbs incident photons of energy $E = hf$, if they exceed the band gap energy E_g, lifting electrons from the valence band into the conduction band. If an external reverse voltage[*] is applied, charge is carried through the material, a photocurrent. Photons with energy below the band gap cannot generate photoelectrons. The cutoff frequency f_c and wavelength λ_c are therefore

$$f_c = \frac{E_g}{h} \quad \text{and} \quad \lambda_c = \frac{hc}{E_g} \tag{3.19}$$

The quantum efficiency η of the optical-to-electrical conversion is the number of charge carrier pairs generated by a flow of photons over a given time period in a given material. In a given time, the number of electron–hole pairs is related to the photocurrent I by the charge e of one electron: I/e; the number of photons is related to the incident light power P by the energy of each photon: $P/(hf)$. The quantum efficiency η is thus

$$\eta = \frac{I/e}{P/hf} \tag{3.20}$$

The ratio I/P is called the responsivity R and is essentially constant for a given material below its cutoff wavelength. R is a key parameter of a photodiode and typically lies in the range 0.5–1 A/W. We can relate R to wavelength by evaluating the term hc/e, which is approximately 1240 nm:

$$R = \frac{I}{P} = \eta \frac{e}{hf} = \eta \frac{e\lambda}{hc} \approx \eta \frac{\lambda(\text{nm})}{1240 \text{ nm}} \tag{3.21}$$

Figure 3.47 shows that responsivity increases with longer wavelengths. It falls off rapidly above the cutoff wavelength as the quantum efficiency drops, when the energy of the incident photons is insufficient to excite electrons from the valence to the conduction band.

Practical photodiodes show good linearity between P and I over hundreds of nanometers. Indium–gallium–arsenide (InGaAs), as used in G-PON receivers, has a responsivity of 0.9 A/W at 1300 nm and 0.95 A/W around 1550 nm.

3.9.1.1 PIN Diodes

Suppose we fabricate a semiconductor structure with an intrinsic (I: undoped) zone between the positive (P) and negative (N) doped regions. A reverse bias voltage applied to the device generates an electric field gradient, which depletes electrical carriers from the intrinsic zone. Photons impinging on the I zone are absorbed and

---

[*]The photoelectric effect, of course, also applies to a forward-biased diode, but the photocurrent is infinitesimal in comparison to the ordinary diode current. The point of reverse bias is that almost all of the diode's current is photocurrent.

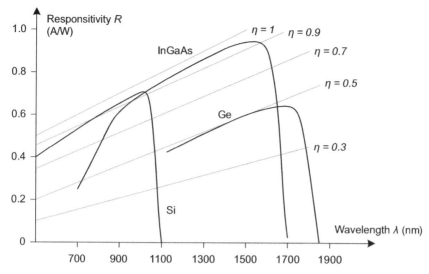

Figure 3.47 Responsivity of silicon (Si), germanium (Ge), and indium–gallium–arsenide (InGaAs).

produce carriers, which are quickly pulled out of the I zone by the electrical field, producing a photocurrent I_{PIN}. Such a structure is called a positive–intrinsic–negative (PIN) diode, shown in Figure 3.48a. The diode is operated in the reverse direction with a bias voltage V_0, which, of course, must be below the breakdown voltage V_B. In the absence of light, surface defects cause a small temperature-dependent dark current $I_{PIN,D}$. Figure 3.48b illustrates the operating curves of the circuit.

A PIN diode is characterized by its conductance G_D and diode capacitance C_D. The capacitance can be varied by altering the length of the depletion zone, that is, the intrinsic length; it greatly affects the bandwidth of the diode and the subsequent electrical amplifier.

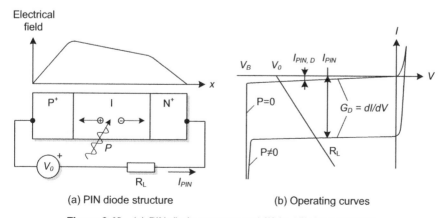

(a) PIN diode structure   (b) Operating curves

Figure 3.48 (*a*) PIN diode structure and (*b*) its V/I characteristics.

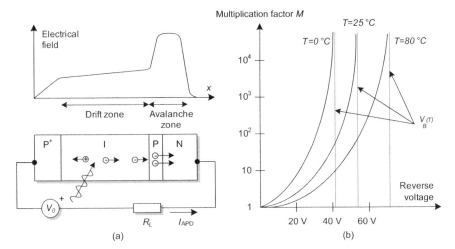

(a)   (b)

Figure 3.49 (a) APD structure and (b) gain over voltage and temperature.

3.9.1.2 Avalanche Photodiodes

It is possible to increase the sensitivity of a photodiode by including a low-noise amplifier directly inside a PIN diode structure. If we introduce another P-doped layer on the N side of the intrinsic zone (Fig. 3.49a), a high electric field can be generated, which enables a process called impact ionization. Photoelectrons are accelerated as they drift through this multiplication zone and release further electron–hole pairs as they collide with the material matrix. This process produces an avalanche of carriers in the material, resulting in a multiplication of the primary PIN-like photocurrent. The multiplication factor M represents avalanche gain:

$$I_{APD} = MI_{PIN} = M.R.P \qquad (3.22)$$

Called an avalanche photodiode, this device is frequently used in G-PON receivers. We see in Figure 3.49b that very substantial avalanche gains can be achieved, but at the cost of a comparatively high-voltage bias supply. An APD must, of course, be operated below its breakdown voltage V_B; the breakdown voltage and multiplication factor both depend on device temperature. To keep the APD gain constant, the bias voltage must be controlled as a function of temperature.

An avalanche photodiode itself is not all that expensive; an APD receiver is costly because of its surrounding components, particularly the bias voltage supply.

The avalanche effect adds multiplication noise because one photoelectron does not always trigger exactly M new carriers. The shot noise in the primary photocurrent increases by a factor $M\sqrt{F}$, where the excess noise factor F is proportional to a power of M. See Figure 3.50.

Because the multiplication factor M affects both signal amplification and the amount of noise generated, it must be optimized to maximize the sensitivity of the receiver. Table 3.8 lists typical parameters for a 2.5-Gb/s APD. As we see, the optimum gain is far below the maximum possible gain suggested by Figure 3.49.

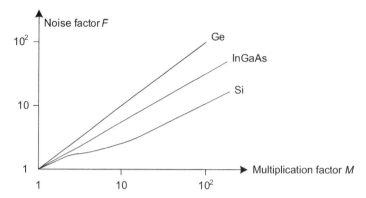

Figure 3.50 Avalanche noise factor versus multiplication factor M.

TABLE 3.8 Typical APD Characteristics

	InGaAs APD	Ge APD
Wavelength range (nm)	1000–1600	700–1800
Quantum efficiency	≥ 0.8	≥ 0.7
Diode capacitance (pF)	≤ 0.5	≤ 1
Time constant (ns)	0.2	0.2
Diameter of active region (μm)	50	30
Typical in-circuit gain (M)	≤ 20	≤ 10

3.9.2 Transimpedance Amplifier

The TIA in a G-PON receiver converts the photocurrent from the PIN or APD detector to a voltage for subsequent clock and data recovery. The TIA must accept a large range of input currents, from several microamperes (20 μA minimum sensitivity) to several milliamperes (2 mA overload).

As illustrated in Figure 3.51, the TIA converts the photocurrent swing ΔI_{PD} flowing through the shunt feedback resistor R_F to a voltage swing ΔV_S. The slope of the transfer curve is the transimpedance Z_T, which is constant for a range of input signals. The average current from the photodetector translates into an average voltage at the output of the TIA, which represents the decision threshold.

The bandwidth of the receiver front end is determined by the photodiode and TIA front-end capacitance as well as the equivalent input resistance, which is $A + 1$ times smaller than R_F due to feedback action (A is the gain of the TIA). Greater bandwidth might seem like a good idea, but greater bandwidth accepts more noise, and, even worse, the noise spectrum turns out to increase at higher frequencies. The choice of bandwidth is thus another optimization tradeoff between signal tracking fidelity and noise performance. In practical designs, the bandwidth of the TIA is chosen to be 10–50% less than the signal symbol rate.

The optical sensitivity of a receiver is mainly determined by the noise performance of the photodetector and the TIA. As suggested in Figure 3.46, TIA noise can

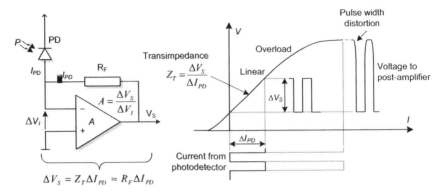

Figure 3.51   TIA transfer function.

be modeled as a frequency-dependent noise current $I_{N,TIA}(f)$ at the input of a noiseless TIA. This is called the input-referred noise current. As we shall shortly see, the sensitivity of a TIA is linearly related to $I_{N,TIA}$, which therefore constitutes one of the most important parameters of a TIA. Table 3.9 lists typical characteristics of the 2.5- and 10-Gb/s TIA circuits used in G-PON receivers.

We are interested in the power spectrum of the noise. Recognizing (but ignoring) the need for a resistance to convert into actual power, we express what we call the power spectrum of the input-referred noise current in squared picoamperes per hertz. The equivalent voltage spectrum has units of squared volts per hertz at the TIA output.

The total noise power (equivalently, the mean-square noise voltage) at the TIA output is derived by integrating the noise spectrum over the largest bandwidth B in the receiver chain, which is usually the bandwidth of the subsequent CDR. In Eq. (3.23), Z_T is the (frequency-dependent) transimpedance whose DC value is the resistance R_F. Assuming no photodiode noise,

$$\overline{V}^2_{N,TIA} = \int^B |Z_T(f)|^2 I^2_{N,TIA}(f)\, df \qquad (3.23)$$

3.9.2.1   Receiver Sensitivity

Returning to the receiver chain of Figure 3.46, the voltage at the output of the TIA contains the desired signal voltage swing V_S and the undesired noise voltage V_N, introduced by the photodetector and the amplifier. In the following analysis, we

TABLE 3.9   Typical 2.5G and 10G PON TIA Specifications

	2.5 Gb/s TIA	10 Gb/s TIA
Small signal bandwidth	2 GHz	10 GHz
Transimpedance (conversion gain)	2000 Ω	500 Ω
Input-referred noise current (RMS)	400 nA	1300 nA
Input overload current	2 mA p-p (0 dBm)	2 mA p-p (0 dBm)
Linear range width	40 µA p-p	40 µA p-p

assume that the postamplifier contributes no additional noise, so that the overall receiver sensitivity is the same as the TIA sensitivity. This approximation is justified by reality.

Equation (3.23) derives TIA output noise as a function of input-referred noise current. So we next need to find the input-referred noise current contributed by the photodetector, which depends on the received data, that is, whether a logical one or a zero is received. Let $I_{N,1}^2(f)$ be the noise spectrum of the photodiode during a 1 bit, and let $I_{N,0}^2(f)$ be the spectrum for a 0 bit. For 1 and 0 bits, respectively, their contribution at the output of the TIA is then given by analogy to Eq. (3.23) as

$$\overline{V}_{N,1}^2 = \int_{}^{B} |Z_T(f)|^2 I_{N,1}^2(f)\, df \qquad (3.24a)$$

$$\overline{V}_{N,0}^2 = \int_{}^{B} |Z_T(f)|^2 I_{N,0}^2(f)\, df \qquad (3.24b)$$

The total RMS (root-mean-square) noise voltage at the TIA output is the contribution of the TIA plus the contribution of the photodiode:

$$\text{One:} \qquad V_{N,1}^{\text{RMS}} = \sqrt{\overline{V}_{N,\text{TIA}}^2 + \overline{V}_{N,1}^2} \qquad (3.25)$$

$$\text{Zero:} \qquad V_{N,0}^{\text{RMS}} = \sqrt{\overline{V}_{N,\text{TIA}}^2 + \overline{V}_{N,0}^2} \qquad (3.26)$$

If the postamplifier adds no noise, the voltage statistics at the decision circuit are the same as at the TIA output, merely scaled up. The detector samples the noisy signal at the center of each bit period, with a threshold that lies halfway between the zero and one values.[*] We therefore have two statistical distributions around the nominal zero and one values of the signal (Fig. 3.52). Both distributions are Gaussian, with the nominal value as the mean, and the zero and one RMS noise voltages as standard deviations. The zero and one variances are not necessarily the same.

The decision circuit compares the sampled signal value against threshold V_{TH}, which is set at the average signal value. A bit error occurs if a zero is interpreted as a one or a one is interpreted as a zero. This BER probability is shown by the shaded areas under the two Gaussian tails in Figure 3.52:

$$\text{BER} = \frac{1}{\sqrt{2\pi}} \int_{Q}^{\infty} e^{-x^2/2}\, dx = \frac{1}{2}\,\text{erfc}\left(\frac{Q}{\sqrt{2}}\right)$$

$$\text{where} \quad Q = \frac{V_S}{V_{N,0}^{\text{RMS}} + V_{N,1}^{\text{RMS}}} \qquad (3.27)$$

---

[*] We discuss the effects of suboptimal sampling time and threshold level below.

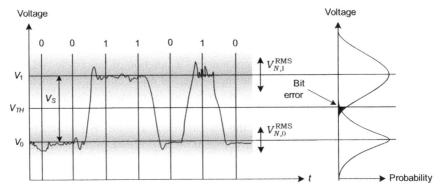

Figure 3.52 Input signal to CDR.

Table 3.10 lists some typical BER and Q values. For a specified BER, the corresponding Q value from Table 3.10 defines the minimum ratio of signal voltage swing to RMS noise voltage at the input to the decision circuit.

The electrical sensitivity of the TIA is now easily calculated from Eq. (3.27) for a given BER. It is defined as the minimum peak-to-peak signal current at the TIA input necessary to achieve a specified BER with an ideal detector circuit.

The signal voltage swing V_S at the output of the TIA is related to the input current swing I_S by Z_0, where Z_0 is the midspan value of the transimpedance transfer function $Z_T(f)$: $V_S = I_S Z_0$. If the input-referred noise currents are $I_{N,1}^{RMS} = V_{N,1}^{RMS}/Z_0$ for a one and $I_{N,0}^{RMS} = V_{N,0}^{RMS}/Z_0$ for a zero, the peak-to-peak minimum electrical sensitivity is

$$I_{sens} = Q\left(I_{N,1}^{RMS} + I_{N,0}^{RMS}\right) \qquad (3.28)$$

The optical sensitivity is defined as the minimum average optical power necessary to achieve a given BER. A photodiode current swing of I_S results in an average current $I_{S,avg} = \frac{1}{2}I_S$, which is generated by an average optical power P_{avg} according to Eq. (3.21) for a PIN diode. For an APD, the current is M times as great, where M is the multiplication factor of the APD. This results in

PIN receiver sensitivity: $$P_{sens,PIN} = \frac{Q\left(I_{N,1}^{RMS} + I_{N,0}^{RMS}\right)}{2R} \qquad (3.29)$$

TABLE 3.10 Relationship between BER and Q

BER	Q	BER	Q
10^{-3}	3.09	10^{-8}	5.61
10^{-4}	3.72	10^{-9}	6.00
10^{-5}	4.27	10^{-10}	6.36
10^{-6}	4.75	10^{-11}	6.71
10^{-7}	5.20	10^{-12}	7.04

APD receiver sensitivity: $\quad P_{\text{sens,APD}} = \dfrac{1}{M} \dfrac{Q\left(I_{N,1}^{\text{RMS}} + I_{N,0}^{\text{RMS}}\right)}{2R}$ $\qquad$ (3.30)

Under the assumption that we can neglect photodiode noise, the input-referred RMS noise currents $I_{N,1}^{\text{RMS}}$ and $I_{N,0}^{\text{RMS}}$ are both equal to the TIA input-referred RMS noise current. The optical sensitivity is therefore determined by:

- Q for the required BER
- Photodiode responsivity R
- Multiplication factor M in the case of an APD
- TIA input-referred RMS noise current

Assuming noiseless photodetectors, then:

PIN receiver sensitivity: $\qquad P_{\text{sens,PIN}} = \dfrac{Q I_{N,\text{TIA}}^{\text{RMS}}}{R}$ $\qquad$ (3.31)

APD receiver sensitivity: $\qquad P_{\text{sens,APD}} = \dfrac{1}{M} \dfrac{Q I_{N,\text{TIA}}^{\text{RMS}}}{R}$ $\qquad$ (3.32)

Table 3.11 lists typical sensitivity values for 2.5- and 10-Gbit/s receivers. Within each block, the sensitivity values in the first column consider only TIA noise; values in the second column also include photodetector noise. We see that the shot noise of a PIN diode is small enough that it does not affect receiver sensitivity, whereas multiplication noise in an APD reduces the sensitivity by about 1 dB.

G.984 G-PON requires ONU receivers with -27 dBm optical sensitivity. As we see, this is achievable only with APD detectors. However, 2.5-Gb/s high-sensitivity TIAs (super-TIAs) have recently become available. Their input-referred noise currents are on the order of 100 nA, providing sufficient optical sensitivity for PIN-based receivers. This development has the potential to affect the cost of ONU optics significantly, especially if the same can be done at 10 Gb/s for XG-PON.

3.9.3 Burst-Mode TIAs

In the previous section, we implicitly assumed continuous-mode operation of the receiver, which is appropriate for downstream flows to an ONU. At the OLT, however, the difference in average power between adjacent upstream bursts from widely separated ONUs is large. In reality, it is far greater than illustrated in Figure 3.53. To support a burst-mode OLT receiver, the TIA design needs to provide a large dynamic range.

The dynamic range of a receiver is the ratio of maximum-to-minimum optical input power that provides the specified BER. The minimum sensitivity P_{sens} determines the minimum power necessary; the overload power P_{OL} defines the limit for large signals. With TIA input overload current values from Table 3.9, a

TABLE 3.11 Typical Sensitivity Values for TIAs

Parameter	Symbol	1.25/2.5 Gb/s TIA	10 Gb/s TIA
Input-referred noise	$I_{N,\text{TIA}}^{\text{RMS}}$	400 nA	1.3 µA
BER $\leq 10^{-10}$			
G-PON downstream			
G-PON upstream			
Input signal swing	I_{sens}	5.1 µA	
Sensitivity PIN ($R = 0.9\,\text{A/W}, M = 1$)	$P_{\text{sens,PIN}}$	-25.5 dBm	-25.5 dBm
Sensitivity APD ($R = 0.9\,\text{A/W}, M = 10$)	$P_{\text{sens,APD}}$	-35.5 dBm	-34.3 dBm
BER $\leq 10^{-4}$			
XG-PON upstream			
Input signal swing	I_{sens}	3 µA	
Sensitivity PIN ($R = 0.9\,\text{A/W}, M = 1$)	$P_{\text{sens,PIN}}$	-27.8 dBm	-27.8 dBm
Sensitivity APD ($R = 0.9\,\text{A/W}, M = 10$)	$P_{\text{sens,APD}}$	-37.8 dBm	-37.1 dBm
BER $\leq 10^{-3}$			
XG-PON downstream			
Input signal swing	I_{sens}	8 µA	
Sensitivity PIN ($R = 0.9\,\text{A/W}, M = 1$)	$P_{\text{sens,PIN}}$	-23.4 dBm	-23.5 dBm
Sensitivity APD ($R = 0.9\,\text{A/W}, M = 20$)	$P_{\text{sens,APD}}$	-33.5 dBm	-32.6 dBm

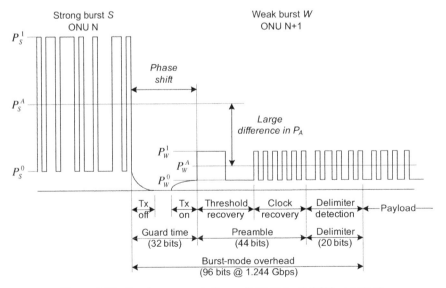

Figure 3.53 Burst recovery challenge (1.25 Gb/s, G-PON upstream).

responsivity of 0.9 A/W, and an APD gain of 10, the overload sensitivity P_{OL} is about -9 dBm. As an example, a 2.5-Gb/s APD-based receiver with BER 10^{-10} supports a dynamic range of 25 dB, between -9 and -34 dBm.

The G-PON recommendations uniformly specify -8 dBm as the overload limit, but keep in mind that this is -8 dBm at the optical connector. It is easy to discover another dB of loss inside the transceiver device (Section 3.10).

The G-PON recommendations specify a dynamic range of up to 20 dB, 15 dB for the ODN and 4.5 or 5 dB for variations in ONU transmit power. So our problem is not with the intrinsic capabilities of the front end but with the transient response of the circuits. We cannot afford extended stabilization times. In the presence of widely varying input signal levels, the receiver must accurately estimate the decision threshold within only a few preamble bits. The difference in clock phase from burst to burst is a related transient response issue, to be discussed in a subsequent section.

Table 3.12 shows that a burst-mode receiver has about 77 ns in G-PON and about 103 ns in XG-PON to reset (guard time), recover the threshold and clock phase (preamble), and detect the start of frame (delimiter).

The simplest way to remove the offset of the average signal level between bursts—which implicitly corresponds to setting the decision threshold—is to use AC coupling between a conventional continuous-mode TIA and the PA, as shown in Figure 3.54.

For each burst, assuming it lasts long enough, capacitance C charges to the burst's average value. With each new burst, the capacitor charges to the new value with time constant RC, where R is some designed or implicit resistance in the circuit. In general, this tends to provide slow settling to the proper data recovery threshold. To avoid threshold offsets, a long burst preamble is needed or the capacitor must be discharged by a reset signal.

A reset decouples one burst from the next. But to discharge the capacitor, we need a reference voltage, which we must choose to be some nominal threshold value. We will do better by choosing a threshold approximately right for weak-signal inputs. This gives weak signals more time to converge on an optimum decision threshold. In contrast, we know that the error rate for strong signals will be small, even if the decision threshold is not quite right.

Another problem occurs if there is an imbalance between ones and zeros in the received data. This causes an incorrect estimate of the average value and a suboptimal threshold.

DC coupling avoids the problems related to AC coupling but requires more sophisticated components. See Figure 3.55. At the beginning of each burst, the

TABLE 3.12 Burst Overhead Allocation

	Tx Enable	Tx Disable	Guard Time	Preamble Time	Delimiter Time	Total
G-PON	16 bits	16 bits	32 bits	44 bits	20 bits	96 bits
1244 Mb/s	12.8 ns	12.8 ns	25.7 ns	35.4 ns	16.1 ns	77.2 ns
XG-PON	32 bits	32 bits	64 bits	160 bits	32 bits	256 bits
2488 Mb/s	12.8 ns	12.8 ns	25.7 ns	64.3 ns	12.8 ns	102.9 ns

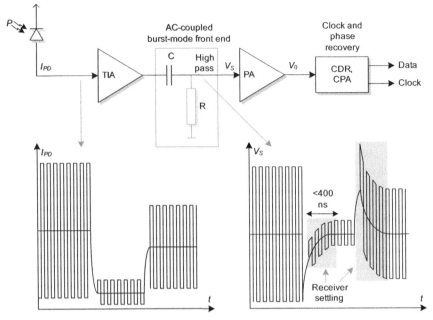

Figure 3.54 AC-coupled receiver performance with burst-mode input.

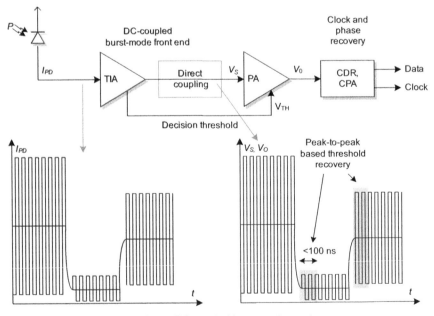

Figure 3.55 DC-coupled burst-mode receiver.

decision threshold for that burst is extracted from the preamble by measuring the amplitude of one and zero bits and setting the threshold halfway between. The threshold extraction process is triggered by a reset signal from the OLT MAC device, which knows the start time of each burst. In principle, the burst preamble for a DC-coupled receiver can be quite short.

The optimum threshold lies in the center of the eye; a suboptimal threshold implies either a wider eye opening or a higher bit error rate. However we choose to think of it, a threshold estimation error results in reduced receiver sensitivity, called a burst-mode penalty. For thresholds within 10–20% of the optimum level, we incur a penalty on the order of 1–2 dB.

3.9.3.1 Automatic Gain Control (AGC)

When operating in burst mode, the TIA must accept consecutive bursts with large amplitude variations. The TIA gain must be adaptively set for every burst—for strong bursts, the TIA must reduce the gain, while higher gain is needed for small signals. The TIA overload current and its sensitivity are both directly related to the feedback resistor R_F. The gain can be adjusted from burst to burst by an adaptive transimpedance.

A simple and fast burst-to-burst automatic gain control mechanism can be implemented by replacing the feedback resistor R_F in Figure 3.51 with the nonlinear passive circuit shown in Figure 3.56. For small signals, the diode isolates the additional resistor R_{F2}. For large-amplitude signals, the voltage drop across R_F is large enough to open the diode, reducing the transimpedance. The change in transfer slope effectively compresses large signals, but as we see, we still need to compensate for differences in average signal value.

3.9.4 Postamplifier

The PA further amplifies the small voltage from the TIA for proper clock and data recovery and further compensates for signal amplitude and offset variations.

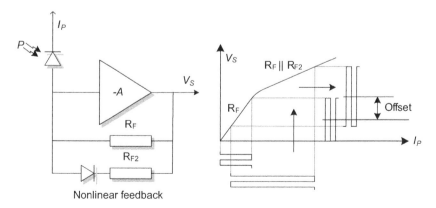

Figure 3.56 Burst-mode TIA with AGC.

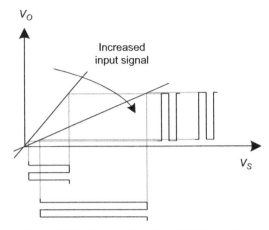

Figure 3.57   Transfer function of AGC amplifier.

For continuous-mode reception, or if the burst-mode TIA already corrects for gain and offset, the PA can be a simple limiting amplifier (LA) without further correction. The transfer function of an LA is independent of the input signal swing. For small signals, the amplifier provides gain in the linear regime; for large signals, the amplifier clips the output signal.

In contrast, an AGC PA adjusts the gain according to the input signal swing, keeping the amplifier always in the linear regime. This avoids distortion of large inputs with low extinction ratio or signals with large offsets. This is apparent when comparing the burst-mode response from an adaptive-gain TIA (Fig. 3.56) with the response of an AGC PA, as shown in Figure 3.57.

The PA needs to provide a large voltage gain. A weak optical burst providing a TIA output swing of 5 mV must be amplified to an output voltage of 1 V—a gain of 46 dB. A 2.5-Gb/s amplifier with 50 dB gain and 3 GHz bandwidth would require a gain–bandwidth (GBW) product of 150 GHz, while a 10-Gb/s amplifier with 50 dB gain and 12 GHz bandwidth would require GBW of 600 GHz—much larger than is achievable with current technology. The solution is through multiple stages of lower GBW amplifiers, each with a gain of around 10 dB and the necessary bandwidth. Except for a slow offset compensation loop, there is no feedback over the amplifier chain, so stability is easy to ensure. Typical PAs have four to eight stages.

3.9.5 Clock and Data Recovery

One way or another, the burst-mode front end provides a constant-amplitude signal. The clock and data recovery unit must recover clock phase and frequency[*] if it is to sample the data at exactly the right instant.

---

[*] Frequency tracking is required in the ONU, whose clock is recovered from the downstream signal and used for upstream transmission. The OLT already knows the upstream frequency.

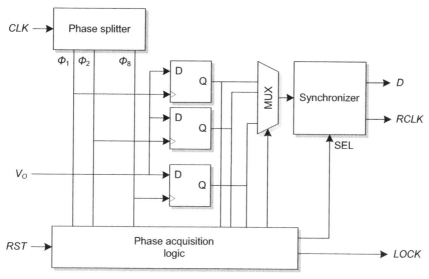

Figure 3.58 Burst-mode CPA.

The ONU's continuous mode receiver uses a conventional PLL-based CDR circuit. The high-speed burst-mode receiver at the OLT commonly uses a CPA, based on temporal or spatial oversampling.

Figure 3.58 is a block diagram of a spatial oversampling CPA. Provided by the MAC just prior to a burst, the reset signal RST informs the CPA that a preamble is arriving and triggers the phase acquisition procedure.

The phase splitter derives eight equally phase-shifted versions Φ_1 to Φ_8 of the reference clock CLK, which in turn is derived from the OLT's downstream bit clock. These edges are used to sample the input data V_O into latches at eight equally spaced instants within each single bit period. Phase acquisition logic then examines the eight separate bit patterns from the eight separately clocked latches and selects the phase closest to the middle of the eye, based on its foreknowledge of the preamble pattern sent by the ONU. Figure 3.59 shows this in the time domain.

The CPA also provides a lock signal LOCK to indicate when the phase has been successfully aligned. A synchronizer delivers the recovered clock RCLK in sync with the recovered data D.

Because the data signal is spatially split, the highest clock frequency in the CPA is that of the bit rate. The same functionality can be achieved by oversampling the input signal at eight times the bit frequency and selecting the best sample. For a 2.5-Gb/s signal, the internal chip clock would be 20 GHz, which is a challenge in CMOS (complementary metal–oxide–semiconductor) processes.

In both G-PON and XG-PON, the CDR and CPA functionality is usually an integral part of the MAC chip set.

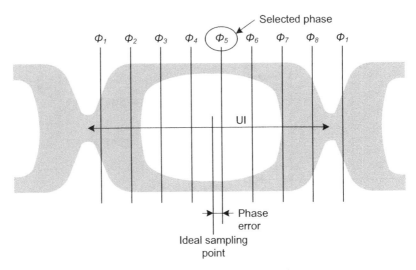

Figure 3.59 Phase selection procedure.

3.10 G-PON TRANSCEIVER MODULES

G-PON transceivers contain a significant number of highly integrated optical and electrical components, as shown in Figure 3.60.

At the OLT, a small-form-factor pluggable (SFP) module contains a dual-wavelength bidirectional optical subassembly (BOSA) and the physical medium-dependent layer integrated circuit (PMD IC). The BOSA comprises a transmitter optical subassembly (TOSA), which contains the laser diode, and a receiver optical subassembly (ROSA), which integrates the photodetector and TIA, as well as a WDM device. The PMD IC contains the bias sources, laser driver, PA, and a microprocessor that controls the other components. The OLT transmitter is designed for continuous-mode operation; the receiver is a burst-mode design.

ONU optics come in different packages. The digital ONU (Internet, IPTV, voice telephony, business services) often uses a fixed-mounted SFF (small-form-factor) optical module that contains a dual-wavelength diplexer BOSA. An ONU supporting RF video contains a three-wavelength triplexer BOSA with an additional ROSA for the video wavelength. The ONU's PMD IC includes a burst-mode laser driver and a continuous-mode PA as well as an APD bias source. Figure 3.61 shows SFP (left) and SFF (right) modules as used in the OLT and ONU, respectively.

The following sections investigate the various components.

3.10.1 Bidirectional Optical Subassembly

In the conventional design shown in Figure 3.61, the optical components are mounted in bulk-optical assemblies that contain stand-alone lasers and photodetectors with the necessary filters and lenses. Optical paths are established in free space by properly

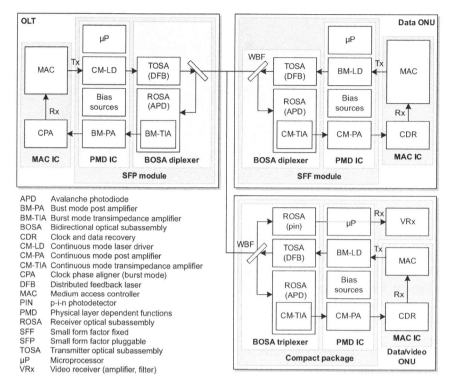

APD   Avalanche photodiode
BM-PA   Bust mode post amplifier
BM-TIA   Burst mode transimpedance amplifier
BOSA   Bidirectional optical subassembly
CDR   Clock and data recovery
CM-LD   Continuous mode laser driver
CM-PA   Continuous mode post amplifier
CM-TIA   Continuous mode transimpedance amplifier
CPA   Clock phase aligner (burst mode)
DFB   Distributed feedback laser
MAC   Medium access controller
PIN   p-i-n photodetector
PMD   Physical layer dependent functions
ROSA   Receiver optical subassembly
SFF   Small form factor fixed
SFP   Small form factor pluggable
TOSA   Transmitter optical subassembly
µP   Microprocessor
VRx   Video receiver (amplifier, filter)

Figure 3.60   G-PON transceiver components in OLT and ONU.

aligning the optical components. Newer transceiver designs are based on PLC technology. Light is routed in waveguides on an optical substrate that is manufactured with the same processing technology used in electronic ICs. We start by looking at the bulk-optics approach.

3.10.1.1 Bulk-Optics Assembly

The ROSA contains a photodetector and TIA, surface mounted in a metallic TO (transistor outline) can (Fig. 3.62). The active diameter of an APD is large

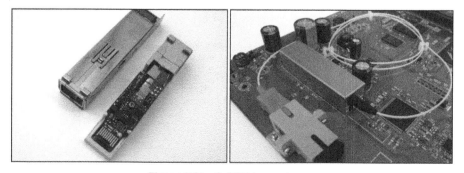

Figure 3.61   G-PON transceivers.

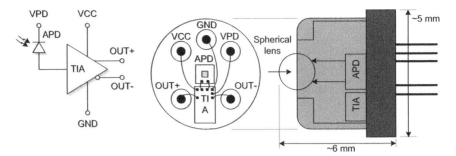

Figure 3.62   ROSA TO-46 APD configuration.

(50 μm), and alignment with a simple spherical microlens provides nearly 100% coupling efficiency. The capacitance of the photodetector and the wire bonds to the TIA greatly affects the noise and therefore the achievable receiver sensitivity. For that reason the TIA is placed as close as possible to the photodiode.

Figure 3.63 shows the TOSA, which contains a DFB laser, together with its monitoring photodiode. It is also typically packaged in a TO-46 can. The monitoring photodiode—usually a low-cost PIN diode—is mounted on the back of the DFB laser. An aspherical microlens reduces aberration and provides a coupling efficiency of typically 60%. An isolator filter may also be present to suppress back reflections into the laser.

In a bulk-optics transceiver, TOSA and ROSA devices are orthogonally mounted in the BOSA package, shown in Figures 3.64 and 3.65. Light flows in free space between the filters and lenses.

In the G.984 G-PON ONU BOSA of Figure 3.64, the 1490-nm downstream wavelength is reflected into the ROSA by a diagonally mounted dichroic TFF. A second TFF at the ROSA implements the wavelength blocking characteristics required by G.984.5. The upstream 1310-nm wavelength passes directly through the diagonal mirror and couples into the fiber. Microlenses in ROSA, TOSA, and

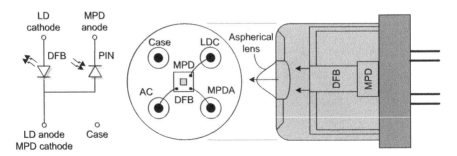

Figure 3.63   TOSA TO-46 DFB configuration.

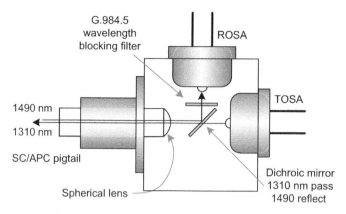

Figure 3.64   G.984 G-PON ONU diplexer BOSA.

fiber pigtail termination support efficient coupling of the light path, and antireflective coating within the enclosure suppresses optical crosstalk.

The diplexer structure can be extended to become a triplexer BOSA, illustrated in Figure 3.65. An additional mirror, blocking filter and video ROSA enable reception of the 1555-nm analog video signal. Figure 3.66*a* shows a triplexer BOSA. Figure 3.66*b* shows an open diplexer BOSA—we clearly see the 90° ROSA reflected in the dichroic mirror.

During device assembly, the components are powered up and manually aligned by monitoring their launched and received power. Temperature gradients in the BOSA can misalign the components during operation, resulting in performance variations. Free-space BOSAs achieve a tracking error in the range of 1–2 dB.

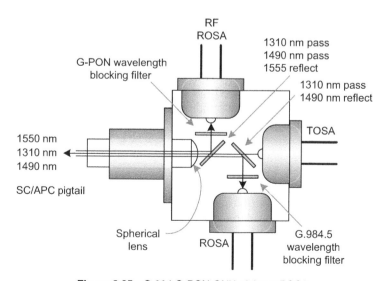

Figure 3.65   G.984 G-PON ONU triplexer BOSA.

(a)

(b)

Figure 3.66 (a) Triplexer and (b) opened diplexer.

3.10.1.2 Planar Lightwave Circuits

Instead of mounting laser, detector, TFFs, lenses, and pigtail into a three-dimensional, free-space mechanical fixture, planar lightwave technology directs optical signals through waveguides on a substrate that can be fabricated in much the same way as conventional electronic ICs. Complex structures can be grown on the substrate itself. High integration and accurate alignment during production eliminate the need for lenses and ease assembly and packaging.

Figure 3.67 shows a PLC triplexer. Waveguides on the surface have been highlighted in white to show the light flow across the device. Observe that the monitoring photodiode is tapped from the front facet of the laser, rather than the more traditional back-facet monitor.

A PLC greatly attenuates the effect of temperature over free-space BOSA designs. PLC-based transceivers operate over the full industrial temperature range with a tracking error less than 0.5 dB and better efficiency. Table 3.13 compares typical bulk

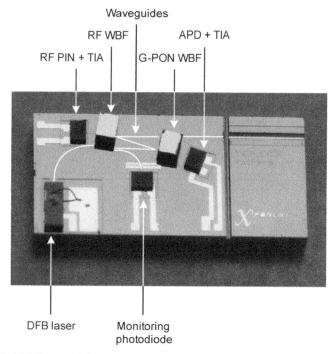

Figure 3.67   PLC-based BOSA triplexer. (Reprinted by permission of Hoya Corporation USA.)

optics and PLC G-PON ONU BOSA performance, assuming APD and DFB components.

3.10.2   ONU Evolution

Optics account for 40–50% of the total cost of the ONU. The industry has introduced several component generations with a strong focus on cost reduction. The main evolutionary steps are:

- *SFF-Based ONU* (original design)   Many of today's popular ONU designs use fixed surface-mounted SFF optical modules, which contain all optical and physical medium-dependent functions, as well as a small control microprocessor. Figure 3.68 illustrates this design.

TABLE 3.13   Comparison of G-PON ONU BOSAs, Free-Space and PLC

Parameter	Bulk-Optics Assembly	PLC
Output power	+2 dBm	+6 dBm
Receiver sensitivity	−28 dBm	−30 dBm
Tracking error	<2 dB	<0.5 dB
Optical return loss	−10 dBm (1310 nm)	−15 dBm (1310 nm)
Size ($W \times H \times D$)	20 mm × 7 mm × 7 mm	2 mm × 2 mm × 1 mm

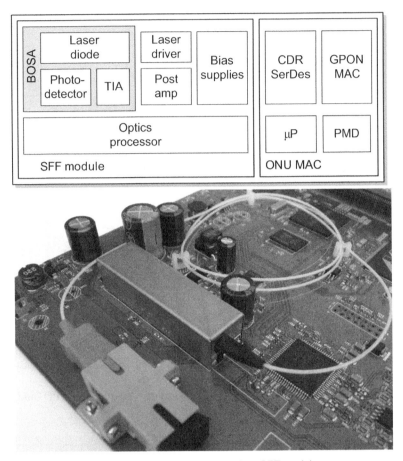

Figure 3.68   ONU design based on SFF module.

- *Discrete Optics ONU* (appeared in 2008)   Some ONU vendors put BOSA and the related analog circuitry directly on the printed circuit board. At first, these ONUs used individually packaged laser driver, limiting amplifier, bias sources, current monitor, processor, and BOSA. Then the laser driver and limiting amplifier were integrated into a single chip, followed by the power supply. Single-chip PMD solutions are now available, interfacing the BOSA and MAC directly. In Figure 3.69, the PMD component is just behind, almost under, the BOSA leads.

- *BOSA on Board* (2011)   Then chip suppliers integrated the optics-related electronics into the MAC chip set, keeping only the BOSA as a separate device. The closely integrated and tightly controlled BOSA assembly with minimal internal power dissipation reduces the effects of temperature and aging, which results in relaxed BOSA specifications and notably lower cost optics. Dual-loop laser drivers and accurate power control circuits eliminate the need for factory calibration. Figure 3.70 shows an example.

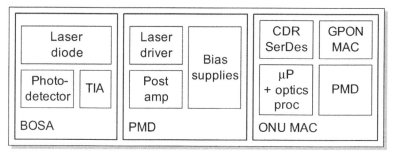

Figure 3.69 Discrete optics ONU.

OLT optics have also changed. When the G-PON market offered only the B+ optical budget option, there was no need for OLT optics to be pluggable, and most OLT designs used SFF fixed-mounted optical modules. With the introduction of the extended C+ budget, the SFP industrial package has become common, so that a given OLT port can be tailored for either budget. An additional benefit is that the port can be repaired individually if the transceiver fails.

3.11   OPTICAL AMPLIFIERS

3.11.1   Semiconductor Optical Amplifiers

A semiconductor optical amplifier (SOA) is similar to a semiconductor laser, but without a resonator. Instead of mirrors, the end faces are angular, with antireflective coating. They reflect as little energy as possible (0.01%) to avoid the risk of oscillation. With no feedback mechanism, light passes through the active region

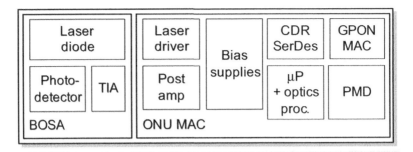

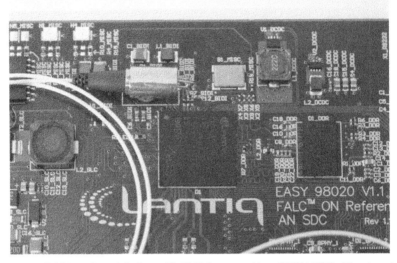

Figure 3.70 BOSA-on-board design, with PMD integrated into GPON MAC. (Courtesy Lantiq North America Inc., 2010.)

only once, amplified by simulated emission. An electrical pump current creates the charge inversion that is the prerequisite for gain. Figure 3.71 shows a simple SOA structure.

A SOA is characterized by a few basic parameters: gain, gain bandwidth, saturation power, noise figure, and polarization sensitivity.

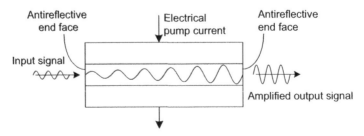

Figure 3.71 Simplified SOA structure.

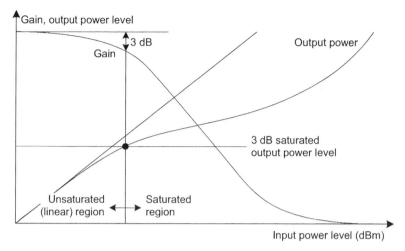

Figure 3.72 Gain and saturation in SOA.

The gain of a SOA depends on the pump current and the input power level. Figure 3.72 shows that the SOA provides high gain for low input signal levels; the SOA operates in its unsaturated region. As input power increases, the signal is still amplified, but the gain declines. For high input power levels, the gain approaches unity, where the output power equals the input power. The output power level at 3 dB gain reduction is called the saturation power. Operating the SOA above the saturated power (outside the linear region) degrades the quality of the signal. This power level, together with the gain of the SOA, defines the maximum allowed input signal level.

Figure 3.73 shows that higher pump current increases both gain and saturated output power. Pump current is therefore the tool to adjust gain and power level to fit the application.

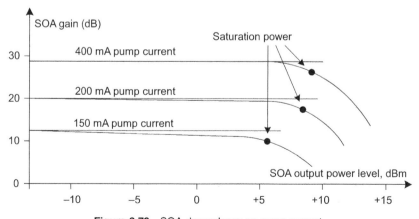

Figure 3.73 SOA dependency on pump current.

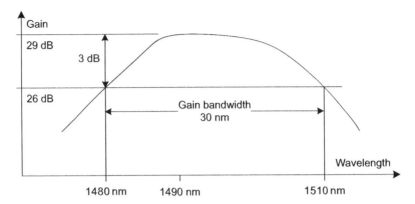

Figure 3.74   Gain versus wavelength for 1490-nm SOA.

SOAs can be tailored to operate at any wavelength supported by the fiber, with a bandwidth span from several nanometers to 140 nm. The gain bandwidth describes the spectrum across which the gain is within 3 dB of its maximum. Figure 3.74 illustrates the gain characteristics of a 1490-nm SOA.

A SOA adds amplified spontaneous emission (ASE) noise to its optical output. Intentional amplification occurs when incident photons cause electrons in the excited state to drop back to the ground state, emitting coordinated photons in the process. But some of the electrons in the excited state drop back to the ground state randomly, and the photons they emit are also amplified as they travel further through the active region. ASE is broadband noise, proportional to the gain of the SOA. To suppress ASE outside the operating spectrum, SOAs are combined with optical filters. With noise figure defined as the ratio of input to output optical SNR (signal-to-noise ratio), practical SOAs achieve a noise figure in the area of 5–10 dB.

In the next section, we discuss fiber amplifiers. To anticipate that discussion, we point out some of the advantages of SOAs over fiber amplifiers:

- SOAs can be tailored to a wide range of wavelength windows.
- SOAs consume less power.
- SOAs are compact, being based on PLC technology.
- SOAs have very short excited-state electron lifetimes, around 200 ps. Amplifiers such as EDFAs (erbium-doped fiber amplifiers) may have excited-state lifetimes of as much as 10 ms. SOAs are therefore perfectly suited for burst-mode operation because SOA gain equilibrates in less than one G-PON bit interval.
- But the property of fast gain dynamics is a drawback when multiple wavelengths (DWDM) are amplified by a single SOA, as cross-gain modulation by high bit-rate signals causes intersymbol interference within and between channels.

Tables 3.14 and 3.15 list typical ratings for SOAs used in G-PON reach extenders and booster amplifiers.

TABLE 3.14   SOA as G-PON Reach Extender

Parameter	SOA 1490 nm	SOA 1310 nm
Peak gain	26 dB	29 dB
Operating wavelength	1490 nm	1310 nm
Gain bandwidth	30 nm	40 nm
Saturation power	+9 dBm	+9 dBm
Noise figure	8 dB	7.5 dB
Polarization sensitivity	1.5 dB	1.9 dB
Typical bias current	390 mA	420 mA
Power consumption	4 W	5 W

TABLE 3.15   SOA as Booster Amplifier (XG-PON OLT)

Parameter	SOA 1570 nm	SOA 1260 nm
Peak gain	9.5 dB	9.5 dB
Operating wavelength	1570 nm	1260 nm
Gain bandwidth	40 nm	40 nm
Saturation power	+12 dBm	+11 dBm
Typical bias current	300 mA	300 mA
Power consumption	3 W	4 W

3.11.2   Erbium-Doped Fiber Amplifiers (EDFAs)

A silica fiber doped with erbium can become an active medium and an optical amplifier. While SOAs use electrical current to excite electrons to higher energy levels, EDFAs use optical pumping. As shown in Figure 3.75, pump photons of short and stable wavelengths raise electrons into a fairly narrow pump band. Electrons decaying from the pump band release some of their energy in the form of phonons (vibrations) as they drop into a metastable band, and then decay from the metastable band back to ground state, amplifying signal photons by stimulated emission.

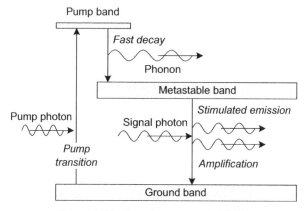

Figure 3.75   Pumping process in EDFA.

TABLE 3.16 Typical EDFA Ratings

Parameter	EDFA 1550 nm (C band)	EDFA 1570 nm (L band)
Peak gain	30 dB	22 dB
Operating wavelength	1550 nm	1570 nm
Gain bandwidth	40 nm	25 nm
Noise figure	5 dB	6 dB
Optical output power	+17 (up to +26) dBm	+13 (up to +17) dBm
Gain flatness	1.5 dB	1.5 dB
Fiber coupled pump power	250–300 mW	300–400 mW

Pump light is injected in the same direction as the signal, via a fused fiber splitter or TFF. This arrangement provides low ASE but limited gain and is appropriate for receiver preamplifiers. Pumping in both directions doubles the gain for use in booster amplifiers but increases ASE.

Originally, EDFAs were limited to the C band (1530–1560 nm) because the erbium gain peak drops off rapidly outside this region. With improvements in doping technology, the availability of high-power pump lasers (optical power levels up to 0.5 W), and longer amplification fibers, L-band EDFAs (1570–1600 nm) with gains around 20 dB became feasible. Table 3.16 lists parameters of typical EDFAs.

3.12 REACH EXTENSION

An RE is an active device placed in an ODN to increase the link budget and thereby the reach or the split ratio of the PON. Reach extension is done at the physical layer only, either by OA or by OEO regeneration, and without changing the G.984.3 or G.987 protocol.

The RE is inserted at the head end of a class B+ or C+ ODN that serves a collection of ONUs, but at the tail end of an optical trunk link (OTL) connected to the OLT. The RE extends the ODN by the length of the OTL. The OTL may be split near the OLT into several fibers, each of which may or may not be served by a reach extender (Fig. 3.76). This option allows the serving area to be tailored to the subscriber distribution, for example, along the length of a valley. The overall split ratio may be increased either through multiple splits ahead of the RE(s) or by designing the RE itself to support several ODNs in parallel, as suggested by RE *b* in Figure 3.76. Splitters may or may not be built into REs, but are in any event likely to be colocated with them.

Recommendation G.984.6 specifies G-PON's physical medium-dependent parameters for the OTL for both OA and OEO REs. See Table 3.17. Recommendation G.987.4 contains the same information for XG-PON.

A reach extender is an active piece of equipment. It requires electrical power and protection against failures of the primary power supply. It requires management access. It requires a protected environment. All of this additional cost makes it

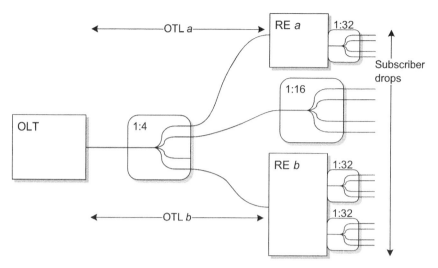

Figure 3.76   Reach extension topology.

TABLE 3.17   Optical Parameters of G.984 G-PON OTL

Parameter	OEO-type RE	OA-type RE
Upstream attenuation	14–27.5 dB	≤28 dB
Downstream attenuation	11–23 dB	≤23 dB
Maximum optical path penalty	1 dB	1 dB
Maximum OTL reach	60 km minus maximum ODN length	60 km minus maximum ODN length

more difficult to justify the business case for G-PON. Further, if a remote active element is really needed, why not a mini-OLT? The demand for G-PON RE equipment has been low.

3.12.1   OEO-Based RE

An OEO-based RE converts the optical signal into an electrical signal and then back into the optical domain with conventional G-PON optics. The process is often designated 3R: retiming, reshaping, regeneration. Figure 3.77 shows an OEO G-PON reach extender. It contains standard ONU optics facing the OLT and slightly modified (resetless) OLT optics facing the ONUs.

At the optical layer, to the OLT, the RE looks approximately[*] like the array of ONUs. To the ONUs, the RE looks like the OLT.

---

[*]*Approximately* because burst header repetition, phase, and power transmitted upstream from the RE are proxies, rather than the real signals from the ONUs. We discuss some of the implications of the difference below and in Chapter 4.

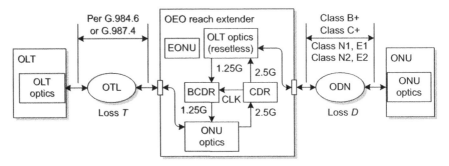

Figure 3.77   OEO reach-extended G-PON.

The operating region of an OEO extender is easily derived from the minimum and maximum insertion loss. Clearly, the OTL and ODN budgets are independent. Figure 3.78 portrays the operating region for a G.984 G-PON OEO reach extender with a class B+ ODN (loss 13–28 dB). If the OTL has symmetric loss, the operating region is bounded by G.984.6 to the 14–23 dB range. The rectangular operating region demonstrates that an OEO reach extender isolates the OTL and the ODN optical budgets. This maximizes network planning flexibility.

For management, the RE includes an embedded ONU[*] (EONU). The EONU may be decoupled from the RE itself by connecting it to the OTL via an optical tap coupler. The EONU is thus still reachable even if some parts of the RE fail. An alternative RE design gains access to the EONU electrically.

The economics of reach extension are dominated by the cost of remote power, management, and environmental enclosure. This cost may be amortized by extending several PONs from the same RE. The realities of geography limit this option to about four PONs, maybe a few more. If multiple REs exist in a single box (multiport RE), a single EONU suffices to manage all REs.

The introduction of an OEO RE into an ODN causes some loss of upstream efficiency due to increase of the total burst overhead. In OLTs, the optics require a reset signal provided by the G-PON MAC to indicate the beginning of a new upstream burst. The reset signal isolates consecutive bursts by normalizing the data recovery decision threshold circuit between bursts. The RE contains no OLT MAC, and it would be nontrivial to adapt an OLT MAC, so OLT optics without the need for reset signals were developed. These optics require additional preamble bits to detect the start of a new burst and stabilize the threshold and gain mechanisms. The OLT must compensate by specifying extra preamble bits at the beginning of each burst. The number of extra preamble bits required by a particular OEO RE implementation is not known to the OLT a priori, so it must be learned during ranging. The OLT follows these steps, which are explained in detail in Chapter 4:

---

[*] G.984.6 calls it an embedded ONT.

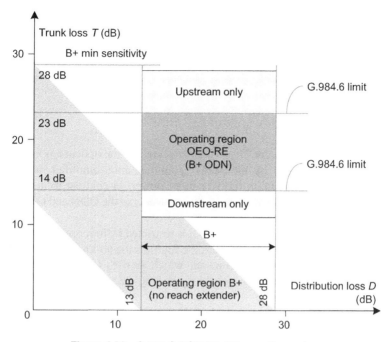

Figure 3.78   G.984 G-PON B+ RE operating region.

1. The OLT broadcasts default G.984 G-PON *overhead* and *extended overhead* messages, or the equivalent *profile* message for G.987 XG-PON.
2. The OLT activates the reach extender by ranging the RE's EONU and establishing an OMCI communications channel with it.
3. Via OMCI, the OLT obtains the reach extender's preamble requirement, as specified by the RE manufacturer.
4. The OLT broadcasts a revised extended overhead message that includes the reach extender's extra preamble. The EONU disregards this message, but ordinary ONUs accept it.
5. Now that the upstream channel is properly aligned, the OLT discovers and activates the ONUs downstream of the reach extender.

3.12.2 OA-Based Reach Extenders

In an OA-based reach extender, two SOAs are used to amplify the optical signals in the downstream and upstream paths separately. The wavelengths are separated by a diplexer filter, individually amplified, then recombined by a second diplexer. Optical bandpass filters restrict the bandwidth of ASE noise generated by the amplifiers.

Because OAs function purely on the optical level, an OA-based RE provides no signal regeneration. ONU and OLT transmitters must support a dispersion reach of the combined length of the OTL and ODN, up to 60 km.

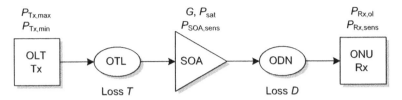

Figure 3.79 Amplified G-PON, downstream.

The link budget varies according to the parameters of the optical amplifier in use. The budget is constrained by the SOAs' amplifier gain, amplifier noise figure, saturated power, and maximum receiver input power. In contrast to an OEO RE, a single optical domain spans the whole fiber network, so the ODN and OTL budgets depend on each other.

The operating region of a SOA-based RE is restricted by four optical limits. Three of them are introduced by the SOA itself; the fourth is the 60-km logical reach limit. Figure 3.79 illustrates the downstream path, which we use to derive the downstream operating area. The upstream analysis is equivalent, *mutatis mutandis*.

1. *SOA Power Limit*   In an unbalanced distribution network with large loss D in the ODN and small loss T in the OTL, the SOA functions as a booster amplifier, improving the OLT transmitter's output power. If the maximum SOA output power is P_{sat}, loss D must be small enough to deliver at least power equal to $P_{Rx,sens}$ to the receiver:

$$D < P_{sat} - P_{Rx,sens} \qquad (3.33)$$

2. *SOA Noise Limit*   In the less likely case that loss T dominates the ODN, and D is small, the SOA functions as a preamplifier. It is necessary that $P_{Tx,min} - T > P_{SOA,sens}$ so

$$T < P_{Tx,min} - P_{SOA,sens} \qquad (3.34)$$

The SOA manufacturer specifies $P_{SOA,sens}$.

3. *SOA Gain Limit*   If the losses are approximately balanced, the SOA functions as an in-line amplifier, providing gain to the optical signal. Called the SOA gain limit, this constraint clearly shows the coupling between OTL and ODN optical domains. Because $P_{Tx,min} - T + G - D > P_{Rx,sens}$,

$$T + D < P_{Tx,min} - P_{Rx,sens} + G \qquad (3.35)$$

To restate Eq. (3.35), the total loss in the optical network must not exceed the normal optical budget plus the SOA gain.

Figure 3.80 illustrates the outer operating boundaries. Considering both directions of transmission, we expect two sets of limits, with the smaller one constraining overall system operation.

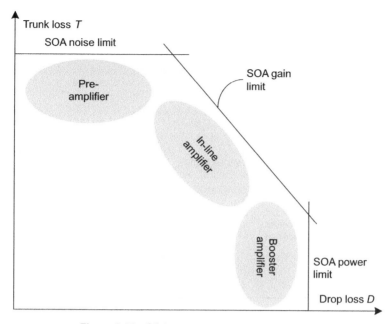

Figure 3.80 SOA operating region—outer limits.

As well as maximum values for losses D and T, there are two constraints on their minimum values.

1. *Receiver Overload Limit* Losses T and D must be large enough to avoid receiver overload:

$$P_{Tx,\max} + G - T - D < P_{Rx,ol} \tag{3.36}$$

2. *SOA Saturation Limit* The SOA output power must be held below its saturation point, also implying a minimum loss T:

$$T > P_{Tx,\max} + G - P_{\text{sat}} \tag{3.37}$$

The most restrictive limits are the saturation limits.

Figure 3.81 illustrates the minimum bounds, showing SOA saturation in both directions and a single-direction R_x overload limit.

As a numerical example, we use the performance values of G-PON SOAs from Table 3.14 and the B+ optical specs from Table 3.6, with a hypothetical SOA noise limit. Combining the above limits into a single graph for the G.984 G-PON upstream and downstream optical path, Figure 3.82 illustrates a typical achievable operating region.

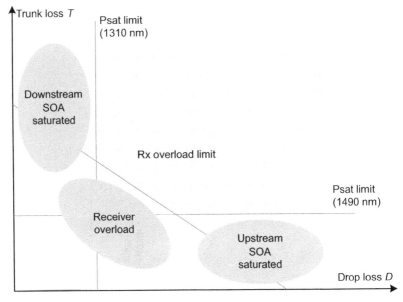

Figure 3.81   SOA operating region—minimum bounds.

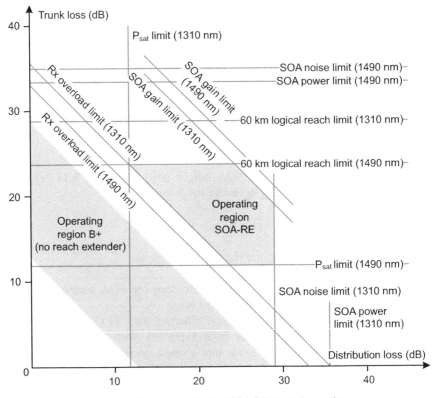

Figure 3.82   Typical G-PON SOA-RE operating region.

From Figure 3.73, we recall that SOA characteristics can be tuned by adjusting the pump current. We can also reposition the operating region with optical attenuators. In either case, we are likely to need special engineering for each OA-based reach extension.

If a reach-extended G.984 G-PON ODN is to be upgraded and cohost an G.987 XG-PON, there are two choices:

1. A separate XG-PON RE can be deployed, with wavelength separators to guide the proper wavelengths to the proper equipments.
2. The G.984 RE can be replaced by a dual-standard RE. Inside the dual-standard RE, we would expect to find much the same components as described in option 1.

4

TRANSMISSION CONVERGENCE LAYER

In this chapter:

- *ONU-ID terminology ambiguity*
- *Payload mapping and framing*
- *FEC, scrambling and encryption*
- *ONU discovery, activation, and delay compensation*
- *Conveying time of day to the ONU*
- *Physical layer OAM (PLOAM) messages; an embedded operations channel whose details appear in Appendix II*
- *Methods by which upstream bandwidth demand is determined and capacity is requested and assigned*
- *Energy conservation*
- *Security*

The transmission convergence (TC) layer is the heart and soul of a PON and, in particular, of a G-PON or an XG-PON. Here is where we find solutions to the unique problems posed by a PON, as well as to the ordinary issues of any network protocol. These unique aspects are largely related to the tree structure, the point to multipoint nature of the PON:

- Discovering hitherto unknown ONUs in a way that does not disrupt existing traffic on the PON

Gigabit-capable Passive Optical Networks, First Edition. Dave Hood and Elmar Trojer.
© 2012 John Wiley & Sons, Inc. Published 2012 by John Wiley & Sons, Inc.

- Recovering previously known ONUs onto the PON after power failures, power down, or other disruptions
- Coping with the fact that the various ONUs are at different distances from the OLT and, hence, have different propagation delays
- Accommodating drift in the propagation delay that could be caused, for example, by daily or seasonal temperature changes in the ODN
- Defining the upstream burst header for fast and accurate optical receiver calibration so that the header can be quickly delimited and parsed
- Orchestrating upstream burst transmit timing among ONUs, such that
  (a) No transmissions collide at the OLT
  (b) The ONUs' varying traffic levels and service-level commitments are honored
- Defining the structure of a payload mapping to preserve efficiency even if upstream payload frames do not fit exactly into the size of the burst
- Protecting the PON from ONUs that might attempt to steal service or invade the privacy of subscriber traffic
- Detecting and diagnosing rogue ONUs, that is, defective ONUs that transmit light at unauthorized times and potentially interfere with normal traffic
- Structuring payload for easy identification of individual flows and grouping of flows into separately manageable traffic classes

Several other topics are addressed by the transmission convergence layer, topics that are not unique to PONs, but are of general interest to any access network:

- Securing the association of OLT and ONU from eavesdropping, tampering, and theft of service
- Adapting (mapping) some set of payload protocols to the transmission medium. The usual G-PON payload mapping is to an Ethernet client layer, but an MPLS mapping is also defined. G-PON also includes a mapping for OMCI, the PON management protocol.
- Measuring the quality of the link, either in terms of optical parameters or in terms of error rate
- Improving the intrinsic quality of the optical link through FEC
- Providing signaling and control channels for the PON link itself, ranging from simple bit-oriented signaling to intermediate-level fast messaging—called PLOAM[*] in G-PON (physical layer operations and maintenance)
- Supporting an embedded high-level operations and management communications channel, OMCC
- Taking steps to actively reduce power demand during periods of low traffic, while minimizing service degradation

---

[*]Appendix II defines the PLOAM message sets in detail.

4.1   FRAMING

G-PON uses straightforward TDM in the downstream direction. The OLT broadcasts all downstream traffic onto the fiber; each ONU selects its own particular traffic in ways described in this chapter. At the physical layer, downstream traffic comprises an unbroken stream of fixed-length 125-μs frames.

In the upstream direction, the OLT authorizes each ONU to transmit bursts of traffic (time division multiple access, TDMA) in such a way that:

- Upstream bursts do not collide at the OLT. We discuss timing compensation in Section 4.3.
- Bursts are sized and spaced appropriately for the traffic and the quality of service committed to each ONU. Section 6.3 covers traffic management.

The upstream flow retains the concept of a 125-μs frame, but it simply marks a repetitive series of ticks that serve as convenient reference points.

Depending on the line rate, different amounts of data can be fit into the physical layer frame. Table 4.1 shows the line rates and physical layer frame sizes for G-PON and XG-PON.

Be aware that the term *frame* is overloaded. Figure 4.1 illustrates several uses of the term in the downstream context. In general, a frame has a header, a payload, and sometimes a trailer. We discuss the various kinds of frame in detail throughout this section:

- Service data units (SDUs) are often called frames, sometimes packets. In this book, particularly for Ethernet, we usually call them frames.
- Each SDU is encapsulated into one or more GEM or XGEM[*] frames for transport over the PON. GEM idle frames are defined for occasions when no other payload is available for transmission.
- A contiguous sequence of downstream GEM frames, together with its overhead, is called a G-PON transmission convergence layer (GTC or XGTC) frame. It repeats at 125-μs intervals.

TABLE 4.1   G-PON and XG-PON Bit Rates and Physical Frame Sizes

Bit Rate, Mb/s	Frame Size, Bytes	Use
1244.16	19,440	G.984 G-PON upstream
2488.32	38,880	G.984 G-PON downstream
		G.987 XG-PON1 upstream
9953.28	155,520	G.987 XG-PON1 downstream

[*] Strictly speaking, the GEM pertains to G.984, while the term XGEM is proper for G.987 XG-PON. Though not identical, they are similar and serve the same purpose. We often use the term *GEM* generically to refer to either. The same is true of the acronyms GTC and XGTC.

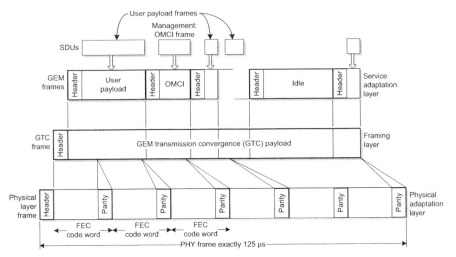

Figure 4.1 Downstream framing, XG-PON.

- The downstream physical layer (PHY) frame extends the GTC frame with a PHY header and FEC. It is a continuous bit sequence at the nominal line rate of 9.953 Gb/s (G.987 XG-PON) or 2.488 Gb/s (G.984 G-PON), with no gaps. The PHY frame header repeats at precise 125-μs intervals. It is used for synchronization as well as for other functions.

In the upstream direction, ONUs transmit bursts, which are normally much shorter than 125 μs. Their size and timing is governed by the bandwidth map (BWmap), which is broadcast to the ONUs from the OLT; timing and addressing in the BWmap determines which ONU transmits a burst, for how long, and when, as measured from the 125-μs upstream frame boundary. Each burst comprises one or more allocations to distinct traffic aggregation entities in an ONU. At the TC layer, each allocation is identified by an alloc-ID; it largely corresponds to a traffic container (T-CONT) at the management layer. We discuss these and their differences in this chapter and in Section 6.3. With the addition of framing overhead, the series of contiguous allocations corresponds to the GTC frame of Figure 4.1.

There is thus no physically observable 125-μs frame in the upstream direction. However, there are well-defined repetitive instants that mark the conceptual boundaries of upstream frames, and the term is meaningful. Section 4.3 on timing and delay equalization defines the concept of the upstream frame more precisely.

The process of building a PHY frame may be understood as the combination of three layers: service adaptation, framing, and physical layer adaptation, which correspond to the three layers shown in Figure 4.1.

Downstream and upstream directions are significantly different, and G.984 G-PON and G.987 XG-PON, although clearly siblings, are by no means identical twins. We therefore discuss them separately, first G.987 XG-PON, then G.984 G-PON. But rather than four distinct expositions, we start by considering the common aspects, and then move into the differences. In the detail of getting payload to and from the PON, the mapping of payload frames into GEM frames (XGEM frames in XG-PON) differs very little.

4.1.1 Service Adaptation Layer

For transport over the PON, G-PON and XG-PON must encapsulate service data units (SDUs). SDUs include user data frames and OMCI management messages. The result of the encapsulation is called a GEM (or XGEM) frame: G-PON encapsulation method. Each GEM frame is tagged with a GEM port ID, which uniquely identifies a particular flow on the PON. GEM frames and GEM ports are visible only between the OLT and ONU; they have no wider significance, either toward the subscriber or toward the core network. There are several reasons for GEM:

- *Traffic Multiplexing*   Particularly in the downstream direction, the GEM port ID facilitates wire-speed filtering of the entire capacity of the PON by each ONU, which need only pick off the traffic destined for itself. A GEM port ID occupies only 12 bits (G.984) or 16 bits (G.987), a small, uniform label that is easy to look up. This also facilitates upstream demultiplexing.
- *Protocol Efficiency*   In contrast to Ethernet, SDUs encapsulated in GEM frames are packed into GTC frames without additional physical layer overhead such as interpacket gaps or preamble-delimiter bytes. GEM frame delineation is accomplished by a short GEM frame header, five bytes in G.984, eight bytes in G.987.
- *Fragmentation*   The service adaptation layer supports SDU fragmentation, which effectively allows maximum utilization of GTC frames. SDUs that will not fit the remaining space in a downstream frame or an upstream allocation are fragmented and transmitted across two or more PHY frames or upstream bursts. This again increases protocol efficiency.
- *SDU Agnosticism*   G-PON and XG-PON are specified in such a way that they can transport a variety of networking protocols over the PON, as well as transporting OMCI, the management protocol of G-PON itself. The provisioned GEM port ID is used to distinguish one protocol type from another. Although Ethernet is the primary protocol in use today, other protocols such as SDH and native IP over GEM were standardized in G.984. G.984 even specified TDM over GEM, though with insufficient detail to support implementation. For lack of interest, these latter three have been dropped from G.987. More recently, multiprotocol label switching (MPLS) over GEM was defined.

4.1.1.1 GEM Frames

Figure 4.2 depicts a G.987 XGEM frame, while Figure 4.3 portrays the structure of a G.984 GEM frame.

Observe that the XGEM frame header contains nothing to identify the protocol of its client. G.984 (no .3) GEM contains a payload-type indicator (PTI) field, but as we shall see, it serves this purpose only in a very minimal sense. The client protocol type is known through configuration and distinguished by GEM port ID.

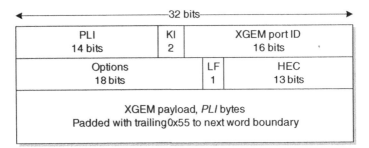

Figure 4.2   XGEM frame, G.987 XG-PON.

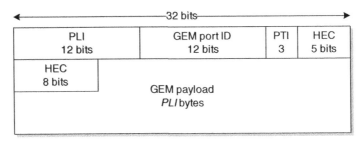

Figure 4.3   GEM frame, G.984 (no .3) G-PON.

Payload in G.987 XGEM frames is always a multiple of 4 bytes, called a word in XG-PON terminology. G.984 GEM frame payloads have single-byte granularity.

As we see, GEM and XGEM frames share features, but they are not the same. We discuss each header field, starting with the common ones.

Common Fields

GEM Port ID   The GEM port identifies a traffic flow and facilitates traffic multiplexing as described above. In G.984 G-PON, GEM port IDs lie in the range 0–4095, while G.987 XG-PON port IDs range up to 65,535. In XG-PON, however, ports 0–1022 are preassigned for the OMCI GEM channels of ONUs with TC-layer ONU-IDs 0–1022, respectively, so they are not available for general use. Further, port 65,535 is reserved to identify idle GEM frames.

PLI   A payload length indicator is essential for variable-length frames. The resolution of PLI is bytes, with a range up to 4095 in G.984 G-PON and 16,383 in G.987 XG-PON. In G.984 G-PON, the GEM frame ends after exactly *PLI* bytes of payload; in XG-PON, the frame is padded with as many as 3 additional bytes of value 0x55 to extend it to the next word boundary, or as many as 7 pad bytes if necessary to create a minimum XGEM payload field of 8 bytes. The actual payload can be as short as 1 byte. No Ethernet SDU is that short—the minimum length is 64 bytes—but it is possible that GEM frame fragmentation could leave a 1-byte orphan.

ONU-ID Terminology Ambiguity

It is important to understand that *ONU-ID* is an ambiguous term.

At the transmission convergence and PLOAM level, where we are in this chapter, ONU-ID is expressed in the form of real bits in real registers. It would be visible, for example, to a snooping device on the PON. Every ONU that has progressed past the first phases of initialization on the PON has a TC-layer ONU-ID.

At the provisioning and management level, ONU-ID is a name, a reference. Service orders, installation orders, maintenance, and provisioning refer to the management-level ONU-ID.

An ONU may be fully operational at the TC level but may have no ONU-ID at the management level, for example, if it is autodiscovered and has not been provisioned for subscriber service. Likewise, a preprovisioned management-level ONU-ID is valid and meaningful, although it may correspond to nothing at the TC level, for example, if its referent ONU has not yet been installed.

The management-level ONU-ID is usually hierarchical:

```
<network element name>
        <slot number>
                <PON port number>
                    <ONU number>
```

whereas the TC-layer ONU-ID exists in a flat name space {0..253} for G.984 G-PON, {0..1023} for G.987 XG-PON. Observe that the range of TC-layer ONU-IDs includes the value 0. At the management level, the final <ONU number> field almost surely starts counting from 1.

The management-level ONU-ID is normally static for the lifetime of a customer subscription, possibly for many years. The TC-layer ONU-ID may be different every time the ONU is activated onto the PON.

During the activation process, the OLT assigns a TC-layer ONU-ID to the ONU. Once assigned, an ONU's TC-layer ONU-ID cannot be reassigned. If it is desired to equate the TC-layer ONU-ID with the management-level <ONU number>, the OLT may deactivate the ONU, which causes it to reactivate. Recognizing the ONU's serial number on the second pass, the OLT can then assign it a different TC-layer ONU-ID. However, it may be simpler for the OLT to maintain a mapping between the TC-layer ONU-ID and the management ONU-ID.

In most cases throughout this book, the interpretation of ONU-ID is apparent from the context. When there is potential for confusion, we disambiguate the identifiers as TC-layer ONU-ID versus management- or OMCI-level ONU-ID. In this chapter, ONU-ID is understood to be at the TC layer unless explicitly stated otherwise.

One way to think of the difference is that a protocol analyzer would see the TC-layer ONU-ID, never the management ONU-ID. In contrast, a management view would never see the TC-layer ONU-ID except perhaps in a troubleshooting or debugging context.

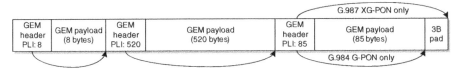

Figure 4.4 Walking from one GEM header to the next.

HEC   Starting with a well-defined synchronization instant, GEM frames are concatenated one after another into the GTC frame (in G.984.3: GTC payload). As illustrated in Figure 4.4, the decoder can only find the next GEM frame if it can properly decode the header of the current GEM frame and in particular, its payload length indicator PLI. So there is an error-checking and correcting code across the GEM frame header. In G.984 G-PON and by tradition, this is called a header error control, but since the same algorithm is used in various places in XG-PON, not always in a header, the acronym was morphed into hybrid error control or hybrid error correction. Appendix I describes HEC.

Figure 4.4 also shows how G.987 XGEM frames are padded to the next word boundary.

Walking the list of GEM frames is a convenient time for the ONU to filter downstream traffic by GEM port, extracting only GEM frames of interest for further processing within that ONU. Each GEM frame represents an individual flow and, except for multicast, therefore goes to (or from) only one ONU.

G.984.3 describes a way to reacquire GEM frame delineation by checking 5-byte candidates in the stream of bits to see if their presumed HEC is correct. Figure 4.5 shows the G.984 state machine that can potentially recover lost GEM frame delineation during the same GTC frame. Each byte is hypothesized as the start of a GEM frame, and the corresponding HEC is computed for that hypothesis.

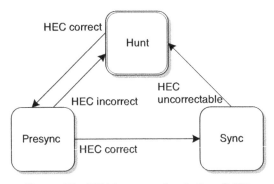

Figure 4.5   GEM frame synchronization, G.984.

Given the extremely low probability of losing GEM frame delineation, and the higher speeds involved, the HEC recovery feature was dropped from G.987. In XG-PON, if the XGEM frame header is corrupted beyond repair, the remainder of the PHY frame is lost.

Fields Unique to G.987 XG-PON

KI, the Key Index   This two-bit field specifies which key, if any, is used to encrypt the current GEM frame. The value 00 indicates an unencrypted payload, and the value 11 is reserved. Values 01 and 10 select one of two possible keys. The reason for two is that during key update, the OLT and ONU may have different views of the validity of a given key. Section 4.7 discusses key negotiation and synchronization in detail.

Options   These 18 bits are unused in G.987 XG-PON. They exist primarily to pad the GEM frame header to a word boundary and are available for future use if needed.

LF   This bit signals the last fragment of an SDU. If it is 0, the GEM frame contains only a leading or intermediate fragment of an SDU. The receiver must buffer the fragment until the next fragment arrives, and the next, and the next, until the final fragment appears, at which time it can reassemble the SDU. In practice, more than two fragments would only occur if large upstream subscriber frames were trying to squeeze through a very small bandwidth assignment. We discuss fragmentation in more detail below.

If $LF = 1$, the GEM frame contains either an entire SDU or the final component of a fragmented SDU.

Fields Unique to G.984 G-PON

PTI   The 3-bit PTI field reflects the ATM heritage of G-PON. Its value is to identify SDU fragmentation.

In theory, PTI also distinguishes user traffic from OAM traffic, meaning OMCI. The definition in G.984.3 says OAM traffic comprises 48-byte ATM-like messages, but G.984.4 subsequently evolved to define an extended OMCI message set with variable-length messages. In any event, since G.984 OMCI is carried over a configured GEM port, the distinction is irrelevant.

4.1.1.2 Payload Mapping to GEM Frames

Now we know what a GEM frame is. How does payload map into a GEM frame? Figure 4.6 illustrates an Ethernet frame mapped into XGEM and GEM frames, respectively.

As we see, the preamble and delimiter are stripped from the Ethernet frame, and the remainder is mapped directly into the payload of a GEM frame. Figure 4.6 shows an untagged Ethernet frame, but tagged frames are encapsulated in the same way: everything after the start-of-frame delimiter to and including the frame check

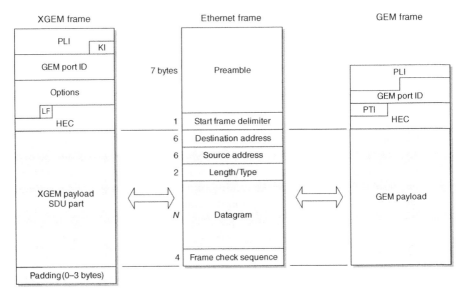

Figure 4.6 Ethernet mapping into GEM.

sequence is copied into the GEM frame. Keeping the Ethernet frame check sequence (FCS) in the GEM frame decouples error detection between the GEM server layer and the Ethernet client layer.

Shown in Figure 4.7, the MPLS mapping is much the same, except of course with none of the Ethernet frame overhead: preamble, delimiter, or FCS.

Ethernet and MPLS are the only external payload mappings defined for G.987 XG-PON. OMCI messages are also mapped into GEM and XGEM frames, also

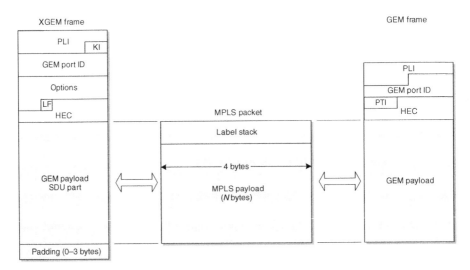

Figure 4.7 MPLS mapping into GEM.

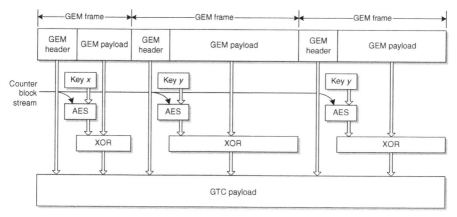

Figure 4.8   GTC payload.

simply by prepending a GEM header, and in XGEM, possibly a trailing pad to the next word boundary. Other mappings exist for G.984 G-PON, but they have not survived into G.987 because they are of little interest to the industry. We omit illustrations of these other mappings; the interested reader is referred to G.984.3.

GTC Payload and Encryption   The GTC payload is the simple concatenation of GEM frames, possibly encrypted. Figure 4.8 illustrates how encryption works.

We do not encrypt the headers of individual GEM frames, nor do we encrypt the additional headers that we will ultimately add onto the structure to build the physical layer frame. The payload of each GEM frame is encrypted (or not) using the advanced encryption system (AES) algorithm in counter mode. The counter block stream is a sequence that can readily be duplicated at both OLT and ONU, while nevertheless including enough complexity to effectively preclude dictionary attacks on the key.

In G.987 XG-PON, the key for each GEM frame may either be a unicast key or a broadcast key, as suggested by keys x and y in Figure 4.8.

G.984 G-PON does not support broadcast keys. If a GEM frame is encrypted, the only choice is the unicast key, of which there is only one at any given time.

In both G-PON standards, encryption is specified with GEM port granularity. Encryption is sufficiently complex that we do not treat it inline here; it is fully described in Section 4.7.

Fragmentation   In principle, one SDU maps into one GEM frame. A GEM frame is never allowed to contain more than one SDU, but the inverse relation is not necessarily true. There are three cases in which an SDU may be fragmented into more than one GEM frame.

- *Insufficient Runway*

  Downstream Every 125 μs without fail, the downstream flow is interrupted for a framing structure that contains synchronization and overhead. If there is not enough time to transmit a GEM frame containing the next complete SDU before the next framing overhead is due, the SDU either has to wait or the GEM layer has to fragment the SDU. We reduce packet delay and improve efficiency by choosing the fragmentation option, unless the remaining time is too short to build a useful fragment.

  Upstream The situation is much the same. An ONU transmits payload according to its upstream capacity allocations. If too little time remains before the end of its allocation to transmit its next SDU, the ONU must either wait or fragment the SDU. If there is more than a negligible amount of remaining space in the allocation, we fragment the SDU.

- *Oversize Payload* If an SDU were simply too long to fit into a GEM frame (G.984 G-PON: 4095 bytes, G.987 XG-PON: 16,383 bytes), it would have to be fragmented.

- *G.984 G-PON Only* It is anticipated that high-priority SDUs might preempt lower priority SDUs. This originally made a certain amount of sense because G.984 was developed with the idea that in the upstream direction, a T-CONT might contain several classes of service, so there could be priority contention between queues within a single T-CONT. However, even a 2000-byte jumbo frame can be transmitted at 1.2 Gb/s in about 13 μs. Further, the industry has moved toward a best-practice service model in which different classes of service are carried in different T-CONTs. There is no preemption between T-CONTs, even within the same ONU. Thus, preemption adds complexity for very little benefit. Preemption is not recognized in G.987.

Once we have determined that our next SDU is to be fragmented, the simple story is that the first GEM frame encapsulates as much of the SDU as will fit. In the first GEM frame header, we indicate that the frame is a fragment. The remainder of the SDU is feedstock for the next GEM frame, which might also contain only a fragment (unlikely, but possible, especially if the upstream allocation were very small). The remaining fragments are at the head of the queue for their particular allocation or (G.984) priority, so they are transmitted as soon as possible. The final GEM frame in the series indicates that it is the last fragment. The receiver must buffer GEM frame fragments until it receives the last one, whereupon it reassembles the original SDU and forwards it.

That is the simple story. Above, we said that we will fragment a frame if we have more than a negligible amount of space into which we can put a fragment. What does *negligible* mean?

In G.984 G-PON, if there remains too little time to send a minimum length GEM frame (6 bytes including GEM header), the transmitter fills the space with all 0 bytes. Otherwise, G.984 will create a fragment with as little as 1 byte of SDU payload.

In XG-PON, if the space remaining before the break in transmission is too small to contain a minimum length GEM frame—8 bytes of payload—no fragment is created, and the transmitter fills the time with either full or short idle frames, the latter defined to be a sequence of 40 bytes.

It is worth mentioning that a sequence of GEM frames, each with 1 byte of payload, would impose considerable stress on the OLT and ONU devices, especially if encryption were active. G.984.3 does not constrain this possibility; G.987 XG-PON reduces this potential overload by specifying a minimum GEM payload length of 8 bytes. In G.987, our 1-byte fragmented SDU trailer would thus be accompanied by 7 bytes of padding.

Idle GEM Frames It can happen that the transmitter has nothing useful to send. The downstream signal is continuous, and with few exceptions, it is also expected that an ONU completely fill out its upstream transmission allocation, rather than simply going silent. We therefore need the concept of an idle GEM frame.

In G.987 XG-PON, an idle frame is a well-formed GEM frame with complete and correct header, of variable length, addressed to GEM port 0xFFFF, or indeed, to any unused GEM port. The payload content of an idle frame is not specified.

Once a GEM frame is committed to transmission, that is, once its PLI is set, it continues to completion. Long idle frames could delay transmission of recently arrived valid traffic frames, so it is encouraged to keep idle frames short, perhaps as short as 64 or 128 bytes, or even 8 bytes—an XGEM header only, no payload.

In G.984 G-PON, an idle frame is simply an all-zeros sequence. Officially it is 5 bytes long, but a trailing idle frame can be truncated to any length necessary to fill the remaining space. Because it includes no way to specify variable payload size, a truncated idle frame it cannot form part of a chain of consecutively linked GEM frames (Fig. 4.4). G.984 also recognizes the possibility of sending arbitrary but well-formed GEM frames to unused GEM ports.

Both G.984.3 and G.987.3 suggest that idle payload be crafted such that its spectral properties assist in the performance of the PON. But to rely on idle payload for PON performance would require a long-term commitment on the part of the operator to remain below some (un)specified traffic load, a constraint that clearly lies beyond the bounds of any operator's ODN engineering rules. Imagine if, as traffic levels increased, the customized idle pattern was no longer possible and PON stopped working!

4.1.2 G.987 XG-PON Downstream Framing

XG-PON framing is based on G.984 G-PON framing, but incorporates several changes, based on experience with G.984 G-PON and expanded requirements, for example, for increased numbers of ONUs. We start by examining G.987 XG-PON and then subset it for G.984 G-PON in the subsequent sections. The additional requirements of XG-PON include:

- *Four-Byte Alignment*   G.984 G-PON aligns fields on byte boundaries. To facilitate parallel processing architectures, particularly in the early days when implementations are likely to be based on FPGAs (field-programmable gate arrays), G.987 XG-PON fields are aligned to 4-byte word boundaries whenever possible.

- *Robustness*   In both G-PON and XG-PON, most fields critical for robust frame interpretation and pattern delineation are equipped with hybrid error correction functionality. At the cost of more overhead—13 bits—the HEC code chosen for G.987 XG-PON can detect (and correct) more errors than the 8-bit cyclic redundancy code (CRC) of G.984 G-PON. G.984 G-PON improves its robustness by transmitting some fields twice and some messages thrice; G.987 XG-PON does not.

- *Increased PLOAM Rate*   Recall that the TC layer includes an embedded message-based overhead channel for PLOAM. When an entire PON initializes, several messages must be exchanged with each ONU, and as the number of ONUs on the PON increases, the G.984 G-PON limit of a single downstream PLOAM message per 125-μs frame was a bottleneck. In G.987 XG-PON, each downstream frame can contain as many as one broadcast PLOAM message plus one unicast PLOAM message to each ONU. A single PLOAM message per upstream burst was retained for simplicity; this does not represent a bottleneck.

- *Reduced PLOAM Rate*   In G.984 G-PON, every downstream frame contains a PLOAM message. In G.987 XG-PON, if the OLT has nothing to say in the PLOAM channel, it does not send a PLOAM message at all.

- *Higher Security*

  The G.987 XG-PON (super)frame counter is 51 bits long, compared to the 32 bits of the G.984 G-PON counter. Together with algorithm enhancements, this improves the security of both physical layer scrambling and payload encryption.

  As well as the downstream unicast encryption of G.984 G-PON, G.987 XG-PON supports upstream encryption and downstream multicast encryption.

- *PON-ID*   The concept of PON-ID was introduced, which may be useful for more robust protection switching.

- *Bandwidth Map Restructure*   The G.984.3 concept of start time and stop time was changed in G.987.3 to start time and duration, which improves the layering structure of upstream allocations.

- *FEC*   Forward error correction is always on in the downstream direction of G.987 XG-PON, contributing to improved link budget and improved error rate. While G.984 G-PON has 10^{-10} as the target bit error rate, G.987 XG-PON is specified at an error rate of 10^{-12}. Because downstream FEC is always on, bit interleaved parity (BIP) adds no value and has been eliminated in the downstream direction. FEC in the upstream direction may be disabled if the optical budget is comfortable, so upstream BIP is retained from its G.984 G-PON ancestor as a way to track errors when FEC is off.

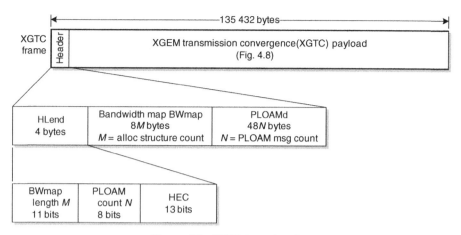

Figure 4.9 XGTC frame header.

We continue to work our way down the stack from user frames (SDUs) to GEM frames, now to the XGTC frame, then to the PHY frame.

4.1.2.1 XGTC Frame Header

Recall from Figure 4.1 that the GTC frame (XGTC frame in G.987 XG-PON) comprises an XGTC header followed by an XGTC payload that comprises a sequence of GEM frames. Figure 4.9 expands the header structure. There are three fields:

1. *HLend*   This 4-byte field tells us the length of the other fields of the header. The BWmap length field is the number of allocation structures in the enclosed bandwidth map. We have 11 bits, but to bound the size and speed required of the algorithms and chips, not more than 512 allocation structures are allowed. Eight bits allow as many as 255 PLOAM messages in the PLOAMd partition, while a 13-bit HEC protects the HLend field from bit errors.

2. *BWmap*   In every downstream frame, the OLT authorizes upstream transmission from zero or more ONUs, the authorization to be effective in the corresponding upstream frame. The bandwidth map BWmap is the vehicle for this authorization. The BWmap field contains zero or more so-called allocation structures, each of which authorizes upstream transmission from a traffic-bearing entity in some ONU, or authorizes an ONU discovery transmission. Section 4.1.2.3 examines the BWmap in considerably more detail.

3. *PLOAMd*   The downstream PLOAM partition can carry as few as zero PLOAM messages, or a maximum of one broadcast PLOAM message, plus one unicast PLOAM message to each ONU on the PON, bounded by the maximum value 255 of the PLOAM count field. Each PLOAM message is 48 bytes long.

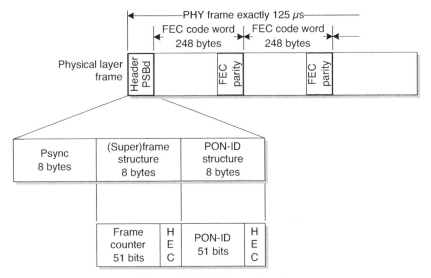

Figure 4.10   PHY frame header.

4.1.2.2 G.987 XG-PON Physical Layer Frame Header

Referring back to Figure 4.1, we see that the downstream XGTC frame is prepended with yet another header to form the physical frame. In XG-PON, this header is called the physical synchronization block downstream (PSBd).

Following the PSBd, the body of the downstream XGTC frame is sliced apart at uniform intervals for the insertion of FEC parity bytes. The original data bytes are not modified in any way. We show how FEC is added and removed in Section 4.1.2.4.

Figure 4.10 illustrates the structure of the physical frame header PSBd. Again, we find three fields:

1. *Psync*   The physical layer sync pattern is a fixed sequence of bits, against which the ONU receiver can match, even if the receiver does not have byte sync. As such, it is important that it have a large Hamming distance[*] from bit-shifted versions of itself—Figure 4.11 challenges us to perform a mental autocorrelation. The PHY frame header is not protected by FEC, and Psync is not even protected by HEC—byte (and word) synchronization is a precondition for both of those. The ONU receiver should be designed to be robust to several bit errors in the Psync field.

2. *Superframe Structure*   This HEC-protected field contains what is universally known as a superframe counter (SFC), even though it actually counts frames, not superframes. With each new downstream frame, the frame count increments by 1, rolling over to 0 if it overflows.

---

[*]The Hamming distance between two binary strings of equal length is the number of bit positions that differ between them.

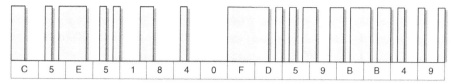

| C | 5 | E | 5 | 1 | 8 | 4 | 0 | F | D | 5 | 9 | B | B | 4 | 9 |

Figure 4.11   Physical synchronization pattern Psync.

3. *PON-ID*   Also protected by HEC, the PON-ID has no equivalent in G.984 G-PON. The concept behind the field is that it could be useful in allowing an ONU to detect protection switches when the ONU is dual homed to two separate OLTs. Its default value is zero.

The physical frame header PSBd is even more robust than is apparent at first glance. The receiver can use several clues to remain in sync.

- The frame structure repeats at exact 125-μs intervals.
- The Psync field is fixed and known.
- The value of the frame counter is always one greater than in the previous frame.
- The PON-ID is also stable, not expected to change over perhaps the lifetime of the OLT.

The 16 bytes of HEC-protected frame counter and PON-ID are exclusive-ORed with 0x0F ... 0F before transmission and again before decoding at the ONU. This anticipates the distinct possibility that at least the PON-ID, and possibly also the frame counter, will be mostly zeros.

The downstream PHY frame is scrambled to reduce its likelihood of containing long sequences of identical bits. Section 4.1.2.5 describes the scrambling process.

At the ONU, the process is reversed. The ONU contains a simple state machine for frame synchronization, depicted in Figure 4.12. The state machine uses the 125-μs interval clue and the SFC clue along with PSync match, and tolerates three (recommended value for *M*) errors before it gives up and declares itself to have lost synchronization (loss of downstream synchronization LODS). Initial synchronization requires an exact match of the 64-bit PSync field and a correctable superframe count field. Subsequently, synchronization is maintained if at least 62 bits of the PSync field match and if the superframe count increments by 1 from the previous frame and remains correctable.

Once the ONU receiver has aligned itself with the physical frame header, the ONU is in a position to process the BWmap, looking for upstream transmission allocations to itself. Then the ONU inspects the downstream PLOAM messages, looking either for a broadcast message or a unicast message addressed to itself, or possibly one of each.

In parallel with processing of the header fields, the ONU can descramble the frame, decode FEC blocks, and correct any errors that may have occurred. Then

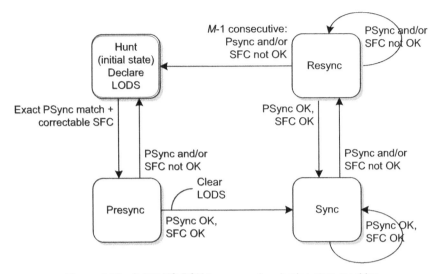

Figure 4.12   G.987 XG-PON frame synchronization state machine.

comes the payload of GEM frames, to be filtered according to their GEM port IDs, and finally the decryption process.

In brief, that is how G.987 XG-PON works, downstream.

4.1.2.3 G.987 XG-PON Bandwidth Map

We skipped over the details of the BWmap because it is a significant topic in its own right. Even now, we only discuss part of the upstream capacity allocation process. The DBA discussion in Section 6.3 explains the factors that the OLT takes into account when it builds the BWmap. For our purposes, we take the BWmap as given.

We first consider the structure and the fields of the BWmap. Figure 4.13 reminds us that the XGTC frame header includes a field that tells us how many allocation structures the BWmap contains.

If several consecutive allocation structures are assigned to the same ONU, without a break in transmission time—that is, to be concatenated into a single upstream burst—we refer to them as a contiguous series. The rules for BWmap construction (Section 6.3.2.2) mean that a contiguous series of allocations is also a contiguous series of allocation structures in the BWmap. A contiguous series uses upstream PON capacity efficiently because it avoids the overhead of separate bursts for multiple allocations to a given ONU.

The fields of an allocation structure are as follows:

1. *Alloc-ID*   Each allocation identifier belongs to not more than one ONU. Alloc-IDs in the range 0–1022 are implicitly bound to the ONUs with the, respectively, identical TC-layer ONU-IDs; alloc-ID 1023 is an invitation for any new ONUs to respond with their serial numbers, making this a serial number grant, as described in Section 4.2. Other alloc-IDs are assigned to ONUs through

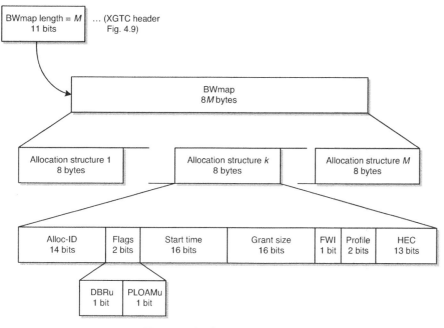

Figure 4.13 BWmap structure.

(PLOAM) provisioning. The alloc-ID represents the fundamental unit of granularity at which the OLT manages traffic from the various contending ONUs on the PON.

Alloc-IDs and T-CONTs

Alloc-IDs are frequently confused with T-CONTs, particularly in discussions of G-PON resource management at the OLT (DBA). They are not the same.

Their identifiers are different. An alloc-ID exists in a flat namespace per PON, while the external identifier of a T-CONT is hierarchical: <management-layer ONU-ID><T-CONT number>. Alloc-IDs and T-CONTs are logically distinct in the layering model. Most importantly, there is not even a one-to-one correspondence: T-CONTs are assigned to alloc-IDs at the OMCI layer for client traffic, but OMCI itself is carried in a default alloc-ID and needs no T-CONT. See also Section 5.1.1.

This is much the same distinction that we saw between the TC-layer ONU-ID (an alloc-ID is visible in the bits flowing across the PON) and the management-level ONU-ID (a T-CONT is a name that refers to an alloc-ID). Think of it this way: subscriber traffic is managed through T-CONTs, while the underlying engine only knows alloc-IDs. Although the ambiguity is mostly harmless, it should be kept in mind whenever there is a need for precision.

2. *Flags*

   DBRu   If this flag is set, the ONU sends a queue occupancy report in the header of the upstream response for that alloc-ID. This is discussed further in the upstream framing section below. The queue occupancy report is one of the primary inputs to the dynamic bandwidth assignment process, DBA.

   PLOAMu   The OLT sets this bit to request the ONU to send an upstream PLOAM message in its response. The bit is meaningful only in the first allocation of a contiguous series addressed to a given ONU. If the ONU has no substantive PLOAM message to send, it just sends a heartbeat acknowledgment message. The ONU is not free to send upstream PLOAM messages at its own discretion, nor can it ignore a PLOAMu request.

3. *Start Time*   In the first grant of a possibly contiguous series, this field specifies the start time of the upstream transmission. Coupled with delay equalization, this is the essential field that allows the OLT to interleave ONU burst responses without collisions. The granularity of the start time is 4 bytes, one word. If it is thought of as a time, rather than as a word count, words flow at the 2.488-Gb/s rate of the upstream PON, not at the 10-Gb/s downstream rate. Start time is an offset relative to the start of the upstream frame, which (Section 4.3.3) is defined to be the first bit following the upstream physical synchronization block, that is, the first bit of the XGTC header.

   In subsequent allocations of a contiguous series, the start time is coded as 0xFFFF, which just indicates continuation without a break from the previous allocation.

4. *Grant Size*   This field specifies the number of words authorized by the grant. The word count includes space for the upstream DBR, if there is one, but does not include space for a possible upstream PLOAM message or other overhead such as the burst header or trailer, or FEC, which exist at lower layers in the client–server model.

5. *Forced Wakeup Indicator (FWI)*   Suppose the ONU supports low-power modes (Section 4.5), and suppose the OLT wishes to awaken it. The ONU may be sleeping at the time the OLT sends the Sleep_allow (off) PLOAM message. In the first allocation of a contiguous series, the FWI bit has the same effect and is more likely to be received by the ONU when the ONU enters its sleep-aware state.

6. *Profile Index*   A burst profile specifies the header of the upstream burst. The OLT defines one or more burst profiles, and communicates them to the ONUs through continuing broadcast Profile PLOAM messages, so that new ONUs can learn them. The two-bit profile field in the BWmap specifies which burst profile is to be used by the ONU in its upstream response. If the ONU does not recognize the profile, it must not respond at all.

7. *HEC*   Hybrid error correction is described briefly in the GEM frame section above and further in Appendix I.

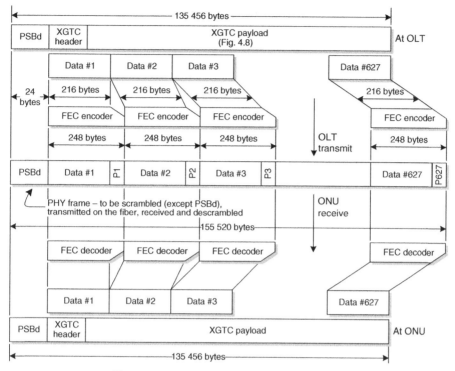

Figure 4.14   Downstream FEC, G.987 XG-PON.

4.1.2.4   *Forward Error Correction*

Appendix I also describes FEC and summarizes the algorithms that pertain to the different directions and G-PON technologies. Here, we take FEC and FEC parity as given and show how parity blocks are mapped into the downstream frame. Figure 4.14 shows framing and FEC processing.

G.987 XG-PON downstream uses an RS(248, 216) code, which can correct up to 16 bytes in error. We add 32 bytes of FEC parity to each 216-byte block of XGTC payload and fit the resulting 248-byte code word into the physical frame structure. The size of the code word is chosen such that an integer number of code words fits into the XGTC frame, excluding the PSBd physical header, which would not benefit from FEC in any event.

We then submit the physical frame to scrambling (next section) and transmit it on the fiber. At the ONU, we reverse the sequence, presenting an XGTC frame to the next layer of the stack for header and GEM frame processing.

Downstream FEC in XG-PON is always on. It costs about 13% of the PON's capacity.

4.1.2.5   *Scrambling*

Scrambling is a technique of pseudorandomizing a data stream to reduce the likelihood of long sequences of identical digits (bits). In non-return-to-zero (NRZ) transmission, as used in G-PON and XG-PON, consecutive identical digits (CID) contain no transitions, and a receiving clock can lose synchronization if the dearth of transitions persists too long. G-PON expects receivers to tolerate strings of up to 72 CID. Shifting the balance of 1 toward 50% also helps the receiver optimize its decision threshold.

A scrambler does not add bits to the flow and therefore imposes no capacity penalty.

It is important to understand that a scrambler does not guarantee to eliminate long, fixed strings of repeated bits. The pseudorandom sequence includes all possible bit sequences up to the order of its generator polynomial, so there necessarily exists a sub-sequence from the scrambler that exactly matches—or complements—any given payload sequence, resulting in a string of consecutive identical zeros (or ones, if it is a complement match).

Scrambling is not to be confused with encryption: a scrambled signal is easy to decipher. However, scrambling does have a security benefit in that it can improve the network's resistance to denial of service (DoS) attacks. If a malicious user uploaded, or downloaded, a file of all 0s or all 1s, the PON might lose synchronization, much to the annoyance of everyone else on the same PON (downstream) or on the same ONU (upstream).

G.984 G-PON uses only 7-bit scramblers, which therefore repeat after 127 bits, and would be comparatively easy to attack with data sequences that complemented the scrambler algorithm. G.987 XG-PON uses a much longer scrambler and initializes it with the (super)frame counter, making it many orders of magnitude harder to predict. Also, the attacker has no way to force transmission of the malicious pattern at any particular instant in the frame, making a deterministic attack even more difficult.

That is the overview. Now for the details:

> The G.987 scrambler applies to the entire downstream frame excluding the PSBd, direct visibility of which is necessary to achieve frame acquisition. Scrambling starts at the first bit following the PSBd and runs to the end of the frame.

> The scrambler polynomial is $x^{58} + x^{39} + 1$. The scrambler is initialized with a different 58-bit value at the beginning of each frame, such that no two nearby frames use the same scrambling sequence. The 58 bits of preload are the 51 bits of the current (super)frame counter, with seven 1 bits at the least significant end.

The scrambler is the same at the OLT and the ONU. Scramblers are conventionally constructed with linear feedback shift registers, as shown in Figure 4.15. The simple and well-defined initialization mechanism allows the ONU receiver to easily generate the same pseudorandom sequence that was used at the OLT transmitter.

4.1.3 G.987 XG-PON Upstream Framing

Figure 4.16 shows how the upstream signal appears at the OLT. Physical layer bursts from a number of ONUs arrive, neatly interleaved in time so that they do not collide.

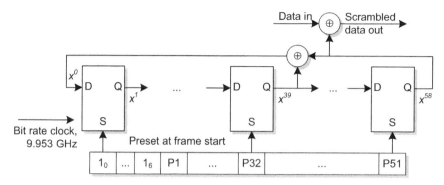

Figure 4.15   G.987 XG-PON scrambler, downstream.

This is the consequence of the OLT's allocation of start times and grant sizes in the downstream BWmap that specified this upstream frame. The boundaries of the upstream physical layer frame are precisely defined—see timing Section 4.3—but they are not directly physically observable.

Figure 4.17 reminds us of the timing and power relationships of the burst header from Section 3.9.3. The 0 and 1 levels define the envelope of normal burst transmission, while the different level shown as *off* recognizes the fact that the ONU transmits some amount of light even at logical level 0 but transmits no light when it is off.

Guard Time   Each burst is separated from other bursts by the guard time t_g, whose recommended minimum value is 8 byte times. The OLT could choose to make it larger, but not smaller. The guard time absorbs slight variations in timing (uncertainty t_u) to guarantee that bursts do not collide. It also allows for

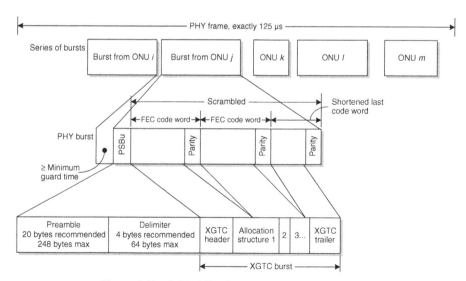

Figure 4.16   G.987 XG-PON upstream frame, OLT view.

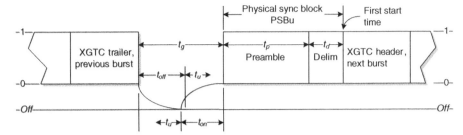

Figure 4.17   Interburst detail.

light decay from the previous ONU as it turns off its transmitter (time t_{off}), and light from the next ONU as it turns its transmitter on (time t_{on}). The t_{off} interval of one ONU may overlap t_{on} from the next.

During the guard interval, the ONU powers up its laser, but is permitted to transmit no more light than that of a logical 0.

The upstream physical sync block PSBu contains:

Preamble   As a sin of omission, Figures 4.16 and 4.17 imply that each burst has the same amplitude. In fact, adjacent bursts can vary by as much as 20 dB, an optical power ratio of 100 : 1. The burst preamble permits the OLT receiver to adjust its gain and thresholds for the amplitude of this particular burst, as well as to acquire clock phase. The preamble length and pattern are specified by the OLT and distributed to the ONUs in the profile PLOAM message. To facilitate cost optimization at the OLT receiver,[*] the profile may contain a very long preamble, although 20 bytes is the recommended length, indicated in Figure 4.17 by t_p.

Delimiter   The same options are available for the delimiter; the recommended length t_d is 4 bytes, but it may be as long as 64 bytes if necessary. Given that bit synchronization is achieved during the preamble phase, the purpose of the delimiter is to synchronize byte boundaries, and especially the boundary between the delimiter and the first byte of the burst header. A longer delimiter is more robust in the presence of high bit error rates, especially if the ONU receiver is designed to tolerate some number of bit errors.

The first byte of the physical layer payload begins the XGTC burst. It also begins the coverage of upstream FEC, assuming that FEC is enabled. Given that FEC is enabled, the remainder of the burst, including the burst trailer, is included in a series of FEC code words, the last of which may be shortened. FEC is discussed in Appendix I; downstream FEC is illustrated in Figure 4.14. Construction of the matching figure for upstream bursts is left as an exercise for the serious student: there are no surprises.

---

[*] In theory, less costly ONU optics could also be accommodated in this way, but, in practice, the OLT may have no way to adaptively discover the need for longer preambles in general or on specific ONUs in particular.

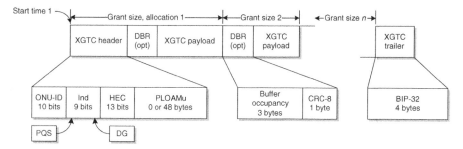

Figure 4.18   XGTC header, upstream

As shown in Figure 4.16, subsequent to FEC parity byte insertion, the burst is scrambled. In a similar way as the downstream frame, the scrambler starts immediately after the PSBu field and runs through the remainder of the burst. It is the same polynomial, with the same preload, as in the downstream direction.

Figure 4.18 expands the XGTC burst—that is, the burst before FEC has been added at the ONU transmitter or after FEC has been removed at the OLT receiver—to show its structure. An XGTC burst comprises an XGTC header and trailer, between which we find a concatenated series of responses to the allocation structures from the BWmap. Each allocation response includes a DBA report (DBR) if it was requested by the BWmap, followed by a series of GEM frames (Fig. 4.8), which may include idle frames. The last frame in each allocation may contain only a fragment of an SDU. Allocation boundaries are respected, even if one allocation is padded with idle frames and the next allocation overflows.

Observe that the first allocation must include space for the XGTC header, optionally including a PLOAM message, as well as whatever payload is intended. All allocations must include space for DBR fields, if they are needed.

The XGTC header comprises four fields, one of which is the ubiquitous HEC. The others are:

ONU-ID   If the ONU is attempting to be activated onto the PON by responding to a serial number grant (Section 4.4), it uses the unassigned ONU-ID value 1023 (0x3FF) for this field. Otherwise, it uses the TC-layer ONU-ID that was assigned to it by the OLT during activation. Strictly speaking, the ONU-ID is unnecessary; it is merely an extra check that everything is working as expected.

Indicators   Only two of the bits in this field are defined. The other 7 bits are reserved.

    PLOAM Queue Status   The most significant bit of the indicators field is a request by the ONU for a PLOAMu allocation in some future BWmap. If the current BWmap already requests an upstream PLOAM message, the PQS bit is set to indicate that, even after transmitting the PLOAM message in the current burst, there still remains at least one PLOAM message in the ONU's queue, awaiting a transmission opportunity.

Dying Gasp   The DG bit is a way for the ONU to signal that it expects to disappear from the PON for its own local reasons, usually loss of power.[*] If subsequent troubleshooting is required, the prior reception of a DG assists the operator in isolating the problem to the ONU itself, rather than to the optical distribution network. DG is a best-efforts indication: the ONU may not be able to send DG before it dies. Further, the ONU may recover power before dying, so it may in fact remain alive on the PON and simply cease signaling DG in future bursts.

PLOAMu   If the first allocation in the BWmap for this ONU specifies an upstream PLOAM message, it appears here. It is not included in the burst header HEC because it contains its own message integrity check (MIC), which includes an error detection function. A secondary reason to exclude the PLOAM message from HEC is to reuse the same HEC algorithm everywhere; PLOAM messages are 48 bytes long, and the error-correcting algorithm tops out at 8 bytes, including the HEC field itself.

Following the XGTC header, we encounter zero or more XGTC responses, one for each allocation in the series specified by the BWmap. Each response begins with a dynamic bandwidth report DBR if and only if the BWmap specifies DBRu for that particular alloc-ID.

The DBR contains two fields:

Buffer Occupancy   This field is a word count of the present backlog in the aggregate of upstream transmit queues for this alloc-ID (T-CONT). It does not subtract off the allocation for the current burst, nor does it attempt to estimate traffic arrival rates. It counts client payload size (Ethernet frames) but not GEM frame headers.

To illustrate, suppose we have 1000 words of upstream Ethernet backlog for this T-CONT, suppose the current burst allocation for this alloc-ID is 100 words, and suppose that traffic is continuing to arrive at the rate of 300 words per unspecified interval—one of the reasons we do not attempt to include arrival rates. Given all of these assumptions, the DBR reports 1000 words. The OLT can subtract the 100-word current allocation, as well as any additional allocations that may already be in the pipeline. Since the OLT governs the rate and size of upstream grants, only the OLT is in a position to act on the estimated traffic arrival rate. Finally, since the reported backlog does not include GEM frame headers, the OLT might choose to grant something larger than 1000 words, perhaps 1050 words.

---

[*] It has been pointed out that ONU reboot would be another valid reason to signal DG.

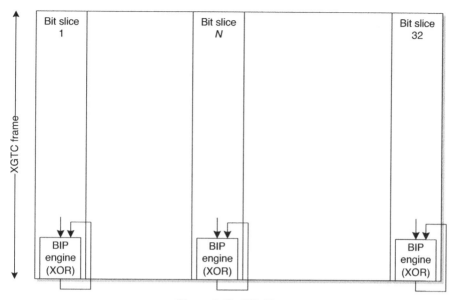

Figure 4.19 BIP-32.

CRC-8 This is just an error check on the buffer occupancy field. Its generator polynomial is $x^8 + x^2 + x + 1$. The standard specifies that cyclic redundancy check failure should cause the OLT to discard the DBR.

To complete our inspection of the XGTC frame, the XGTC trailer is a BIP-32 block (Fig. 4.19). What BIP—bit interleaved parity—means in this context is that the BIP engine looks at the XGTC burst, starting with the XGTC header, as if it were a stream of 32 independent parallel bitwise slices. The BIP engine sets each of its individual bits to produce an even number of 1s in each flow, an exclusive-OR function, counting the BIP itself as the final bit of the checked sequence. At the OLT, the same check detects the case when an odd number of errors occurs in the bit slice. BIP errors are intended to be counted as a performance monitoring parameter.

For a high-quality channel, the probability of multiple errors in a given bit position across a burst is negligible, and BIP is a good estimate of the total error rate. At higher raw error rates, the probability of two errors in a given bit slice increases, and BIP saturates; the saturation point depends on the length of the burst, but typically lies in the BER range 10^{-4} to 10^{-5}. In practice, it does not much matter where the BIP saturates: if the BIP count exceeds some quite small threshold, it is advisable to turn on upstream FEC. At high raw error rates, BIP errors can still be collected, but the count does not mean much. Even when the BIP count is not saturated, it must be compared with some independent count of the total number of bytes received before the quality of the link can be evaluated.

The OLT also collects FEC statistics: if the number of corrected code words were consistently zero for a given ONU—which is quite possible for ONUs with comfortable optical budgets—we could gain some additional upstream traffic capacity by disabling upstream FEC for that ONU.

4.1.4 G.984 G-PON Downstream Framing

Though not an identical twin of G.987 XG-PON, G.984 G-PON framing is clearly a sibling. We recognize many family characteristics as we look into its details.

- G.984 G-PON framing differs in part because it was not intended to operate at high raw bit error rates, so it did not need the same level of inherent robustness. FEC is optional in both directions.
- The speed of G.984 G-PON electronics did not warrant 4-byte alignment.
- The fiber was understood to be secure physically, so that upstream content did not need to be encrypted. Downstream multicast was (correctly) expected to be encrypted by the middleware, so that further multicast encryption by the PON layer was thought to be unnecessary.
- With fewer ONUs on the PON—32 envisioned initially, 64 in the current view—there was less need to send several PLOAM messages in a single downstream frame.

Downstream physical layer frames of the G.984 G-PON transmission convergence (GTC) layer occupy 125 μs with continuous transmission, as shown in Figure 4.20.

4.1.4.1 GTC Frame Header

The downstream GTC frame header is called PCBd, the physical control block. It contains seven fields, one of them a repetition.

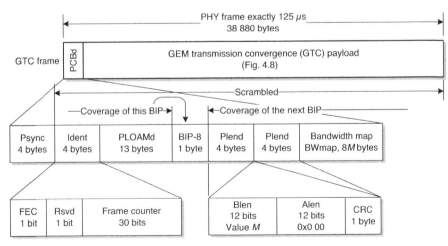

Figure 4.20 Downstream GTC frame.

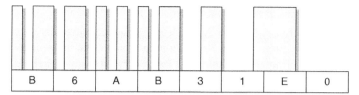

B	6	A	B	3	1	E	0

Figure 4.21 G.984 G-PON Physical sync pattern PSync.

PSync As with G.987 XG-PON, the purpose of the physical synchronization pattern is to provide a well-defined bit sequence against which the ONU receiver can acquire byte delineation. As we would expect, Psync is not scrambled. Figure 4.21 invites another mental autocorrelation exercise to convince ourselves that bit-shifted versions of the Psync pattern are reasonably immune to misinterpretation.

Figure 4.22 shows the state machine by which a G.984 G-PON ONU acquires and maintains frame sync. Examining every bit as a candidate for start of frame, the machine goes from hunt to presync state upon a single match. It then waits exactly $125 \, \mu s$ and checks again. If it sees $M_1 - 1$ additional consecutive Psync patterns at the expected intervals, it goes to sync state, clears the loss of signal/loss of frame indication LOS/LOF, and begins to decode the rest of the downstream frame; but a single Psync error during the presync state puts the state machine back into hunt mode. Once synchronized, the machine remains in sync state until it sees M_2 consecutive incorrect Psync fields. The recommended value for M_1 is 2, and for M_2 is 5.

Ident

- *FEC* If this bit is set, the downstream frame is protected with FEC. By comparison, downstream FEC is *always* active in G.987 XG-PON. To protect against bit errors, the ONU requires four consecutive frames with the same FEC bit before it recognizes a change. This presents no service disruption problem in practice because FEC is not intended to be changed dynamically.

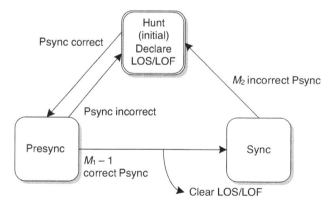

Figure 4.22 G.984 G-PON Psync state machine.

- *Rsvd*   The reserved bit is not used.
- *Frame Counter*   As in G.987 XG-PON, the recommendation calls this a superframe counter. It is a 30-bit circular counter that increments with each downstream frame. Its primary use is in encryption key switching and time-of-day synchronization, both of which are described later in this chapter. Just as in G.987 XG-PON, the predictably incrementing frame counter is available to provide additional robustness in synchronizing to the downstream frame.

  This option is not mentioned in G.984.3; but then, G.984 G-PON was not originally intended to operate at the same high error rates as XG-PON. The introduction of the C+ optical budget class changes that assumption (10^{-4} raw error rate), and it may be wise for G.984 implementations to consider robustness enhancements.

PLOAMd   In G.984 G-PON, PLOAM messages are 13 bytes long. Every downstream frame contains one; if the OLT has nothing to say, it sends a *no_message* message. Appendix II discusses PLOAM messages.

BIP-8   This field computes bit-interleaved parity on each of 8 parallel bit slices in the bytes of the frame. Its scope includes all bytes, excluding possible FEC parity bytes, transmitted since the BIP-8 field of the previous frame. At the ONU, BIP is computed *after* a possible FEC correction step. BIP counts can be used for performance monitoring and, in theory, as the input to SDH-like signal fail and signal degrade conditions.

Plend   The payload length field does not indicate the length of the payload.

- In its 12-bit field *Blen*, it specifies the number of 8-byte allocations in the bandwidth map BWmap.
- As well as the GEM partition, the original G.984 G-PON specification included an ATM partition, and a similar 12-bit field *Alen* was defined for it. ATM was subsequently removed from the GTC layer, in fact completely removed from G-PON, so the Alen field is always set to zero on transmit and ignored on receive.
- The *CRC* field checks and corrects errors. The specified code, $x^8 + x^2 + x + 1$, can correct one error and detect multiple errors.

The downstream frame includes two copies of Plend for robustness, so the ONU makes an additional check of the CRC results to determine whether the BWmap and the downstream frame are usable. Refer to Table 4.2.

BWmap   The bandwidth map specifies the assignment of upstream capacity to the alloc-IDs of the various ONUs on the PON.

4.1.4.2   G.984 G-PON Bandwidth Map

Figure 4.23 portrays the BWmap in G.984 G-PON. As in G.987 XG-PON, it comprises a series of 8-byte allocation structures, each of which grants permission to transmit an upstream burst. As in XG-PON, several allocations to a given ONU

TABLE 4.2 PLend Error Processing

First Plend	Second PLend	Corrected Results	Decision
Uncorrectable	Uncorrectable		Discard
Correctable	Correctable	Different	Discard
No errors	No errors	Different	Discard
No errors	No errors	Same	Either PLENd
Correctable	Correctable	Same	Either PLENd
No errors	>0 errors		First PLENd
Correctable	Uncorrectable		First PLENd
>0 errors	No errors		Second PLENd
Uncorrectable	Correctable		Second PLENd

may be concatenated into a single burst. We describe the rules for start time and stop time pointers below.

There are five fields in a G.984 G-PON allocation structure:

Alloc-ID   The alloc-ID specifies the owner of the upstream grant. Values less than 254 match the corresponding TC-layer ONU-ID directly and authorize upstream OMCI transmission (it is allowed but not encouraged to carry subscriber traffic in this, the so-called default alloc-ID). Alloc-ID 254 invites

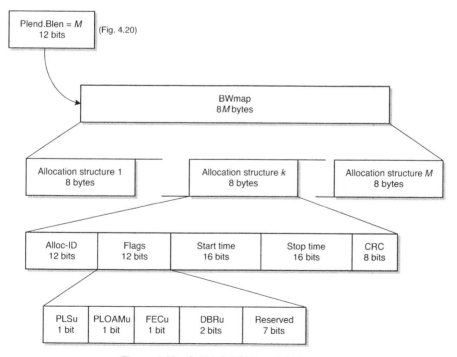

Figure 4.23 G.984 G-PON bandwidth map.

a serial_number PLOAM response from undiscovered ONUs, those that do not yet have an assigned TC-layer ONU-ID. Alloc-ID 255 is reserved, and the remainder of the space is available for alloc-IDs that are mapped to T-CONTs, to carry subscriber traffic.

Flags The flags field comprises 12 bits, of which 5 are used.

- *PLSu* G.984 G-PON originally contemplated power leveling. This was a feature whereby the OLT could reduce the dynamic range of the upstream power received from the various ONUs by commanding nearby ONUs to reduce their transmit power levels. The feature has been deprecated; this bit is set to 0 on transmit and ignored on receive.

- *PLOAMu* If this bit is set, the ONU sends a PLOAM message in the upstream allocation. This bit is only meaningful in grants sent to the default alloc-ID. The PLOAM message is often shown pictorially as the first of a contiguous series of allocations, but the default alloc-ID, and its PLOAM message, may in fact occur in any position of the burst.

- *FECu* This bit requests that FEC be enabled in the upstream burst. Unlike G.987 XG-PON, G.984 G-PON ONUs are allowed to not support FEC, in which case they ignore this bit and transmit ordinary data, just as if FEC had not been requested. A bit in the upstream burst header tells the OLT whether FEC is in fact present or not.

- *DBRu* DBRu is a 2-bit field that instructs the ONU to send a queue occupancy report in the upstream burst, or not, and which mode to use. G.984 G-PON specifies two modes of DBA reporting (even more, if we go back far enough into history), mode 0 and mode 1. Mode 0 is the default, and is the only mode defined in G.987 XG-PON. In G.984, the mode 0 report is a single byte containing a nonlinear code (Table 4.3) that approximates the total queue occupancy in bytes. Mode 1 contains two similarly encoded bytes, one for green traffic and one for yellow (conformant traffic and traffic beyond the guarantees of its contract, as described in Section 6.3).

Start Time This 16-bit field specifies the offset of the allocation's start time from the beginning of the upstream frame. It is measured in bytes at the upstream

TABLE 4.3 DBA Report Encoding

Queue Length	Binary Input (ONU)	Coding of Octet	Binary Output (OLT)
0–127	00000000abcdefg	0abcdefg	00000000abcdefg
128–255	00000001abcdefx	10abcdef	00000001abcdef1
256–511	0000001abcdexxx	110abcde	0000001abcde111
512–1023	000001abcdxxxxx	1110abcd	000001abcd11111
1024–2047	00001abcxxxxxxx	11110abc	00001abc1111111
2048–4095	0001abxxxxxxxxx	111110ab	0001ab111111111
4096–8191	001axxxxxxxxxxx	1111110a	001a11111111111
>8191	01xxxxxxxxxxxxx	11111110	011111111111111
Invalid	N/A	11111111	N/A

rate, 1.2 Gb/s. The burst header occurs prior to the first start time, and must be taken into account by the OLT when it plans the BWmap.

Stop Time The stop time also refers to a byte offset from the start of the upstream frame.

CRC The allocation structure is protected by a CRC-8 that can correct one error and can reliably detect two. If the CRC detects more errors than it can correct, the ONU discards the allocation.

As with G.987 XG-PON, the concept of contiguous grants also pertains to G.984 G-PON—that is, several grants in a row, all of them directed to the same ONU. An ONU is expected to be able to support up to eight allocation structures in a given BWmap, although it is not specified that they need be contiguous. The way in which contiguous grants are indicated in G.984 is that the start time of the subsequent grant is exactly one greater than the stop time of the current grant. If this condition is not met, the OLT must allocate the second grant as a completely separate burst, complete with burst overhead.

Upstream FEC is either on or off for the entire burst. This means the FECu bit should theoretically be the same in each allocation. In practice, the ONU would configure the burst according to the first allocation and ignore the others.

When it is planning the BWmap, the OLT must consider that an upstream PLOAM message consumes capacity from a grant to the default alloc-ID, that a DBA report consumes capacity from each allocation for which it is specified, and that FEC is an overhead of 16 bytes added onto each 239-byte chunk of payload content, starting with the BIP field[*] of the upstream burst header, running continuously across allocation boundaries, and ending with a possibly shortened last code word.

4.1.4.3 Assembling the Downstream G.984 G-PON Frame
The BWmap completes the header of the downstream frame. We now append the GTC payload, which comprises a series of GEM frames, possibly including idle frames. Before transmitting the frame, we optionally insert FEC parity bytes, and we always scramble it.

FEC FEC is discussed in detail in Appendix I and in the downstream XG-PON framing Section 4.1.2. Here we describe the details of G.984 G-PON downstream FEC, as shown in Figure 4.24.

If FEC is disabled, the G.984 G-PON downstream physical frame is identical to the GTC layer frame (but scrambled). If FEC is enabled, the entire frame, including PCBd, is protected by a series of FEC code words. The code is RS (255, 239), which means that 239 data bytes are followed by 16 parity bytes. The first FEC code word begins with the first byte of the PCBd.

---

[*] Although FEC starts with the BIP field, which is the first byte of the burst header, the start time and the allocation interval refer to the first byte *after* the burst header. This byte is either the first byte of a PLOAM message, the first byte of a DBA report, or the first byte of a GEM payload frame. Refer to Figure 4.27.

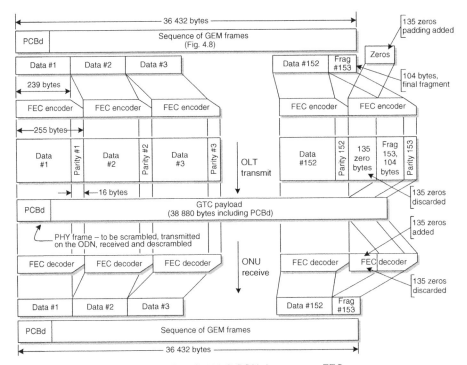

Figure 4.24 G.984 G-PON downstream FEC.

The physical frame is 38,880 bytes long, enough for slightly more than one hundred fifty-two 255-byte FEC code words. We therefore have a shortened last code word 153 in each frame, 104 bytes of data with the usual 16 bytes of FEC parity. After subtracting FEC parity overhead, 36,432 bytes remain in the downstream frame; the cost of downstream FEC in G.984 G-PON is 6.3%.

A G.984 G-PON ONU is not required to support FEC decoding, but it is required to be able to detect that the downstream FEC bit is set, and if so, skip past the FEC parity bytes and properly receive the payload. In fact, all current G-PON ONUs support FEC.

Scrambling Scrambling is described in Section 4.1.2.5. In G.984 G-PON, the downstream frame is scrambled with a frame-synchronous polynomial, $x^7 + x^6 + 1$, whose output repeats every 127 bits. The scrambler is initialized to all 1s on the first bit following the Psync field, and as shown in Figure 4.20, runs continuously across the frame, including FEC bytes. The Psync field itself is not scrambled.

Figure 4.25 illustrates the G.984 G-PON scrambler.

The short polynomial and fixed initial state of G.984 G-PON conceivably pose a risk of DoS attacks, supposing that a malicious user were able to transfer data that caused the scrambler to generate a long sequence of identical digits. This risk was addressed in the much longer scrambler and varying preset states of the G.987 XG-PON scrambler.

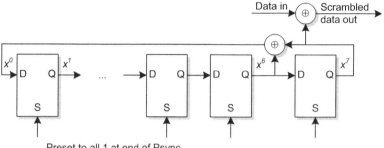

Figure 4.25　G.984 G-PON scrambler.

4.1.5　G.984 G-PON Upstream Framing

A G.984 G-PON upstream burst comprises a burst header and a series of allocation intervals. We discuss the burst header after the following explanation of the framing details.

A generalized upstream allocation interval comprises a GTC overhead block, followed by a GTC payload section that contains a sequence of GEM frames. An allocation to the default alloc-ID may include an upstream PLOAM message, as shown in Figure 4.26, while any allocation interval may include a dynamic bandwidth report DBR:

PLOAMu　An upstream PLOAM message is present:

- Only if the allocation was directed to the ONU's default alloc-ID or to alloc-ID 254 for ONUs that are attempting to be activated onto the PON,
- Only if the PLOAMu flag was set in the BWmap allocation.

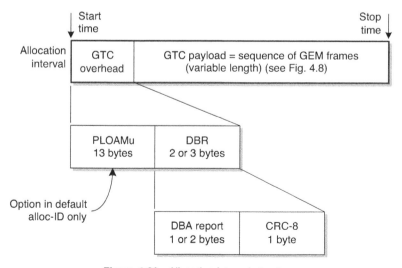

Figure 4.26　Allocation interval structure.

If the ONU has no substantive message to send, it sends a *no_message* message.

DBR   The DBA report is likewise present only if requested by the OLT, but DBR can occur in any allocation structure, reporting on the queue backlog for that particular alloc-ID. The granularity of the report defaults to 48 bytes, for reasons that have to do with the ATM history of ITU-T PONs. It may be provisioned by the OLT through OMCI as the GEM block length attribute of the ANI-G managed entity.

The DBA report comprises 1 byte reporting the entire queue backlog, (mode 0) or 2 bytes reporting green and yellow backlog separately (mode 1). Section 6.3 discusses colored traffic.

It is not specified whether the DBA report should include or exclude the current allocation. The ambiguity is recognized in G.984.3, which requires the ONU to be consistent in its reporting practice, and requires the OLT to tolerate both, on an ONU by ONU basis.

Table 4.3 shows the coding of the DBA report octets. The letters a, b, ... g represent bits that express the next-to-most significant part of the queue length (except for backlog <127, the most significant bit is always 1). Bits shown as x are unspecified: they may be either 0 or 1.

The idea of this compact logarithmic representation is that fine-grained reporting is less important as the queue backlog grows larger. This is appropriate because we do not know exactly how much overhead will be needed for GEM framing or DBR or for new traffic that may arrive before we get a grant for the existing backlog.

To illustrate, suppose the ONU has a backlog of 753 bytes (below); the OLT will interpret our backlog report to be 767. This 2% overestimation is typical of the accuracy to be expected. Because we need extra overhead for GEM headers and such, it is also appropriate that the OLT overestimate the backlog.

	Pattern	Example Value	Decimal
Backlog at ONU	000 001a bcdx xxxx	000 0010 1111 0001	753
Coding of octet	1 110a bcd	1 1100 111	
Binary output (OLT)	000 001a bcd1 1111	000 0010 1111 1111	767

Upstream Burst Structure   Figure 4.27 shows the concatenation of the burst overhead and burst header with the allocation intervals to form the complete burst, ready for FEC and scrambling. The burst overhead exists to permit the OLT receiver to set the sampling threshold and phase correctly and establish byte delineation; the physical layer overhead PLOu is the complete set of fields transmitted by the ONU prior to the individual allocations. The guard time is not part of PLOu because it does not represent a transmission, merely the laser turn-on interval.

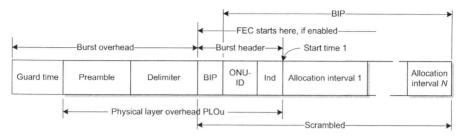

Figure 4.27 G.984 G-PON burst structure.

Yes, all these overlaps are confusing. Let us first discuss the burst header, then the burst overhead:

BIP   This bytewide field represents the bit-interleaved parity of the burst, excluding preamble, delimiter, and FEC (about which, more below). The BIP computed in this current burst appears in the header of the next burst from the same ONU. This is clumsy because we do not know the quality of the current burst until we receive the next burst, which may be some arbitrary time in the future. It was revised in G.987 XG-PON, whose BIP appears in the burst trailer.

ONU   The unique 8-bit identifier of the ONU on the PON (TC-layer ONU-ID).

Ind   The bits of this byte are assigned as follows:

   7   When set, this bit requests a PLOAMu grant from the OLT: the ONU has something to send. Originally, this bit was defined to indicate that an urgent PLOAM message was waiting, but there *are* no urgent PLOAM messages—or from another viewpoint, there are no *non*urgent PLOAM messages—so urgency was removed from the definition.

   6   When set, this bit indicates that upstream FEC is on. Recall that the ONU may disregard the OLT's request to use upstream FEC; this bit represents the reality. Observe that this bit is (or is not) itself protected by FEC.

   5   RDI, remote defect indication. This bit informs the OLT of a signal failure in the downstream direction. Since most downstream signal failures prevent the ONU from transmitting, this bit is of limited value. Conceivably, it could signal high BER even after FEC correction; the criteria for setting the bit are not defined.

The other bits are reserved.

G.984 G-PON limits bursts to lie entirely within their respective upstream frames. In comparison, G.987 an XG-PON burst need only have a start time within the upstream frame interval.

Forward Error Correction   When FEC is enabled, the first code word begins with the BIP field. Every 239 bytes of data are followed by 16 bytes of FEC parity, if

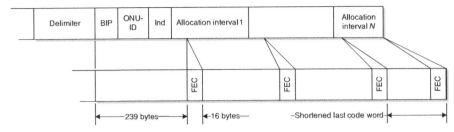

Figure 4.28   G.984 G-PON upstream FEC.

necessary with a shortened last code word at the end of the burst. Figure 4.28 illustrates the operation.

In G.984 G-PON, the FEC bytes are part of the allocation as defined by start and stop time fields in the BWmap. This implies certain rules in the construction of the BWmap. If the start time pointer of a contiguous allocation series were to point to one of the parity byte locations of a previous allocation, it could be interpreted as an attempt to preempt FEC parity, which is not permitted. It could also be interpreted as a request to start the next allocation on the first byte after the parity block or as a request for a shortened FEC code word. G.984 G-PON therefore requires the start time not to point to a parity byte.

Recall that in contiguous allocations, the stop time pointer is exactly one less than the subsequent start time. The restriction on start time value therefore implies that embedded stop time pointers are forbidden to point to the last byte before a parity block, or to any of the first 15 parity bytes.

The stop time rule is irrelevant at the end of the burst because there *is* no subsequent start time. The ONU creates a shortened last code word that includes the full complement of 16 parity bytes. The well-behaved OLT avoids a stop time that implies a shortened last code word of fewer than 16 bytes.

Scrambling   After applying FEC, the upstream burst is scrambled. The scrambler is the same as is used downstream, with generator polynomial $x^7 + x^6 + 1$. It is preset to all ones on the first bit of the BIP field and runs continuously throughout the burst.

G.984 G-PON Burst Overhead   Figure 4.29 reminds us of the burst overhead shown in Figure 4.27. The burst overhead is determined by the OLT, which periodically broadcasts upstream_overhead PLOAM messages and may also broadcast extended_burst_length messages as well. The extended burst length allows for a longer burst header for ONUs that are attempting to be activated

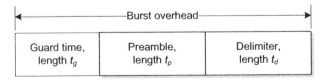

Figure 4.29   G.984 G-PON burst overhead.

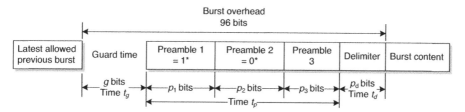

Figure 4.30   Definition of burst overhead fields.

onto the PON, specifically a longer type 3 preamble (Fig. 4.30). Once they have been discovered and brought into service, ONUs use a different type 3 preamble byte count, which is also specified in the extended_burst_length message.

G.984.2 defines the length of the various fields, as shown in Table 4.4. The idea of guard time and the enable and disable times is the same as is illustrated in Figure 4.17.

The upstream_overhead PLOAM message specifies several fields that combine to form the burst overhead. The results are shown in Figure 4.30. Asterisks indicate repetition of a bit value.

g—Number of Guard Bits   Guard time t_g is provided to allow time for transmitter turn-off and turn-on, for drift in equalization delay, and for tolerances in measuring and setting equalization delay. The ONU cares about guard time because it represents the earliest instant at which the ONU is allowed to turn its laser on in preparation for an upcoming burst.

p_1 and p_2

p_1 Number of type 1 preamble bits. Preamble 1 bits are all 1.

p_2 Number of type 2 preamble bits. Preamble 2 bits are all 0.

The p_1 and p_2 intervals are intended to give the OLT receiver a chance to set its decision threshold halfway between 1 and 0 values. Because the preamble 1 and 2 bits contain no transitions, they are useful only in setting the receiver gain and its decision threshold, but not in recovering timing. It is allowed to set either or both of p_1 and p_2 to zero, appropriate if the receiver prefers a preamble type 3 pattern instead.

TABLE 4.4   G.984 G-PON Burst Overhead Components

Component	Symbol	Units	Value
Guard time	t_g	Bits, min	32
Preamble time	t_p	Bits, suggested	44
Delimiter time	t_d	Bits, suggested	20
Total burst overhead	—	Bits, mandatory	96
Transmit enable	—	Bits, max	16
Transmit disable	—	Bits, max	16

Pattern for Type 3 Preamble Bits   one byte. This is likely to be a bit pattern such as 10101010, with a maximum transition density that allows the OLT to recover clock phase.

The length of preamble 3 is computed as

$$p_3 = 96 - g - p_1 - p_2 - p_d \text{ bits} \qquad (4.1)$$

where

96 is the length of the burst overhead, as specified in G.984.2.

g, p_1, p_2, and p_d are as defined here.

Delimiter, p_d   The delimiter, which has length p_d, is the OLT's choice of a pattern that provides robust byte delineation and frame start indication in the presence of bit errors. The upstream_overhead PLOAM message has space for 3 bytes of delimiter. If fewer than 3 bytes are needed, as suggested in Table 4.4, they may be provisioned to extend the preamble type 3 pattern.

As well as burst header information, the upstream_overhead PLOAM message contains various additional information, not pertinent to this discussion. Appendix II contains the details.

The extended_burst_length PLOAM message, also expanded in Appendix II, is intended to allow the OLT to specify a longer burst overhead for ONUs in the discovery process. The primary content of the extended_burst_length message is a specification of the number of type 3 preamble bytes to transmit while the ONU is in its serial number and ranging states. As a secondary feature, the extended_burst_length message also allows the specification of type 3 preamble repetition for the ONU in operation state. This is an escape route from Eq. (4.1), which may impose an unduly restrictive upstream_overhead message format, especially for ONUs behind a reach extender.

4.2   ONU ACTIVATION

We now have a good view of how payload is mapped into frames, along with the various chunks of necessary overhead, and how the frames are placed onto the fiber. Timing is an important aspect of upstream burst transmission. We discuss timing in Section 4.3, but it may help to understand the details of timing if we first consider the process of activating an ONU that newly appears on the PON. The activation process requires timing measurement and delay adjustment.

We first outline the basic sequence of steps to bring an ONU onto a working PON, then go back and discuss the variations. States and transitions for G.987 XG-PON are shown in Figure 4.31, with G.984 G-PON in Figure 4.32. We consider G.987 XG-PON first, then G.984 G-PON.

4.2.1   G.987 XG-PON Activation and Operation

1. When the ONU powers up and initializes, it must remain silent. Anything it transmitted on the PON could disrupt existing traffic. In initial state O1, the

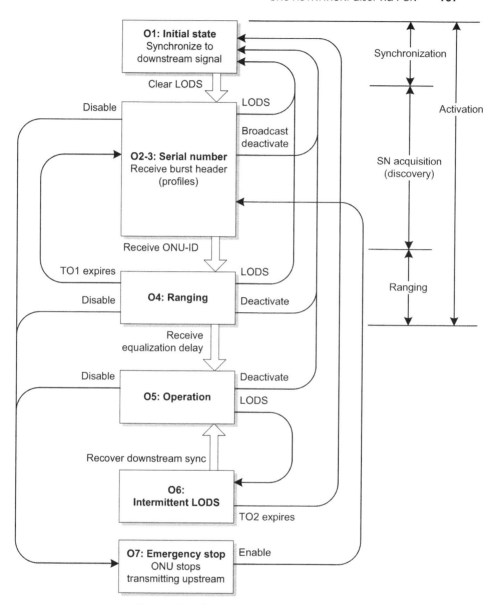

Figure 4.31 G.987 XG-PON ONU states.

ONU synchronizes itself to the downstream signal, acquiring bit and frame synchronization and frame count. Then it enters serial_number state O2-3 to listen quietly for the burst header information it will need before it can properly sign on.

The ONU also clears, or discards, all TC-layer information it may have had: its ONU-ID, equalization delay, burst profiles, encryption keys, and so forth.

2. From time to time, the OLT broadcasts header information to define the parameters that ONUs must have before they can attempt to join the PON. In G.987 XG-PON, this information takes the form of a profile PLOAM message, discussed in detail in Appendix II.

3. Once it has acquired one or more profiles, the ONU waits for a serial_number grant that specifies one of the profiles in its repertoire. The ONU is never permitted to transmit if it does not know the specified profile.

4. From time to time, the OLT withholds transmit permission from all known ONUs, thereby opening a so-called quiet window in the upstream direction. Section 4.3 discusses the factors affecting the size and placement of the quiet window. For this window, the OLT generates a so-called serial_ number grant, an invitation for any ONU in serial_number state to respond. A serial number grant is a BWmap allocation directed to alloc-ID 1023, requesting a PLOAM message upstream, and allocating zero XGTC payload capacity.

5. More than one ONU can be waiting to sign on at any given time, and they can be at nearly the same distance from the OLT. If they all transmitted at the same time, their transmissions would collide at the OLT (see Fig. 4.38) and no one would be recognized. Worse, this deterministic collision would recur forever. Therefore, the ONU delays its response by some random value; transmissions may collide during this particular discovery cycle, but sooner or later they will be distinct and the OLT will recognize one of them, then the other.

   The random delay takes on values from 0 to 48 μs; the size of a serial number burst varies depending on the profile settings, but is typically less than 1 μs, so collisions are likely to be resolved reasonably soon, even if several ONUs simultaneously contend for recognition (e.g., when the entire PON is being initialized).

6. The OLT recognizes the serial_number_ONU response from the ONU and assigns a TC-layer ONU-ID to that serial number. If the ONU sees another serial number grant before it gets an ONU-ID, it deduces that it has not been recognized. It generates a new random delay and tries again.

7. The OLT can, and probably does, measure the round-trip delay from the serial_number response PLOAM message. Since the response conveniently includes the value of random delay used by the ONU, round-trip delay is easy to derive.

8. The ONU accepts the ONU-ID as its own and henceforth uses this value for its PLOAM communications as well as for its default alloc-ID for bandwidth grants and OMCI. Gaining a TC-layer ONU-ID allows the ONU to enter ranging state O4.

9. According to the script, the OLT should now measure the propagation and processing delay of the newly assigned ONU. The OLT opens another quiet window and invites the newly assigned ONU to transmit

again. The ranging window may be as wide as the serial number window, less an allowance for random delay, which is not present in this step. In practice, if the OLT has even an approximate value for the delay, and merely wishes to make a second, precise measurement, the OLT need not open a full window.

This grant is called a ranging grant, primarily distinguished because the ONU is in ranging state O4 when it happens. The ranging grant is directed to the newly assigned alloc-ID (ONU-ID); it contains no allocation for XGTC payload, only for an upstream PLOAM message. An ONU in ranging state O4 responds with its registration ID, and with no random delay.

10. Here or at serial number acquisition, the OLT measures the delay precisely, computes an offsetting value called equalization delay, and transmits the equalization delay value to the ONU. Henceforth, the ONU delays all of its transmissions by the equalization delay value. Section 4.3 explains equalization delay in detail.

11. Reception of an equalization delay puts the ONU into operation state O5, where it is now ready for service. The OLT is free to audit the ONU's MIB and provision it for service.

Consider a variation on this theme. If the OLT already knew a lot about the ONU—if, for example, the ONU fell off the PON because the subscriber powered it down—the OLT could blindly issue PLOAM messages to assign a TC-layer ONU-ID to the known serial number and set the known equalization delay, whereupon the ONU could go into operation state O5 without ever having responded to either a serial number or a ranging grant.

But see also the discussion in Section 4.6.1 on the PLOAM MIC, which requires the registration ID to be known to the OLT. So unless the registration ID is known, the OLT must send a ranging grant, and the ONU must respond to it. In the case of a previously known ONU, the OLT may well know the ONU's registration ID, but building OLT software around the corner cases may be more trouble than just issuing a ranging grant.

States O6 and O7 are not part of the normal progression. The *intermittent loss of downstream sync* state O6 occurs when the ONU's frame synchronization state machine declares LODS (Fig. 4.12). The ONU starts timer TO2. If TO2 expires, the ONU recognizes that it is no longer on the PON and restarts the activation sequence from scratch. When the ONU returns to initial state O1, it discards its ONU-ID and all other TC-layer provisioning.

If a G.987 ONU recovers downstream synchronization before TO2 expires, it simply returns to operation state O5 on its own initiative and carries on. There is an assumption that, if the equalization delay is suddenly wrong because of a protection switch, the OLT knows about it and will take whatever action is appropriate. The PON ID field (Fig. 4.10) may also be useful for the ONU itself to detect a protection switch and avoid disruptive responses until it has been redirected by the OLT. Details of how this might work are for further study.[*]

O7 is called the emergency stop state. It is intended as a holding pen for rogue ONUs. Assuming that the ONU is capable of responding, the broadcast downstream disable_serial_number PLOAM message puts it into state O7, where it is forbidden to transmit anything upstream. If the ONU receives a subsequent disable message whose action code point is set to enable, then and only then may it return to the serial number state O2-3 and go through the normal discovery and ranging process.

More formally, the states are defined as follows:

G.987 State O1, Initial

Behavior   The ONU transmits nothing. In this state, the ONU waits for the presence of a downstream signal and then synchronizes to that signal. The ONU discards any TC-layer parameters that it may have known from its previous session: ONU-ID, default alloc-ID, profiles, equalization delay, encryption keys.

G.987.3 states that when the ONU enters O1, including entry from some other state, it should reset its downstream synchronization machine (Fig. 4.12). An ONU that already has downstream sync may well choose to skip this detail.

Exit Criterion   Bit, byte, frame, and superframe synchronization achieved: LODS cleared $\rightarrow$ O2-3.

G.987 State O2-3, Serial Number

Behavior   The ONU is never permitted to transmit if it does not know the profile specified in the bandwidth grant. When it sees a serial number grant that specifies a known profile, the ONU transmits a serial_number_ONU PLOAM response, delayed by a locally generated random value. In this state, the ONU continues to respond to serial number grants, each time with a different random delay.

This is the only state in which the ONU is permitted to respond to global bandwidth grants, also the only state in which it uses a random delay.

Exit Criteria
- Assign_ONU-ID PLOAM message $\rightarrow$ O4.
- LODS $\rightarrow$ O1.

G.987 State O4, Ranging

Behavior   When it enters state O4, the ONU starts timer TO1, whose initial value is recommended to be 10 s.

In this state, the ONU regards any grant directed to itself as a ranging grant. The grant is expected to request an upstream PLOAM message and provide no XGTC payload allocation. The ONU responds with the

---

[*] The phrase *for further study* is ITU speak for an admission that an issue has been recognized but that no one wants to take the time and trouble of solving it now. Later, maybe, but only if it becomes important.

registration PLOAM message. There is no random delay in this message; it is a response to a directed grant, so there is no possibility of collision with autonomous transmitters.

Whether or not it has received and responded to a ranging grant, the ONU waits for a ranging_time PLOAM message from the OLT, which establishes its equalization delay.

Exit Criteria

- Ranging_time PLOAM message $\rightarrow$ O5.
- Timer TO1 expires $\rightarrow$ O2-3.
- LODS $\rightarrow$ O1.

G.987 State O5, Operation

Behavior   This is the normal state of ONU operation. The ONU responds to grants directed to it, delaying its response by the value of the equalization delay.

Exit Criterion   LODS $\rightarrow$ O6.

G.987 State O6, Intermittent LODS

Behavior   When it enters state O6, the ONU starts timer TO2, whose initial value is recommended to be 100 ms. The ONU transmits nothing. In this state, it has lost sync with the downstream signal.

Exit Criteria

- Timer TO2 expires $\rightarrow$ O1.
- Recovery of downstream sync, LODS cleared $\rightarrow$ O5.

G.987 State O7, Emergency Stop

Behavior   The ONU transmits nothing. It remains in this state until explicitly reenabled, even when power cycled.

Exit Criterion   Reception of a disable PLOAM message with enable code point set $\rightarrow$ O2-3.

Further Notes

- In any state, reception of a disable PLOAM message with a disable code point causes a transition into emergency stop state O7.
- In any state except O7, reception of an assign_ONU-ID PLOAM message that matches one, but not both, of the ONU's serial number and TC-layer ONU-ID causes a transition to O1.
- In any state except O7, reception of a deactivate PLOAM message causes a transition into initial state O1.
- As we went to press, G.987.3 amendment 1 had been consented in ITU-T. Among other things, it includes an optional suspension state O8 that may help to mitigate rogue ONU behavior.

4.2.2 G.984 G-PON Activation and Operation

Comparing Figures 4.31 and 4.32, we see that the XG-PON and G-PON state machines are similar but not identical. The loss of downstream sync LODS of G.987 (Fig. 4.12) is effectively the same as the LOS/LOF of G.984 (Fig. 4.22) and is declared by the downstream synchronization state machine.

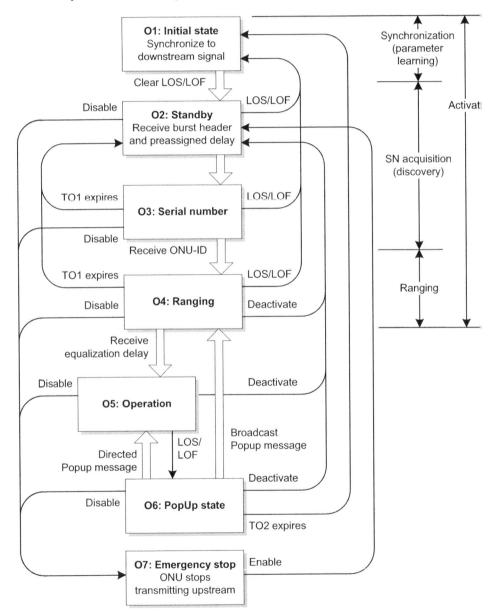

Figure 4.32 G.984 G-PON ONU states.

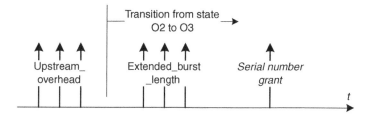

Figure 4.33 ONU discovery sequence, G.984 G-PON.

Because of the similarities, we abbreviate the discussion of G.984, focusing on the differences.

1. In initial state O1, the ONU synchronizes itself to the downstream signal, acquiring bit and frame synchronization and frame count. Then it enters standby state O2 to listen quietly for the burst header information it will need before it can properly sign on. The ONU also clears, or discards, all TC-layer information it may have had: its ONU-ID, equalization delay, burst profiles, and the like.

2. The OLT broadcasts header information to define the parameters that ONUs must have before they can attempt to join the PON. In G.984 G-PON, they are defined by the upstream_overhead and optional extended_burst_length PLOAM messages, discussed in Appendix II. The logical time to do this is just before the OLT issues a serial_number grant (Fig. 4.33). One of the parameters of the upstream_overhead message is a preassigned delay, which allows the OLT to adjust the relative position of the quiet window. G.987 XG-PON has no concept of a preassigned delay; the OLT positions the quiet window through its management of start time pointers.

3. When it acquires the burst header parameters, the ONU waits for a serial_number grant in state O3. By comparison with Figure 4.31, we see that states O2 and O3 were merged in XG-PON because the division served no externally visible purpose.

4. The OLT opens a quiet window during which it generates a serial_number grant, an invitation for any ONU in serial_number state to respond. A serial number grant is a BWmap allocation directed to alloc-ID 254, requesting a PLOAM message upstream, and allocating zero GTC payload capacity.

5. The ONU delays its response by the preassigned delay value specified in the upstream_overhead PLOAM message plus a random delay in the range from 0 to 48 μs.

6. The OLT recognizes the serial_number_ONU response from the ONU, and assigns a TC-layer ONU-ID to that serial number. If the ONU sees another serial number grant before it gets an ONU-ID, it generates a new random delay and tries again.

7. The OLT can measure the round-trip delay from the serial_number response PLOAM message. Since the response conveniently includes the value of random delay used by the ONU, round-trip delay is easy to derive.

8. The ONU accepts the ONU-ID as its own and, henceforth, uses this value for its PLOAM communications as well as for its default alloc-ID for bandwidth grants. Gaining a TC-layer ONU-ID allows the ONU to enter ranging state O4.

9. The OLT now opens another quiet window and issues a ranging grant, primarily distinguished because the ONU is in ranging state O4 when it happens. The ranging grant is directed to the default alloc-ID of the newly assigned ONU; it contains no allocation for GTC payload, only for an upstream PLOAM message. The G.984 ONU in ranging state O4 repeats the serial_ number message, delayed only by the preassigned delay, with no random delay. The G.984 ONU uses the preassigned delay only in the serial_number and ranging states.

10. Here or at serial number acquisition, the OLT measures the delay precisely, computes an offsetting equalization delay, and transmits the equalization delay value to the ONU. Henceforth, the ONU delays all transmissions by the equalization delay value.

11. Reception of an equalization delay puts the ONU into operation state O5, where it is now ready for service.

If the OLT already knew a lot about the ONU—if, for example, the ONU fell off the PON because the subscriber powered it down—the OLT could blindly issue PLOAM messages to assign a TC-layer ONU-ID to the known serial number and set the known equalization delay, whereupon the ONU could go into operation state O5 without ever having responded to either a serial number or a ranging grant. The discussion in Section 4.6.1 on the PLOAM MIC does not pertain to G.984 G-PON, which has no MIC; it is a feature only of G.987 XG-PON.

The PopUp state O6 occurs when the ONU's downstream synchronization state machine (Fig. 4.22) declares downstream LOS/LOF. The ONU starts timer TO2. If TO2 expires, the ONU recognizes that it is no longer on the PON and restarts the activation sequence from scratch. When the ONU returns to initial state O1, it discards its ONU-ID and all other TC-layer provisioning.[*]

Assuming that the ONU recovers synchronization and can receive PLOAM messages, the G.984 OLT can return the ONU to operation state O5 by sending a PopUp message to that specific ONU—it may not help, but it does not hurt anything. The G.984 OLT can also send a broadcast PopUp message, which causes all ONUs in state O6 to enter ranging state O4. This was intended to allow fast recovery in case of protection switches. The PopUp state changed its name and behavior in G.987 XG-PON, and the PopUp PLOAM message was dropped in favor of automatic recovery.

State O7, the emergency stop state, behaves much the same in G.984 as in G.987.

---

[*] G.984.3 is less explicit than G.987 about details such as resetting the TC-layer parameters. The principles make sense, however, in G-PON as well as in XG-PON.

The states are more formally defined as follows, still focusing on differences from G.987 XG-PON:

G.984 State O1, Initial

Behavior The ONU transmits nothing. In this state, the ONU waits for the presence of a downstream signal and then synchronizes to that signal. The ONU discards any TC-layer parameters that it may have known from its previous session: ONU-ID, default alloc-ID, burst header parameters, equalization delay. Although this discard action is only spelled out explicitly in G.987.3, it also makes sense in G.984 G-PON ONUs.

Exit Criterion Bit, byte, frame, and superframe synchronization achieved, LOS/LOF cleared → O2.

G.984 State O2, Standby

Behavior The ONU transmits nothing. In this state, the ONU listens for the broadcast upstream_overhead PLOAM message, which defines the basic parameters of the bursts it will subsequently transmit, including a value for preassigned delay.

If the OLT also broadcasts an extended_burst_length PLOAM message (described in Appendix II), the ONU modifies its response parameters accordingly. This message allows for a longer burst header during ONU discovery—the unranged type 3 preamble bytes field—and also after the ONU is operational—the ranged type 3 preamble bytes field. The idea is that a marginal optical budget may become usable if the ONU's preamble is extended; in particular, a reach-extended ONU is almost certain to require an extended burst overhead, for reasons explained in Section 3.12. The ONU is permitted to ignore this message if it does not support the feature, so in theory, the OLT's effort could be in vain. In practice, current ONUs are expected to support extended burst length capability.

If a newly initialized ONU does not see an extended_burst_length message before it responds to a serial number grant, it ignores subsequent extended_burst_length messages that may appear later on. The intended state and message sequence of G.984 G-PON ONU discovery is quite deterministic, as shown in Figure 4.33.

The OLT is expected to transmit three upstream_overhead messages, followed optionally by three extended_burst_length messages, and only then to issue one or more serial number grants. If a new ONU appears on the PON in time to receive at least one of the upstream_overhead messages, it can be assured of receiving an extended_burst_length message before having the opportunity to respond to a serial number grant.

Exit Criterion Upstream overhead PLOAM received → O3. State transition does *not* depend on having received an extended_burst_length PLOAM message.

G.984 State O3, Serial Number

Behavior   The ONU responds to a serial number grant by generating a serial_ number_ONU response PLOAM message, delayed by the preassigned delay plus a random locally generated delay. In state O3, the ONU continues to respond to serial number grants, each time with a different random delay. This is the only state in which the ONU is permitted to respond to global bandwidth grants, also the only state in which it uses a random delay.

Exit Criterion   Assign_ONU-ID PLOAM message received → O4.

G.984 State O4, Ranging

Behavior   When it enters state O4, the ONU starts timer TO1, whose initial value is recommended to be 10 s. In G.984 G-PON, the response to a ranging grant is another serial_number PLOAM message, delayed by the preassigned delay value. There is no random delay in this message because it is a response to a directed grant.

Exit Criteria

- Ranging_time PLOAM message received → O5.
- Timer TO1 expires → O2.

G.984 State O5, Operation

Behavior   This is the normal state of ONU operation. The ONU responds to grants directed to it, delaying its response by the value of the equalization delay.

Exit Criterion   Loss of downstream signal/frame LOS/LOF → O6.

G.984 State O6, PopUp

Behavior   When it enters PopUp state O6, the ONU starts timer TO2, whose initial value is recommended to be 100 ms. When it enters this state, the ONU has lost sync with the downstream signal. If it remains unsynchronized with the downstream flow, the only exit from state O6 is through expiration of timer TO2. However, while in state O6, the ONU may clear LOS/LOF, whereupon it resumes downstream processing, and can then recognize downstream PLOAM messages.

Whether or not the ONU has downstream sync, it transmits nothing while in state O6.

Exit Criteria

- Timer TO2 expires → O1.
- Directed PopUp PLOAM message received → O5.
- Broadcast PopUp PLOAM message received → O4.

G.984 State O7, Emergency Stop

Behavior   The ONU transmits nothing. It remains in this state until explicitly reenabled.

Exit Criterion   Disable PLOAM received, with enable code point set → O2.

Further Notes

- In any state, reception of a disable PLOAM message with a disable code point causes a transition into emergency stop state O7.
- In any state except O7, reception of a deactivate PLOAM message causes a transition into standby state O2.

4.3 ONU TRANSMISSION TIMING AND EQUALIZATION DELAY

This section describes the timing relationships of G.987 XG-PON in detail, and discusses the differences between XG-PON and G.984 G-PON. It is based on the following definitions.

1. The start time of the downstream frame is the moment of transmission/reception of the first bit of the PSync field.
2. The start time of the upstream PHY frame is the moment of transmission/reception (either actual or calculated) of the first bit of the word identified by a StartTime pointer of zero value.
3. The start time of an upstream PHY burst is the moment of transmission/reception of the first bit of the word identified by the StartTime of the corresponding bandwidth allocation structure. This is the first bit of the XGTC burst header in G.987 XG-PON (Fig. 4.18). In G.984 G-PON, it is the first bit following the physical layer overhead PLOu (Fig. 4.27), which may be an upstream PLOAM message, a DBA queue occupancy report DBR, or a GEM frame header, depending on the flags in the BWmap that authorized the burst.

4.3.1 Fundamentals

We start with some fundamental concepts and notation. The terminology—PSBd, for example—is preferentially taken from G.987. The concepts are common to both G-PON and XG-PON.

Consider Figure 4.34.

Suppose that the OLT transmits frame N onto the fiber, starting at time t_{sendN}. If we care about time of day (ToD), we require that t_{sendN} be an absolute time, that is, some specific date and time. Otherwise, t_{sendN} is just a convenient reference. In either case, all of the other times are delays, most of them relative to t_{sendN}.

Frame N propagates down the fiber to various ONUs, located at varying distances L from the OLT. When we engineer the optical distribution network ODN, we need to specify the fiber distance of the farthest possible ONU that will need to be accommodated. Designate this hypothetical ONU as ONU_{max}, at distance L_{max}. After propagation delay t_{max}^D, frame N reaches ONU_{max}. If there were a real ONU here at L_{max}, it would perceive frame N to start at time $t_{sendN} + t_{max}^D$.

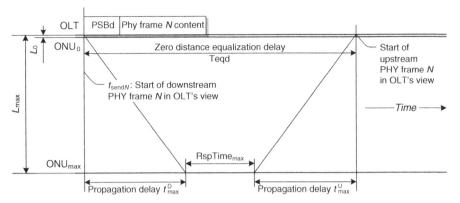

Figure 4.34 Basic timing concepts.

Now, a real physical ONU cannot recognize a downstream frame, interpret its BWmap, and generate a corresponding upstream frame in zero time—the content of upstream frame N depends on the BWmap information received in downstream frame N. We define an ONU response time RspTime to account for this delay. Further, we standardize its value to be $35 \pm 1\,\mu s$, meaning that all real ONUs are expected to be able to respond within $36\,\mu s$, and if an ONU is capable of responding in less than $34\,\mu s$, it must introduce additional delay.

The earliest possible burst (start time zero) from ONU_{max} is guaranteed to start not more than $RspTime_{max} = 36\,\mu s$ later. After upstream propagation delay t_{max}^{U}, the OLT receives the burst. The OLT can safely position its upstream frame N timer at this instant (or later). Let Teqd be the offset in time between the OLT's view of the downstream frame and the latest possible start of the corresponding upstream frame. Consider a second hypothetical ONU_0 located at zero distance L_0. It is apparent that, if ONU_0 were to generate upstream frame N based only on its own RspTime, the frame would arrive at the OLT far too early. The OLT therefore compensates by provisioning an additional delay into each ONU, a delay known as equalization delay, EqD. The total delay for an ONU at zero distance is called the zero-distance equalization delay, and is equal to Teqd. The ONU's actual RspTime is implicit in the delay measurement and compensation.

The equalization delay for our hypothetical ONU_{max} with $36\,\mu s$ RspTime would be 0, and we see from Figure 4.35 that for any ONU i, at intermediate distance L_i, there will be some equalization delay EqD_i that lies between 0 and Teqd—more precisely, between 0 and $(Teqd - RspTime_{min})$. By the way, since the OLT knows the propagation delay, it is trivially easy for the OLT to compute and display the ONU's distance, as long as we do not expect really precise results (1% accuracy is the objective stated in the recommendations).

To summarize: the OLT derives the zero-distance equalization delay Teqd based on the provisioned maximum reach of the PON. As it activates ONUs onto the PON, the OLT compensates each ONU such that its upstream frames arrive at the OLT aligned to a common reference, which could be later than $t_{sendN} + $ Teqd, but not earlier. Strictly speaking, we should therefore show $Teqd_{min}$ and $Teqd_{actual}$ as two distinct delays; but to remain consistent with the description in the standards, we

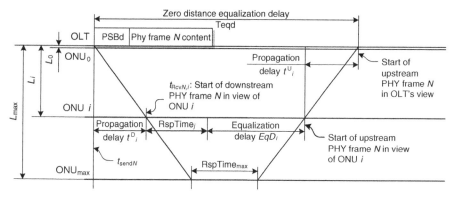

Figure 4.35   Values for ONU i.

show them as a single value Teqd, with the understanding that a constant additional delay does not affect the ultimate result. Another way to think of it is that L_{max} can be chosen arbitrarily, as long as it is at least as great as the actual maximum reach.

We have defined separate designations for downstream and upstream propagation delays. Although they are indeed approximately the same, it is also true that the group velocity of light in fiber is a function of wavelength, and since the downstream and upstream signals on a G-PON are at different wavelengths, their propagation velocity differs. Not by very much, but if we care about nanoseconds over a reach of 20 km or more—and we do—it matters. To briefly reprise the discussion in Section 3.1: the index of refraction n is a measure of the speed of light v in a particular medium, in comparison to the speed of light c in free space:

$$n = \frac{c}{v} \tag{4.2}$$

Let the downstream group index of refraction be n_D, and the upstream index be n_U. To illustrate the concept, suppose that we sent a signal downstream through a fiber whose refractive index n_D was 1.5, and that we returned the signal upstream through a radio link, through the air, whose refractive index n_U is essentially 1. We would know that 60% of the total round-trip delay was consumed in the downstream direction, 40% upstream. Denote the fraction of round-trip delay claimed by downstream propagation as η, where $\eta = n_D/(n_D + n_U)$. Table 4.5 shows the numbers for G.652 fiber at the G-PON and XG-PON wavelengths. The results are derived in appendices of G.987.3 and G.984.3, respectively, along with estimates of the error.

At 20 km, the difference is about 30 ns (G-PON), about 55 ns (XG-PON), with an expected tolerance of 3 or 4 ns. We shall see η again, as a factor in the fine detail of the time-of-day feature.

4.3.2   Time of Day

By virtue of the physical layer itself, an ONU is always frequency synchronized to the OLT. Historically, it was not necessary for an ONU to know date or time. The

TABLE 4.5 Propagation Velocity Differences

	G.987 XG-PON	G.984 G-PON
Downstream wavelength λ_D	1577 nm	1490 nm
Upstream wavelength λ_U	1270 nm	1310 nm
Downstream fraction η	50.0134%	50.0065%[a]

[a]The body of G.984.3 amendment 2 suggests 50.0085%, but Appendix VII recommends 50.0065% as the value to be assumed for compatible calculations across the industry.

OLT took care of time stamps on performance monitoring (PM) or alarm messages. However, G-PON has been extended to backhaul cellular radio traffic, an application that may require ToD to be available at the ONU, depending on the wireless protocol and whether the radio base station contains a separate reference such as GPS. The accuracy requirement on ToD information is the consequence of an allocation of impairments, and the ONU provisionally claimed ± 1 µs. Continuing analysis may require that this tolerance be tightened up, perhaps to as little as 100 ns.

IEEE 1588 is a packet timing protocol with the potential to deliver accurate time and frequency information over a packet network. However, the straightforward packet timing algorithms measure round-trip delay and compensate the far end under the assumption that the delay is symmetric and stable. On a PON, upstream bandwidth allocation is driven by the DBA algorithm and tends to run independently of the downstream flow. It is not inconceivable that downstream queuing and the DBA algorithm could be aligned to achieve symmetric and stable delay, but the industry has chosen instead to distribute ToD as a separable feature of the PON TC layer. Here is how it works:

The OLT is assumed to have its own source of accurate ToD information—IEEE 1588, for example, or GPS. The ToD distribution mechanism uses an attribute of the OLT-G managed entity of OMCI. Whenever it activates an ONU that needs ToD, and from time to time at its own discretion, the OLT sends down the value of a reference frame count N (in G.987 XG-PON, the 32 least significant bits of the 51-bit frame counter), along with a time stamp T_{stampN}, expressed in IEEE 1588 format and with nanosecond resolution. Refer to Figure 4.36.

During normal operation, ONU i has an equalization delay EqD_i set by the OLT in such a way that upstream bursts from ONU i align with bursts from all other ONUs when they reach the OLT. For ToD distribution, assume that ONU i is properly compensated and operating normally.

As an aid to the following derivation, extend the diagonal time lines of Figure 4.34 past ONU_{max} until they meet at a hypothetical reflector at time T_{stampN}, corresponding to an imaginary ONU with zero RspTime. The downstream propagation delay can be allocated from the total delay according to η (Table 4.5). Teqd and t_{sendN} are known to the OLT, so it can compute T_{stampN}:

$$\begin{aligned} T_{stampN} &= t_{sendN} + t_{Refl}^{D} \\ &= t_{sendN} + \Delta_{OLT} \\ &= t_{sendN} + \eta \cdot \text{Teqd} \end{aligned} \tag{4.3}$$

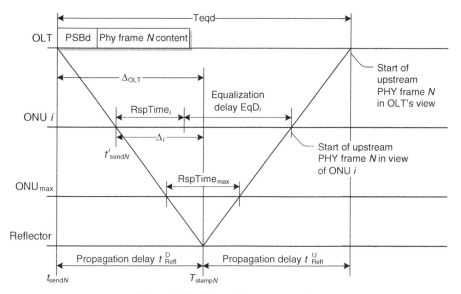

Figure 4.36 Time-of-day computation.

In this application, t_{sendN} is an absolute time, so T_{stampN} is also absolute.

The OLT now sends the pair $\{N, T_{stampN}\}$ to any or all ONUs via OMCI.

Our objective is to have an accurate value for T_{stampN} everywhere throughout the PON. At our imaginary reflector, T_{stampN} marks the time of arrival of frame N. But if ONU i were to make the same assumption at time t^i_{sendN}, it would be too early. So ONU i needs to derive an offset Δ_i from t^i_{sendN}, the instant of arrival of frame N, that corresponds to T_{stampN}.

ONU i does not know the propagation delays, but it does not need to. Since all triangles in Figure 4.36 are geometrically similar, the ONU need only apply the factor η to the delays it *does* know, namely $RspTime_i$ and EqD_i. From the perspective of ONU i, frame N arrives at t^i_{sendN}, and

$$
\begin{aligned}
T_{stampN} &= t^i_{sendN} + \Delta_i \\
&= t^i_{sendN} + \eta(RspTime_i + EqD_i)
\end{aligned}
\tag{4.4}
$$

Observe that frame N need not be in the future or even in the recent past. The ONU can calculate the correct time of day by offsetting the current frame number M by $(M - N) \times 125\,\mu s$.

If the OLT changes the equalization delay EqD_i at some time, due to drift or possibly a protection switch, ONU i retains the time of day derived from its local counter. Changing the delay does not by itself reset the clock. When the OLT someday sends a subsequent ToD calibration update, it would be expected that the altered physical delay would be exactly compensated by the η-weighted amount of the new correction, and the ToD clock would not incur a phase hit.

Once the ONU knows what time it is, it needs to convey this information to the radio base station. A new IEEE 1588 secondary clock could be instantiated on the ONU; if the radio base station interface is Ethernet, this is probably the most straightforward solution. For non-Ethernet backhaul, no standard interface existed at the time of writing, although the question was under study. A number of existing devices, especially GPS receivers, implement a so-called 1 pulse per second (1 PPS) interface. This interface uses the edge of an RS-422 signal to designate the start of each second and a message to carry time. It has been called out as a requirement by some operators and is a likely candidate for standardization.

Multiple clock domains—that is, the ONU's ability to honor multiple opinions about date and time—are out of scope. Various options could be pursued if a network client were to require a ToD distinct from that of the network provider, but the question has not arisen in practice.

Having described the standard solution, we mention that other mechanisms have also been proposed for ToD transfer, such as a transparent clock, in which corrections are added by each equipment through which the IEEE 1588 message passes.

4.3.3 ONU Upstream Burst Timing

As we have seen, ONU transmissions are defined relative to the start of the downstream PHY frame carrying the bandwidth map that authorizes the upstream burst. That is, the bandwidth map in downstream frame N governs the burst structure of upstream frame N. The response time RspTime gives the ONU time to decode the bandwidth map and prepare a response burst accordingly.

Is There Enough Time to Decode the BWmap?

A G.987 XG-PON downstream frame can contain a maximum of 512 eight-byte bandwidth allocation structures, which implies a transmission time of up to 3.3 μs. The BWmap is transmitted before any possible PLOAM messages to give the ONU the maximum possible headroom in preparing a response. The G.987 BWmap is structured in increasing order of start time, giving the ONU further time in which to decode the earliest allocations.

In the G.984 G-PON downstream frame, PLOAM comes before the BWmap, along with several other fields (Fig. 4.20), but there is only one PLOAM message, of only 13 bytes, the other fields are small, and the BWmap is limited to 256 allocation structures. The latest allocation in a BWmap would be transmitted in a bit more than 6.6 μs.

These up-front delays do not seriously impinge on the ONU's 36-μs maximum response time limit.

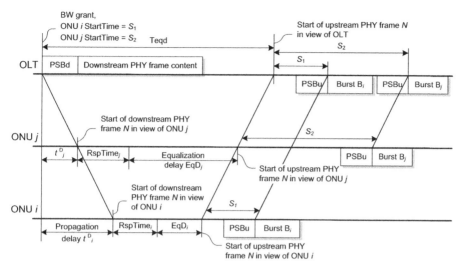

Figure 4.37   ONU timing diagram, general case.

Each ONU maintains a running upstream frame counter that is synchronized to the downstream frame count, but delayed by a precise amount, the sum of RspTime and EqD, as shown in Figure 4.37. As a rule of thumb, 10 km of reach corresponds to 100 μs of round-trip delay, so if the PON is configured for its maximum logical reach of 60 km, the OLT will have launched two more downstream frames $N + 1$, $N + 2$, before it begins to receive upstream frame N.

In each allocation, the OLT specifies a start time, which is an offset from the defined start of the upstream frame. The ONU sends the first byte of the burst content at the start time instant (meaning that burst overhead PSBu (G.987) or PLOu (G.984) *precedes* the start instant—see Figs. 4.17, 4.18, 4.27). The OLT's bandwidth allocation algorithm is responsible to leave sufficient guard time between bursts to avoid collisions. This means the OLT must know the length of the burst header, the length of the payload allocations, and the number of FEC bytes that will be added to the burst. There are enough variables that the OLT could conceivably use an approximation, rather than exact values, as long as it rounds the errors in the conservative direction.

The OLT expects to receive the upstream burst body—the start of the GTC overhead—at an offset of StartTime from Teqd. Figure 4.37 illustrates how differing StartTimes from different ONUs arrive as expected at the OLT. The ranging and equalization delay process removes the effect of the relative delays at the OLT, so the OLT's bandwidth allocation algorithm can do its work in the domain of a single time reference, that is, $t_{\text{start}N} + \text{Teqd}$.

From time to time, or continuously, the OLT measures the actual arrival instant of an upstream burst from ONU i against the expected instant and tests the deviation against a threshold, which is recommended to be eight bit times at the G.987 XG-PON upstream bit rate, four bit times in G.984 G-PON. If the difference in delay exceeds the threshold, the OLT sends a ranging_time PLOAM message to

ONU i with a new value for EqD$_i$. Exceeding the drift threshold is called *drift of window* (DoW) in both G.984 G-PON and G.987 XG-PON. It calls for a correction but is not an exceptional event.

Both G-PON technologies also specify a second drift threshold, twice as large as the DoW value, violation of which suggests a serious problem. Either the delay is changing more rapidly than expected or the ONU is not responding properly to equalization delay correction commands. The OLT recognizes a defect known as transmission interference warning (TIW).

According to G.984.3, the OLT should deactivate the offending G-PON ONU upon TIW. This response is somewhat inappropriate. Deactivation merely resets the ONU, causing it to vie for reactivation. If the propagation delay is in fact changing rapidly for some reason, resetting the ONU does not correct the situation; likewise if the ONU is unable to correctly respond to equalization delay adjustments. G.987.3 specifies deactivation as one option but also suggests disabling the ONU (emergency stop state O7), which may be more appropriate for an uncontrollable ONU.

4.3.4 Timing Relationships During ONU Discovery

We have seen how timing works during an ONU's normal operation, with equalization delay that compensates for the differences between the distances and response times of the various ONUs on the PON. But when an ONU first announces itself on the PON, its distance and delay are unknown and must be discovered. Discovered, moreover, without disrupting existing traffic on the PON.

From time to time, the OLT opens a quiet window for ONU discovery, an interval during which it suspends upstream transmission from all known ONUs. As we shall see, this window is several frames long—250–450 µs—so the OLT needs to plan it in advance.

The OLT transmits a serial number grant, which invites any ONU that has not already been activated onto the PON to send a serial number PLOAM message. The OLT times the serial number grant such that all possible responses will occur during the quiet window, given the full range of physical distances, ONU response times, random delay, plus the (G.984 G-PON only) preassigned delay. If it sees a serial number response from an ONU, the OLT assigns it a TC-layer ONU-ID in a downstream PLOAM message, which advances the ONU's state (Section 4.2) so that it does not respond to further serial number grants.

It can also happen that the OLT does not receive the ONU's response. In that case, the ONU remains in serial number state and responds again when it detects the next serial number grant.

The usual reason for the OLT's failure to see an ONU is contention among several ONUs, each of which desires activation onto the PON. This may take the form of a collision in time, in which case neither contender succeeds, or it may simply be that the OLT recognizes the first burst received and does not have processing depth to recognize a second ONU during the same quiet window.

The second case would eventually resolve itself, as the OLT recognized ONUs one by one in order of appearance. ONUs mitigate the first possibility by adding a random

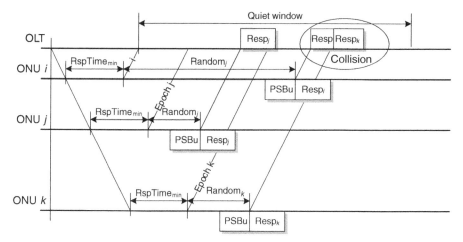

Figure 4.38 Discovery failure.

delay to each serial number response. Although bursts from two or more ONUs may collide during any given quiet window, sooner or later their random delays will relocate their responses so that they can be distinguished. The range of random delay is 0–48 μs.

Figure 4.38 illustrates the possibilities. The epoch lines show the earliest possible responses from the various ONUs. The actual responses are offset by random delays. During the illustrated discovery cycle, ONU j succeeds in being registered, while, due to unfortunate choices of random delay, burst responses i and k collide.

Now, how does the OLT determine the quiet window? The quiet window must span the time between the earliest conceivable ONU response and the latest possible ONU response. See Figure 4.39.

The earliest possible response is determined by:

- The minimum round-trip delay, determined by the nearest ONU expected on the PON. L_{min} is a network design parameter; if it is not provisioned, it is implicitly taken as 0.

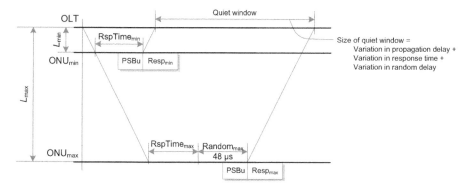

Figure 4.39 Size of quiet window.

- The minimum ONU RspTime, 34 μs.
- The minimum random delay generated by candidate ONUs, namely 0.
- The preassigned delay, G.984 only (not shown in the figure).

The latest possible response is determined by corresponding factors:

- The maximum round-trip delay, determined by the maximum reach of the PON, L_{max}, which is likewise a network design parameter. The G-PON standards allow the maximum reach to be as much as 60 km.
- The maximum ONU RspTime, 36 μs.
- The maximum random delay generated by candidate ONUs, 48 μs.
- The preassigned delay, G.984 only.

The default assumption about an ODN is that it spans 0–20 km. As mentioned in Chapter 2, some operators require reach and differential reach of 40 km. In real deployments either maximum reach or differential reach, or both, are potentially provisionable, to optimize the size and placement of quiet windows and thereby gain a bit of efficiency in use of the PON's upstream capacity.

For 20-km differential reach, the variation in round-trip propagation delay is 200 μs. For 40-km differential reach, the variation in round-trip propagation delay is 400 μs. When the differential reach is 20 km, the suggested duration of the serial number quiet window is therefore 250 μs, and for 40-km differential reach the suggested duration is 450 μs.

Observe also that the OLT must extend the quiet window on the leading edge to include the PSBu of the earliest possible response burst, and it must extend the trailing edge to include the actual body of the burst containing the latest possible response. Both of these are small factors but are not to be ignored.

The OLT can position the ranging window according to its own convenience. The quiet window typically spans several 125-μs frames, so the OLT must coordinate several bandwidth maps to create the window. G.984 G-PON has two ways to control the position of ONU responses:

- The preassigned delay, broadcast to all prospective ONUs. The ONU adds the preassigned delay to its own delays before generating a serial number response. In G.987, there is no preassigned delay, so this term is 0.
- The serial number grant startTime.

Expanded differential reach is not free. ONUs at the near end of a wide differential reach must be able to store more BWmaps because they incur greater internal delay before the start of the upstream frame.

As to extending the maximum reach, one of the difficulties is the responsiveness of the DBA algorithm to varying traffic rates. The objectives for DBA response are on the order of 2 ms, while 60-km reach would consume 600 μs of that budget just in propagation delay. In Section 6.3, we explore other artifacts of increased differential reach in the DBA algorithm.

4.4   ONU REGISTRATION

In traditional copper pair telecommunications, there is a direct relationship between a given pair of wires and the terminal at the subscriber's premises. The twisted pair may be cross-connected from one cable to another along its route, maybe more than once, which makes it important to keep accurate records of cable and pair information on each segment. In principle, the same relationship is true in point-to-point optical fiber connections.

In North America, access loop inventory is often maintained in the venerable loop facility assignment and control system (LFACS), a Telcordia product. In LFACS, hence by tradition across the industry, the first cable, the feeder cable from the central office, contains so-called F1 pairs (or fibers), which are cross-connected to F2 pairs (fibers) at a cross-box of some kind in the field, and so on. Tracking outside plant inventory and connectivity is a nontrivial operations expense, but at least in principle, the operator knows from the database exactly where a one-to-one subscriber termination appears at the central office, as well as the exact details of the connection all the way to the subscriber.

In a PON, there is no directly observable physical relationship between a given subscriber terminal and the signal on the PON as viewed from the OLT side of the splitter. That is, at the physical level, the signal at splitter port 1 is indistinguishable from the signal at splitter port 2 or splitter port 32.

It is clearly necessary to identify each individual ONU: a phone number, for example, must be provisioned to the right subscriber, and Internet and IPTV subscriptions must be matched to the subscriber who will be paying the bill. In the world of G-PON, ONU identification is usually done through the ONU's composite serial number,[*] which the OLT learns during the discovery process.

The (composite) serial number gives us a unique identifier for each ONU on the PON, but we still do not know which ONU is installed at which subscriber's premises. There are several ways to make this association, which is referred to as registering the ONU.

If we can, we want to preprovision service on the ONU, that is, set up all of the necessary information on the OLT and further up in the network, before we actually install the ONU. We want the OLT to bring up service as soon as the ONU is installed, so the installer can verify that all services are working properly before leaving the site. Additional truck rolls are expensive, to be avoided if at all possible.

---

[*] The serial number itself is not necessarily unique until combined with the vendor ID. The industry—and this book—frequently uses the term serial number to designate the combination of vendor ID and the serial number assigned by the ONU vendor.

If we could guarantee that only a single ONU was to be installed on a given PON at any given time, it would be easy: we're only expecting one ONU, so the ONU that appears must be the one we're expecting. This is a standard premise for theatrical comedy, but it is hardly feasible when installing ONUs.

A second alternative: all of the provisioning information can be preestablished in the OLT, except for the ONU's serial number. When field craft install the new ONU or replace a defective ONU with a new one, they phone the network operations center (NOC) with the address of the subscriber (more likely, the installation or repair work order number) and read off the ONU's serial number from its sticker. The NOC operator then associates the serial number with the particular subscriber and provisions it into the OLT. Until the NOC operator does this, service on the ONU does not come up, and craft in the field must wait, if they are to verify it. We might designate this the **learned serial number** registration method. Unfortunately, it violates our wish to avoid real-time coordination between field craft and NOC operators.

As described, this is unsuitable. But there are meaningful use cases that work in essentially the same way, from the viewpoints of OLT and ONU. So-called *auto-discovery* represents the case in which a new ONU is simply discovered by the OLT, without having been preprovisioned. The ONU's serial number is reported in a message to the management system, whereupon the ONU and OLT merely await further provisioning. The ONU sits harmlessly on the PON; the OLT refrains from provisioning it and grants it no upstream capacity except perhaps an occasional query to see if it is still there.

Even with preprovisioning, ONU autodiscovery could occur if there were an error of some kind, and the ONU were, for example, installed on the wrong PON.

A similar use case is the possibility that a newly built subdivision might be preequipped with ONUs, few of which were actually in service. The ONU's serial number and the corresponding street address might be recorded in a database, awaiting a service order. When the service order was subsequently placed, the OLT could be provisioned accordingly. It is also possible that we might choose not to record the correlation of ONU serial number to address in our database. In that case, a variation of the registration ID method (below) could be invoked at service order time.

Any G-PON system would naturally support autodiscovery. We would surely not want an orphan ONU to sit out there, undiscovered and unreported to the management system!

We avoid the real-time coordination problem in a second form of registration in which we preassociate the ONU's serial number with the given service order and thereby with a given subscriber. From the management system, we preprovision the serial number into the OLT along with the rest of the service parameters. When the physical ONU appears on the PON, the OLT can immediately activate the subscriber's services. The installer verifies service, closes the order, and goes on to the next job. This is known as the *serial number registration* method.

Serial number registration is used when it makes sense. It requires no special support from either the ONU or the OLT. If the OLT does not recognize the ONU's serial number, we have an autodiscovery situation.

Making Learned Serial Number Registration Work

It would be possible for the installer to have a wireless terminal, much as we find at rental car check-ins, but simpler and smaller, because it would not need paper storage or printers. The installer could enter the particulars of an installed ONU on a keyboard or preferably with a bar code or radio frequency identification (RFID) reader, along with subscriber or work order information. The information could then flow through the operator's back office software and back down into the OLT and ONU, turning up preprovisioned service without the need for human coordination at the NOC. All of this except perhaps the RFID reader could be built into an app on a cell phone. If this were put in place, the learned serial number registration method could become the preferred installation procedure.

As far as we know, no one is doing this. Yet.

The complication with the serial number registration method is that it requires coordination of physical ONUs between service order, warehouse inventory, and the correct truck on the correct day. And what if the preselected ONU turns out to be damaged or defective?

Logistics considerations make this alternative undesirable for single-family ONUs, where we just want to select the next of many identical ONUs from the warehouse, or from the truck. But it may be quite appropriate for the installation of an MDU or a reach extender, devices that are regarded as network equipment and installed under engineering work orders rather than subscriber installation or repair orders. For these, it may well be practical to assign a particular equipment unit to a particular site in advance.

Another approach is called the *registration ID* method. This registration model assumes that the ONU has some kind of local interface that lets an installer—or the subscriber, for that matter—enter an identifier string, designated the registration ID. Not infrequently, the local interface takes the form of a simple web page available to whomever connects a PC to one of the ONU's Ethernet ports, but it can be as simple as a craft butt set connected to a POTS port.

In the registration ID method, the ONU's serial number is never provisioned into the OLT. Instead, the OLT learns the ONU's serial number. Either during—recall the G.987 XG-PON registration PLOAM message in Section 4.2.1—or after ONU activation, the OLT retrieves the registration ID from the ONU. The OLT then associates the serial number with a particular preprovisioned management-layer ONU-ID (subscriber) through the registration ID.

In G.987 XG-PON, the registration ID is a string of up to 36 bytes, conveyed from ONU to OLT in the registration PLOAM message. In G.984 G-PON, the registration ID is conveyed in a confusingly named 10-byte variable called *password* that is conveyed in the password PLOAM message. The name came from the early days of B-PON, when it was expected that some sort of identifier would be useful, but the registration concepts had not yet been fully worked out. The G.984 registration ID is usually a character string, but it could be anything that can fit into 10 bytes, for example, a 20-nibble hex string or a cryptographic hash of some other string such as a passphrase.

In the ONU-G managed entity (Chapter 5), OMCI has been adapted to allow for 36-character credentials for operators who cannot fit the registration ID into the 10 bytes of G.984. The same managed entity includes an attribute through which the OLT can signal the status of the ONU's credentials back to the subscriber or installer. This feedback facilitates a second try if things go wrong, for example, because of typing errors.

The registration ID method enables the OLT to turn up preprovisioned services immediately. No personnel coordination is needed, and no logistics are involved in getting the right physical ONU to the right subscriber. For good reason, this is the most popular method, used in one form or another by most operators.

The registration ID is at least potentially accessible to the subscriber, so the long-term security of this technique must be addressed. One of the ways in which the registration ID process can be secured is the so-called lock mechanism. When the ONU appears initially on the PON, the OLT learns its serial number. The OLT associates the serial number with the given subscriber through the registration ID. The OLT then locks the serial number association, either immediately, or after a predetermined time that allows for the possibility of discovering and replacing a defective ONU or upon command from the network operations center. The OLT subsequently recognizes the ONU only by its serial number and ignores its registration ID.[*] If the ONU later fails and needs to be replaced, the network operations center must unlock the association—reenabling registration ID recognition—before the replacement ONU can be brought into service.

It has not become a formal registration method in the standards, but it would also be possible for a subscriber to do the complete service order process directly from an autodiscovered ONU, with no hands-on involvement by the operator at all. This would require the network to set up a promiscuous IP association for any auto-discovered ONU, an association that would only accept HTTP(S) (hypertext transfer protocol secure), and would be redirected to the operator's own service order process. The ONU used by the subscriber to order the service would be implicitly recognized by its serial number in the process—a variant of autodiscovery—and provisioning would flow in real time to the necessary network elements, so that service could be instantly available. No preassigned registration ID would be needed.

This is done with DSL service; it can be done with PON.

4.5 ONU ENERGY CONSERVATION

After the discussion of the problems of power in Chapter 2—and we hardly mentioned heat dissipation—we trust it is clear that power is a major concern of

---

[*] But registration ID also plays a rôle in security. See the further discussion in Section 4.6.

the industry. Existing power converters may be 85–90% efficient; if it is possible to gain another 1 or 2%, it is well worth doing. Inexpensive batteries that could tolerate temperatures down to −40°C, with indefinite lifetimes, with more stored energy and less energy loss during full charge and trickle charge—of such stuff are dreams made.

Reducing the resistance of copper wires would be nice, but it will not happen. What does happen, has happened for years, and shows no signs of stopping is Moore's law reduction of semiconductor size, along with a reduction of the power needed to perform, say, one floating-point multiplication. The progress in power reduction is more impressive than is apparent at first glance because each new generation of semiconductors multiplies the functionality of its predecessor.

But good, solid engineering, the kind we have been doing for years, looking for excess milliwatts and killing them, is not glamorous. We do not do press releases about that 1% gain in power converter efficiency, installing the ONU closer to the power converter, or going to 48 V instead of 12. And it is important to be seen to be doing something about energy consumption.

If we could save 1 W per ONU, and if we had a billion subscribers, we could avoid building two 500-MW power plants! That is pretty exciting. It even has the merit of being true, given the assumptions, which include the questionable ideas that our whole billion subscribers reside within one or two time zones, and that power can be reduced during peak demand hours.

The maximum power reduction at an ONU, of course, occurs when it is switched off, for example, when the subscriber goes on holiday or possibly even overnight. But intrusion monitoring, smart meter reading, the intelligent home, machine-to-machine communications—these services require access even when the homeowner is away for a month. The upside is that the power-down option places the choice on the subscriber, who knows best what his needs are.

Unwilling to rely on subscribers making their own choices—there would be no press releases, and the great unwashed might actually make the wrong choice!—the first standards proposals for saving ONU power contemplated shutting down the entire ONU overnight, without benefit of subscriber consent. Arguably, the ONU could be awakened at any time by subscriber action, but there are times when an emergency call needs to be directed *toward* the ONU, not just *from* the ONU. A sound asleep ONU could not receive any sort of wakeup signal from the OLT. Long periods of unilateral unavailability guaranteed customer dissatisfaction, not to mention liability lawsuits.

The discussion then turned to the idea of brief intervals of sleep, after each of which the ONU and OLT would check with each other to see if anything interesting was going on. That model is formalized in G.987.3, XG-PON. The durations have never been tied down, but a sleep cycle on the order of 10–100 ms is the likely range.

Two modes are defined: sleep (also known as cyclic sleep) and doze. In sleep mode, the ONU's PON interface is entirely powered down; in doze mode, the downstream direction remains operational. Through OMCI, the OLT learns some interval values that are built into the ONU, sets some interval values of its own choice, and then grants permission for the ONU to enter one or either of these low-power modes at its own discretion.

At run time, the OLT and ONU exchange PLOAM messages that keep them in approximate synchronization with regard to low-power states. The ONU may exercise either doze or sleep modes at will, assuming that the OLT has authorized both, but it must coordinate with the OLT if it wishes to change mode from doze to sleep or vice versa. We discuss the state machines below.

4.5.1 Sleep Mode

While the ONU is sleeping, its optical transceiver is powered down. Traffic arriving in either direction is discarded by default, or possibly—if there is not very much—buffered for later delivery. A timer in the ONU periodically triggers the transceiver to power up, whereupon it regains synchronization to the PON and exchanges handshakes with the OLT to confirm that it is still alive and well. If necessary, the OLT can bring the ONU into full active mode (active held state—see Fig. 4.40) during one of these handshake episodes.

It is important to understand that the handshakes take the form of bandwidth grants and responses, with the ONU granted the opportunity to send at least one upstream PLOAM message, but *without* a mandatory PLOAM message exchange.[*] This relieves the message processing load, especially from the OLT, which would otherwise have to process several exchanges per second from perhaps 100 or more ONUs. Only if the ONU does not respond to a grant during the expected interval does a timer in the OLT bring its processor into the picture or, of course, if the ONU signals that it wishes to wake up.

To expedite the process of bringing the ONU into active mode, the OLT can set the FWI bit in the first allocation structure of every burst authorization directed to that ONU—refer back to Figure 4.13. The ONU will see this bit in the first frame it decodes after recovering downstream synchronization, in contrast to the sleep_allow (off) PLOAM message, which serves the same purpose, but which the OLT may send less often because it may involve processor overhead.

Management communications between OLT and ONU are generally expected to be reliable. But when the ONU is asleep, messages may fail to be delivered, including, for example, broadcast PLOAMs that might redefine the burst profile or request a new encryption key from all ONUs. The OLT is responsible to deal with such possibilities, possibly by forcing the ONU awake before beginning a management transaction.

The ONU may also respond to a local stimulus such as a subscriber off-hook event—called a local wakeup indication (LWI)—to awaken itself spontaneously. Whether awakened by a local indication or a forced wakeup from the OLT, the ONU remains awake until released by the OLT to resume low-power operation.

The criteria for what constitutes an LWI are not specified. Examples include the aforementioned off-hook event at an analog telephone interface, upstream messages such as IGMP join requests, heartbeat or alarm messages generated by applications

---

[*] A no-op heartbeat acknowledgment PLOAM message from the ONU can presumably be discarded before it reaches the processor.

such as premises surveillance, or timed events such as 802.1ag CCMs initiated by the ONU itself. Guided by market requirements, it is the vendor's choice how much of the ONU's functionality to retain during PON interface sleep.

4.5.2 Doze Mode

While the ONU is dozing, its optical transmitter is powered down, but its receiver remains active, and it is expected to forward downstream traffic. Doze mode admittedly saves less power than full sleep mode; on the other hand, doze mode can be used during the many long hours of TV watching—almost entirely downstream traffic, with only occasional upstream IGMP messages—while sleep mode cannot.

As with sleep mode, the dozing ONU periodically powers up its transmitter and shakes hands with the OLT. As with sleep mode, local wakeup indications can bring the ONU into full active mode spontaneously. Unlike sleep mode, the OLT can force the ONU awake at any time, albeit with a start-up delay while the ONU powers up its transmitter.

The question of management transactions remains but is less constrained in doze mode. The ONU does receive downstream PLOAM and OMCI messages; it just cannot respond instantly. But the ONU is allowed $750\,\mu s$ to respond to a PLOAM message, and $1\,s^*$ for OMCI. If the ONU regards these messages as remote wakeup indicators, it might be prepared to respond to PLOAM messages within the PLOAM window and can certainly respond to OMCI messages within the OMCI window. However, the standard does not specify that the ONU treat these events as remote wakeup indicators, so the OLT may be well advised to bring even a dozing ONU into full wakefulness before beginning a management transaction.

Doze mode is mandatory in G.987 ONUs; sleep mode is optional.

4.5.3 Tools

Two PLOAM messages are defined for power saving coordination, one in each direction.

- The OLT issues the sleep_allow PLOAM message either to a single ONU or as a broadcast message to all ONUs. The message contains only a single parameter: sleep_allow (on) or (off). As would be predicted, the value *off* brings the ONU into full active held state. The value *on* represents permission to the ONU to enter one of its enabled low-power modes—or not—based on its own best judgment. The ONU ignores the message if the designated low-power modes have not been enabled through OMCI.

- The ONU issues the sleep_request message, which also has only one parameter, chosen from the set {sleep, doze, awake}. The values *sleep* and *doze* indicate

---

* G.984.4 allows as much as 3 s for low-priority OMCI responses, but low priority is undefined. In G.988, all OMCI responses are expected within 1 s.

which mode the ONU intends to enter, while the value *awake* tells the OLT that the ONU is now fully active and will remain in active held state until permitted by the OLT to reenter one of its low-power modes.

If the ONU supports both sleep and doze modes, the ONU vendor defines two times: for sleep mode, the expected time required for the entire transceiver to power up and regain synchronization to the downstream PON; for doze mode, the time required for only the transmitter to power up and be ready to respond to bandwidth grants. This read-only information is available to the OLT via OMCI to assist it in setting up a satisfactory timing contract with the given ONU.

The OLT determines three intervals that it sends to the ONU via the ONU dynamic power management control managed entity in OMCI.

I_{sleep} Measured as a count of 125-μs frames, this interval is the maximum duration that the ONU is permitted to remain offline during a low-power state. That is, before the interval I_{sleep} elapses, the ONU is expected to have powered itself up, regained synchronization to the downstream PON if necessary, and be fully prepared to respond to grants. The ONU runs a timer T_{sleep} to guarantee this performance. Observe that, if the initial value of T_{sleep} were the same as I_{sleep}, the ONU would have to come fully awake in zero time. It is the ONU's responsibility to appropriately offset the initial value of T_{sleep}, depending on how long it needs to recover. If the OLT's timing requirements were too onerous, the ONU would be expected to remain fully active.

I_{aware} In the absence of external stimuli, the ONU returns to awareness every I_{sleep} frames and remains aware for at least I_{aware} frames. During this time, it responds to bandwidth grants and may exchange PLOAM messages or user traffic with the OLT. Once the interval I_{aware} elapses, in the absence of external stimuli, the ONU returns to the low power state. The ONU runs a timer T_{aware}, whose purpose is to guarantee that the constraints of I_{aware} are satisfied.

I_{hold} When the ONU enters active held state, either because of a local or remote wakeup event, it remains there for a minimum duration, I_{hold} frames. The purpose of I_{hold} is to avoid the race that might occur if the OLT were to send a downstream sleep_allow (on) PLOAM message that crossed the ONU's upstream sleep_request (awake) message. Not surprisingly, we find that the ONU runs an analogous timer T_{hold}.

The OLT maintains its own timers, whose values are derived by the OLT itself:

$T_{alerted}$ When the OLT attempts to rouse a sleeping or dozing ONU, it either sets the FWI bit or sends sleep_allow (off), or both. The OLT must allow time for the ONU to receive this indication, including delay in sending the PLOAM message, round-trip time, ONU recovery time (this is why the ONU declares its power-up times), and delays in granting upstream transmission opportunities. If the OLT does not receive a response from the ONU before $T_{alerted}$

expires, the OLT begins counting missed response events toward the declaration of an ONU loss of bursts (LOB) defect.

C_{lob} This count defines the number of consecutive bursts whose absence triggers a loss of bursts declaration by the OLT. G.987.3 specifies that C_{lob} is 4. The relationship between grant count and elapsed time is not specified; the OLT would be expected to exercise good judgment.

T_{er} The expected response timer T_{er} is similar in a way to $T_{alerted}$, inasmuch as it defines the latest instant at which a response is expected from the ONU. It differs from $T_{alerted}$ in that it runs when the OLT believes the ONU is in one of its unresponsive low-power states. After I_{sleep} elapses, and subject to delays in granting upstream transmit opportunities, the OLT expects to receive a response. Each time it receives a response, the OLT resets timer T_{er}; if T_{er} times out, the OLT attempts to force the ONU awake to confirm that it is still present and healthy.

4.5.4   State Machines

The low-power state machines in ONU and OLT only exist when the ONU is in its operation state O5 (Section 4.2). The OLT has no direct visibility of ONU state but has strong indications when it is safe to believe that the ONU is in state O5.

Either the OLT or the ONU may be incapable of supporting sleep mode (doze mode is mandatory in G.987), and either OLT or ONU may be provisioned to avoid modes that they are actually capable of supporting. In these cases, the states for unsupported modes simply do not exist; the associated messages are never transmitted and are silently discarded if received. In the discussion below, we assume that both OLT and ONU support both modes.

During its aware states, and when it awakens spontaneously, the ONU responds to bandwidth grants from the OLT. These grants occur at intervals wholly under the control of the OLT. The pertinent interval I_{aware} is also wholly controlled by the OLT. It is therefore the OLT's responsibility to issue grants with sufficient frequency to permit the ONU to respond in a timely manner. If the OLT were to permit I_{aware} to elapse without having granted the ONU an upstream transmission opportunity, it could hardly blame the ONU for the timeout!

Transitions between full-power and low-power state groups are signaled between ONU and OLT with PLOAM messages. However, upstream PLOAM messages rely on the PLOAMu grant bit in an upstream bandwidth allocation, which is only available at the OLT's discretion. Where the ONU state machine shows the emission of a sleep_request (SR) message on a transition, the associated timers do not start running until the PLOAM message is actually transmitted.

Figures 4.40 and 4.41 portray the state machines.

4.5.4.1 ONU State Machine

Figure 4.40 illustrates the ONU power-saving state machine. At the time of power-saving provisioning by the OLT, if not before, the ONU creates the state machine and initializes it into active held state. The OLT provisions the necessary modes and

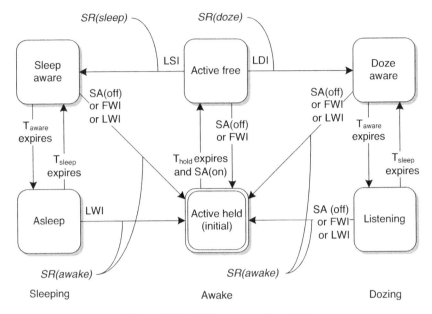

Figure 4.40   ONU power-saving states.

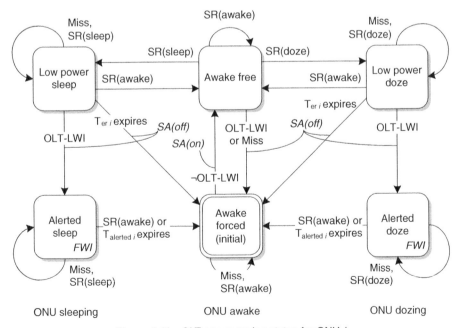

Figure 4.41   OLT power-saving states for ONU *i*.

intervals into the ONU via OMCI, after which it issues a sleep_allow (on) PLOAM message. When timer T_{hold} expires, the ONU enters active free state, in which it has permission to enter a low-power state according to its provisioning, its capabilities, and its assessment of the local traffic environment.

Notation

- Italic text designates the transmission of a PLOAM message. (No-op acknowledgment messages are not shown.)
- SR means sleep_request; SA means sleep_allow. These are the PLOAM messages whose exchange coordinates ONU and OLT state machines. The OLT sends only the SA message; the ONU sends only SR.
- LWI is the local wakeup indication.
- LDI and LSI are local doze and sleep indications, respectively. Criteria for declaring LDI and LSI are left to the ONU vendor, just as with LWI.
- FWI is the presence of the forced wakeup bit in the first allocation structure of the BWmap.

Depending on circumstances, active traffic, for example the ONU may remain in active free state for long periods of time. Suppose that the ONU has remained active free and, for its own reasons, has not entered a low-power state. Now suppose that an event at the OLT causes the OLT to want to keep the ONU active. The OLT can force the ONU back into active held state by sending a sleep_allow (off) PLOAM message, by setting the FWI bit in a BWmap, or both.

Left to its own devices, the ONU evaluates its environment and sooner or later determines that it is appropriate either to sleep (LSI) or doze (LDI). The criteria for making this decision are not specified but would include at least the level of upstream traffic and the need to send management messages such as OMCI responses.

Suppose that our ONU has decided to doze. The ONU informs the OLT about its decision by sending a Sleep_request (doze) PLOAM message upstream. The ONU enters doze aware state, in which it continues to respond to upstream grants while counting down its local timer T_{aware}, a timer based on I_{aware} as set by the OLT. The aware state keeps the ONU responsive long enough to be sure that the OLT was not in the process of forcing it awake just as the ONU decided to doze.

Assuming that there was no race with the OLT, timer T_{aware} expires after an interval at least as large as I_{aware}. At this point, the ONU enters listening state, stops responding to upstream grants, and powers down its transmitter. The listening ONU continues to process and deliver downstream traffic.

Because the ONU's receiver is still alive, the ONU can respond not only to LWI but also to requests from the OLT, be they delivered via the FWI bit or the Sleep_allow (off) PLOAM message, or both. These events cause the ONU to power up its transmitter; after it is capable of responding normally, the ONU signals sleep_request (awake) and returns to active held state.

But if nothing else is going on—the most likely case for 3 AM—timer T_{sleep} expires. After an interval not greater than I_{sleep}, and after having powered up its transmitter, the ONU returns to doze aware state. When it enters doze aware state, the ONU must be fully capable of responding to bandwidth grants.

In doze aware state, the ONU responds to grants from the OLT, primarily as a keep-alive mechanism. It is not forbidden to send real traffic upstream during this time, for example, a brief heartbeat message from our premises surveillance application, as long as the grant specifies the correct alloc-ID for the traffic. It is also possible that the ONU will regard even a small amount of real traffic as an LWI, and will no longer be in doze aware state. It is also possible that the OLT will grant bandwidth only to the default alloc-ID. Observe that there is no PLOAM exchange beyond a heartbeat acknowledgment message, and therefore little or no CPU involvement at the OLT, in the frequent transitions between doze aware and listening states.

Sleep mode operation is much the same, except that downstream messages cannot be delivered while the ONU is asleep.

4.5.4.2 OLT State Machine

In Figure 4.41, we see the matching state machine, an instance of which is maintained by the OLT for each ONU i that has a low-power mode enabled. The state names differ slightly from those of the ONU state machine, avoiding the possibility of ambiguous interpretation.

Although the states do not map one for one—beware!—the OLT state machine is synchronized to that of the ONU, subject to delays in propagation, message generation, and bandwidth grants.

The notation is the same as in Figure 4.40, with the following additions:

- *Miss* indicates that the OLT issued a bandwidth grant and received no response.
- *FWI* in the alerted states indicates that the OLT sets the forced wakeup indicator bit in the first allocation structure of each BWmap directed to ONU i.
- $\neg OLT$-LWI designates the cessation of the local wakeup indicator at the OLT.

The OLT instantiates the state machine for ONU i in awake forced state. The OLT is free to hold the machine in awake forced state indefinitely. Of course, if it wished the ONU never to sleep or doze, the OLT would simply not create the state machine at all. At the ONU, the corresponding initial state would be active held, which would prevent the ONU from ever going into a low-power mode.

Having created its local state machine, having provisioned ONU i with suitable interval values, having enabled one or both low-power modes via OMCI, and having determined that traffic levels are low, the OLT emits a Sleep_allow (on) message to permit the ONU to exercise at least one of the low-power modes at its own discretion. This puts the OLT into awake free state.

If a local wakeup indicator LWI gives the OLT subsequent reason to want the ONU to remain awake, it can send Sleep_allow (off) or FWI and return to awake forced state. One reason for this action could be the OLT's failure to receive a response to an upstream grant (a miss event) without having first received a Sleep_request PLOAM message. This is a safeguard against the possibility that the Sleep_request message from the ONU was lost in transit.

Suppose that the OLT is now in awake free state for ONU i, and that it receives a Sleep_request (doze) PLOAM message in response to one of its bandwidth grants (necessarily a grant to the default alloc-ID with the PLOAMu bit set). Having no objection, the OLT enters low-power doze state and starts timer $T_{er\,i}$. The OLT does *not* acknowledge the SR message from ONU i.

The OLT continues to send bandwidth grants periodically. Whenever it receives a response from ONU i, it restarts timer $T_{er\,i}$.

Having given notice that it intends to doze, the ONU stops responding to upstream grants after an interval not less than I_{aware}, whereupon OLT timer $T_{er\,i}$ begins to count down. The OLT continues to send bandwidth grants, in case the ONU awakens spontaneously, but when it receives no response (miss), it just remains in the same state, as long as timer $T_{er\,i}$ is still running.

Under normal circumstances, the ONU responds again to upstream grants after an interval not longer than I_{sleep}, T_{er} resets again without expiring, and the cycle repeats indefinitely. Observe that no explicit signaling or state changes occur at the OLT during this repetitive cycle. This is an important feature to offload unnecessary processing from the OLT state machine, an instance of which exists for each of perhaps 100 ONUs on the PON, and perhaps 8 or 16 PONs on an OLT blade.

In comparing state machines, observe also that a dozing ONU alternates between listening state and doze aware state, during both of which the OLT's state machine remains in a single state, namely low-power doze. The OLT and ONU state machines are synchronized but do not match state for state.

If ONU i dies in its sleep, $T_{er\,i}$ expires. The OLT sends Sleep_allow (off), probably sets the FWI bit, and enters awake forced state, where it starts deciding whether to declare loss of bursts (LOB) against ONU i.

The OLT may have some less dramatic reason—traffic or management, for example, shown as LWI—to want the ONU to return to full wakefulness. It again sends Sleep_allow (off), probably sets the FWI bit in the first allocation structure of each bandwidth grant to ONU i, and enters alerted doze state, where it waits for the ONU to power up its transmitter and respond. In comparing ONU and OLT state machines, observe that the ONU has no states that match the OLT's alerted states.

When it sees a response from ONU i—the proper response being a sleep_ request (awake) PLOAM message—the OLT enters awake forced state. It does the same if $T_{alerted\,i}$ expires, but in this case, it is considering whether to declare loss of bursts.

The state machine for sleep mode is exactly the same but subject to slightly different timing and possibly a different policy to force wakefulness.

G.987.3 defines the state machines in more detail, specifically covering the corner cases. There is no equivalent power-saving model for G.984 G-PON.

4.6 SECURITY

We encourage a skeptical attitude toward security in general and toward G-PON security in particular. There are at least two perspectives for the skeptical eye:

- Is the threat model complete? What did we miss? Of the threats we listed, are they real?
- If the threat were actually to materialize, how effective would the security barriers be?

The security features of G.987 XG-PON go considerably beyond those of G.984 G-PON. In G.984 G-PON, downstream unicast transmissions can be encrypted. Full stop. By default,[*] unicast keys are generated by the ONU and sent upstream to the OLT in the clear, in the expectation that the PON is physically secure in that direction. Control and management traffic is subject to CRC error checks but not to validation.

G.987 XG-PON, in contrast, allows for encryption of unicast traffic in both directions and encryption of downstream multicast traffic. G.987 defines two mechanisms for strong mutual authentication of ONU and OLT—OMCI and IEEE 802.1X—where G.984 defines only the OMCI method. G.987 key exchange is protected by encryption, and PLOAM and OMCI traffic is cryptographically protected from forgery. There is no error correction capability in this layer.

As we have done before, we start with G.987 XG-PON, following which G.984 G-PON is readily to be understood.

4.6.1 Security in G.987 XG-PON

4.6.1.1 Threat Model

G.987.3 lists four threats:

Theft of Service   Anyone capable of receiving the downstream PON signal can, in principle, intercept all downstream traffic. This threat is realistic. While we, as subscribers, may have nothing to hide, we would certainly object to someone eavesdropping on our services, even in only the downstream direction. As to multicast, it does matter! The ethically challenged may consider it a venial sin to steal video streams, while on the other hand, operators expect to derive a substantial portion of their G-PON-related revenue from IPTV. Theft of multicast service is discouraged by middleware encryption, but it might be possible for a knowledgeable group of users to purchase one legitimate subscription, snoop the keys from a legitimate—but hacked—set-top box, and distribute the keys to other members of the group.

---

[*] G.984 G-PON includes an option for OMCI strong authentication. If this option is exercised, the ONU encrypts upstream keys with the resulting master session key.

ONU Impersonation This threat is a refinement of the theft of service model. Since the serial number of a single-family ONU is readily available—marked right there on the outside of the box—someone capable of counterfeiting that serial number could impersonate a legitimate ONU, and the OLT would happily deliver that subscriber's complete service package to the imposter. If a forged serial number appeared on a PON that differed from its home PON, the OLT would simply fail to recognize the serial number (see the autodiscovery discussion in Section 4.4), and no harm would be done. It is not specified what should happen if two ONUs appear on the same PON with the same serial number, but surely the OLT ought to take notice. Of course, if it happens that the legitimate ONU has simply been stolen and relocated by the neighbors while the owner is away, there is little the OLT can do about it.

OLT Impersonation The third threat listed in G.987.3 is that an attacker could gain access to the optical network. ONU impersonation and theft of service are yet once again cited as risks. The only new risk in this category is that an attacker could impersonate the entire OLT. This threat definition came from those who presumably have intelligence connections into the underworld, but one has to ask what the legitimate OLT is doing while its functions are being usurped by an imposter, and why the operator is not out there responding to critical alarms. Be that as it may, this is the justification for *mutual* authentication of ONU and OLT. The ONU is expected not to speak to strangers.

Replay Attacks The final threat listed is that an attacker could record packets and replay them later, either intact or modified, to cause something bad to happen. The discussion typically refers to PLOAM messages as vulnerable, but no one has ever brought forward a detailed example of specific messages or message sequences whose compromise would cause a problem. Replay attacks with subscriber traffic are potentially serious, so we intend to encrypt it. Given the security of the keys, it is fair to say that subscriber traffic is safe from replay.

Observation: the agreed security mechanism does not protect against replay attacks of PLOAM or OMCI messages.

OMCI can carry information such as session initiation protocol (SIP) (RFC 3261) and remote authentication dial in user service (RADIUS) (RFC 2865) registration, information that admittedly ought to be protected from interception. The OMCI channel can be encrypted.

There is no question that downstream encryption is necessary. It has been part of ITU-T PON since the early days and, in particular, is a feature of G.984 G-PON. A case can be made for multicast encryption, which at least restricts the group of service thieves to the scope of a single PON. It is harder to be convinced of the dangers of counterfeit OLTs and intercepted PLOAM traffic.

Is excessive security a problem? Only inasmuch as it increases complexity and brittleness and delays implementation and deployment. There is also the risk of complacency if the security turns out not to be as robust as might have been desired.

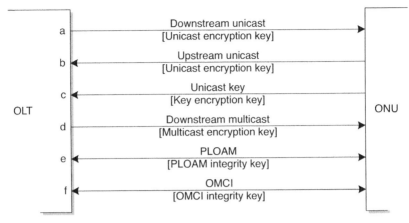

Figure 4.42 G.987 XG-PON security overview.

4.6.1.2 What We Want to Achieve

G.987 XG-PON security appears in any number of contexts and can be confusing. Figure 4.42 is an overview of which information is encrypted and with which key.

(a) Downstream unicast traffic may be encrypted with the unicast encryption key. The key is generated by the ONU. This is the only form of cryptographic security in G.984 G-PON.

(b) Upstream unicast traffic may also be encrypted, also with the unicast key.

(c) When a G.987 XG-PON unicast key is generated or updated, it is conveyed to the OLT under the protection of the key encryption key (KEK). There is also an option for this in G.984 G-PON.

(d) Downstream G.987 XG-PON multicast traffic may be encrypted. The multicast encryption key is generated by the OLT and conveyed to all concerned ONUs via OMCI, protected by the KEK.

(e) G.987 XG-PON PLOAM messages in both directions are sent in cleartext form, but they are protected against alteration by a MIC. The MIC employs the PLOAM integrity key (PLOAM_IK). G.987.3 specifies that the receiver discard a PLOAM message whose MIC fails.

> The editor of G.987.3 points out that the safety rules on construction sites specify that heavy equipment operators accept emergency stop commands from anyone under all circumstances, even from passers-by on the street. At the time of writing, the same policy was under consideration for PLOAM messages such as deactivate_ONU-ID.

(f) G.987 XG-PON OMCI messages in both directions are sent in cleartext form, but possibly in an encrypted GEM port. They are protected against alteration by a message integrity check. The MIC employs the OMCI integrity key (OMCI_IK). G.987.3 specifies that the receiver discard an OMCI message whose MIC fails.

In the following sections, we investigate how all of this works. Initialization and update are topics of special interest, and we offer parenthetic observations on the actual degree of security achieved.

4.6.1.3 Security Toolkit

We need to digress at this point to fill in some of the fundamentals and notation. All XG-PON security is based on the advanced encryption standard AES, defined in National Institute of Science and Technology (NIST) Federal Information Processing Standard (FIPS) 197. The underlying AES encryption system can be used in several modes, three of which are applicable to XG-PON. G.987 XG-PON specifies 128-bit keys, so in the recommendations, we sometimes see the algorithms designated explicitly with the trailing term -128 is the trailing term. We also find the same 128-bit qualifier as an explicit argument in some of the equations.

CMAC A cipher-based message authentication code (NIST special publication 800-38B) does not conceal the original message but instead produces a digest of the message that is sent along with the message itself. The digest is easy for the receiver to regenerate if the message has not been altered and if the CMAC is generated with the same key at each end. The algorithm is designed such that it is computationally difficult to generate a valid digest if either of these conditions is not satisfied, that is, if the message has been altered or if the receiver does not know the key.

We express an AES-CMAC derivation in the following notation:

$$\text{Result} = \text{AES-CMAC}(\text{key, operand, 128}) \qquad (4.5)$$

That is, the 128-bit result is derived by applying the AES-CMAC algorithm to a key and an operand.

ECB An electronic code book (NIST special publication 800-38A) is a way of transforming cleartext input to ciphertext through the use of a key. The characteristic of ECB mode is that the same input, with the same key, always produces the same output. Large samples of ECB output could therefore be subject to dictionary attack, especially if the corresponding cleartext could reasonably be deduced. Key encryption is a good application for ECB mode because the output sample space is small and also because it would not be instantly obvious whether the output of an attack algorithm was or was not the cleartext.

CTR In contrast, AES counter mode continually changes a seed value, such that the output of a given input with a given key is never[*] the same twice, at least never within whatever time scale is regarded to be significant. Successive counter initialization values must be known or predictable by both sides, which is why it is a counter, a simple function that changes continually and is readily predictable by the engines at both ends. In G.987 XG-PON (also in G.984 G-PON), counter mode is used for payload encryption, with material derived from the frame counter and the cleartext's position within the frame or upstream burst as the initial counter value.

4.6.1.4 System Keys

It is apparent from Figure 4.42 that G.987 XG-PON relies on an abundance of keys. The keys used for XG-PON message exchanges are based on two additional keys. Their purpose is to assist with initialization and to make each OLT-ONU association unique, so that there is very low probability of two associations generating the same keying material.

- The session key (SK) is used to generate the key encryption key (KEK), PLOAM_IK and OMCI_IK.
- A master session key (MSK) helps generate the session key.

At every stage of initialization and operation, we need keys that are derived with the same algorithms at OLT and ONU, from raw material that is preferably supplied jointly by the OLT and the ONU but is, of course, known to both. Starting with initialization, let us trace the derivation of the various keying material needed for XG-PON. Refer to Figure 4.43.

Notation

- The vertical bar | designates bit or byte concatenation.
- A hex pair followed by a subscript designates a repeated byte value. For example, $0xFF_{16}$ indicates 16 repeated bytes of all-one values.

The sequence of causality is as follows: before the ONU can attempt to be activated onto the PON, it must learn the burst profile. The OLT periodically broadcasts the profile, a PLOAM message that, like all PLOAM messages, includes a message integrity check. G.987.3 specifies that the ONU discard PLOAM messages whose MIC is not valid, but MIC validation requires a PLOAM integrity key PLOAM_IK. The PLOAM integrity key for the profile message[*] is the default key $0x55_{16}$. This

---

[*] At the time of writing, a vulnerability in the XG-PON application had been identified in which the counter value is duplicated in two successive 125-µs frames once every 4000 years. The proposed solution is to initialize the counter to zero at OLT boot-up and defer the issue for further study for the subsequent 2000 years.

[*] All broadcast PLOAM messages use the default PLOAM_IK, $0x55_{16}$ and thus have no security whatever.

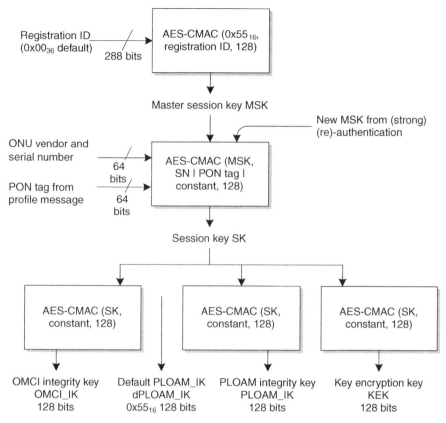

Figure 4.43 Key derivation sequence.

default PLOAM_IK,† which we designate dPLOAM_IK, permits the ONU to perform error checking, but not error correction, on the profile message.

The profile PLOAM message contains an 8-byte PON tag field. It is expected that the PON tag remain constant across all profiles that may be broadcast by the OLT, and that it remain constant over extended periods of time. Now that the ONU recognizes the profile, the OLT and ONU share common knowledge of the PON tag.

Observation: anyone snooping the downstream PON also knows the PON tag.

When the ONU responds to a serial number grant, it supplies its vendor ID and serial number in an upstream serial_number_ONU PLOAM message. Since we have no other key, the PLOAM message MIC is again based on dPLOAM_IK = $0x55_{16}$. The OLT and ONU now share common knowledge of the ONU's vendor ID and serial number.

---

† Be aware that dPLOAM_IK is *not* the same as a PLOAM_IK computed on the basis of the default registration ID.

Observation: anyone capable of snooping the upstream PON or reading the sticker on the ONU enclosure also knows the ONU's vendor ID and serial number.

The next step is for the OLT to assign a TC-layer ONU-ID, using the dPLOAM_IK key to generate the PLOAM MIC. In the state progression of Section 4.2.1, the ONU is now in ranging state, where it responds to a ranging grant by sending its registration ID, again with MIC based on dPLOAM_IK.

The OLT and ONU now share knowledge of the registration ID and can derive the master session key (MSK), the first step in the sequence of Figure 4.43.

Observation: anyone snooping the upstream PON also knows the registration ID and can generate the same MSK.

Based on common knowledge, information known to both ONU and OLT, with part supplied by each, the OLT and ONU agree when they separately compute the SK and the integrity keys for OMCI and PLOAM, as well as the KEK. We are in business.

Observation: anyone snooping the PON is also in business.

At any time, the OLT may reauthenticate, either by requesting the registration ID (weak)—always computing MIC with dPLOAM_IK—or through the OMCI or 802.1X authentication mechanisms (strong). Any strong reauthentication event causes recomputation of all of the keys at both ONU and OLT.

Observation: if a PON snooper can derive the system keys, the payload keys are equally vulnerable. It appears that we must rely on strong authentication. If strong authentication fails, the recovery mechanism is not well specified.

Corner Case The initialization model of Section 4.2.1 allows the OLT to usher the ONU directly through ranging state O4 and into operation state O5 by simply sending a ranging_time PLOAM message, and without retrieving the registration ID.

In theory, this is feasible. By factory default, the registration ID is a sequence of null octets, $0x00_{36}$. If the operator does not use the registration ID, the keys computed from the default are valid. The other possibility is that the ONU has already registered on some previous occasion, and its registration ID is already known.

However, if the OLT fails to guess the correct registration ID, the ranging_time PLOAM message fails because it is required to be signed with a MIC that is derived from a real PLOAM_IK, not the default. If the ONU discards the ranging_time PLOAM message because of a MIC mismatch, as specified by G.987.3, the OLT has no choice but to send a ranging grant. The OLT only knows whether the equalization delay was successfully set by granting the ONU an upstream transmission opportunity, so this particular corner case could affect service.

Rather than elaborate software to deal with this corner case—and with negligible benefit—a real-world OLT can be expected to always retrieve the ONU's registration ID.

4.6.1.5 Strong Authentication

We have seen that the simple default approach introduces complexity but hardly security. The alternative is strong encryption, which may be supported in either (or both!) of two ways, one based on OMCI, the other built on IEEE 802.1X. Both methods presuppose some kind of shared secret, or in the latter case, possibly a certificate that can be verified. Even the registration ID is, in some sense, a shared secret. For that matter, even the ONU's MAC address could be considered a shared secret under certain circumstances—almost anything that can be known to both sides, that has at least some degree of shielding from casual observation, and whose complexity minimizes the odds of a lucky guess.

Whatever it is, the shared secret must be known to both OLT and ONU. And thereby hangs a tale.

Authentication Options Even within a single operator's network, authentication methods may vary. For example, an MDU that is regarded as telecommunications equipment may need nothing more than serial number identification, while ONUs that are CPE, easy to move and easy to compromise, may be subject to 802.1X or OMCI strong authentication. The OLT therefore needs to know what level of authentication is necessary with which ONU. This policy can be one of the parameters preprovisioned into the OLT as part of the equipment or installation order, presumably looked up in an EMS table that correlates the operator's policy options with the available and approved ONU types.

The problem is a bit more difficult at the ONU. For many good reasons, vendors want to deliver the same ONUs, and the same software, to all customers worldwide. This may mean that the ONU has to support all of the authentication options, including no authentication. So be it. In theory, an ONU practicing mutual authentication should refuse to talk to an illegitimate OLT—but how would it know?

Strong authentication presupposes not only that both sides have the same shared secret information but that both sides know which particular information is the shared secret. If one side uses the MAC address as its shared secret and the other side uses a passphrase, they are not likely to succeed in their negotiation. Nothing in the standards allows the OLT and ONU to discover each other's opinions about the nature of the shared secret.

We hypothesize some out-of-band mechanism to set the value of the shared secret on the ONU, for example, a web page served by the ONU to the subscriber or installer. This mechanism also needs to be able to define which information is to be regarded as the shared secret, and which authentication algorithms the ONU should accept. Sounds a lot like a simple registration ID, does it not!

Keying Material Updates, Proper Treatment of MIC Errors Clearly, both OLT and ONU must have the necessary raw material for key computation. When one of the input values changes, both recompute their keys. But a change in raw material originates either at the ONU or at the OLT, and the other does not know about the change for a certain amount of time. Each side must be prepared to continue to accept messages with the old key for a certain amount of time. For management transactions, G.987.3 specifies 10 ms for regeneration of the keys of Figure 4.43. Alternatively, the OLT may simply refrain from issuing management transactions for 10 ms. (But see newly-consented G.987.3 amendment 1.)

The standards do not specify it, but the suggested 10-ms interval would naturally begin when the information change was communicated by one side to the other.

- If the registration ID changes, the timer would begin when the ONU sent an upstream registration PLOAM message with the new value. This might not happen until the next time the ONU reinitialized, which could be many years.
- The PON tag is intended to remain stable indefinitely, but if it were to change, the timer would begin when the OLT transmitted an updated profile PLOAM message.
- In OMCI strong authentication, the timer would begin when the ONU transmitted an attribute value change (AVC) message that confirmed the creation of a new MSK.
- In 802.1X authentication, the timer would begin when the OLT forwarded an extended authentication protocol (EAP) success message (RFC 3748) to the ONU.

The alternative, to recompute keys unilaterally, exposes the OLT-ONU association to difficult corner cases, which we need not explore.

By the way, if the ONU homes onto two OLTs for protection, it is assumed that the OLTs have a way to apprise each other of key updates.

The algorithm for key generation is not specified, except for the admonition to use a cryptographically adequate source of entropy in generating random numbers. We may remark that this is a difficult requirement to satisfy, also difficult to test.

It is also not specified how often to change keys. The strength of the AES algorithm with 128-bit keys is such that it probably suffices to change keys once per month. On the other hand, if the key derivation or the key storage is not secure, no update rate is good enough to ensure security. However, a key update every few seconds or minutes might discourage an attacker.

Consider a use case. If the neighborhood IPTV theft of service club is capable of hacking the middleware key out of a set-top box, its technology expert can also presumably find the broadcast key by delving into a G-PON ONU. But if coordinating a key update with the other members of the club—whose usefulness is limited to the 100 ONUs on the same PON—causes seconds or minutes of disruption to the smooth flow of stolen video, it may simply not be worth the trouble. Instead, the thief with the legitimate subscription will just retransmit the decoded video back up the PON to his friends and neighbors, possibly concealed within his own privately arranged encryption stream.

MIC Errors   There is always the possibility that, somehow, OLT and ONU keys get out of sync. It could expedite recovery from keying errors to define a code point in the acknowledgment PLOAM message (or possibly a new message) to flag a PLOAM message MIC error. When it received an acknowledgment PLOAM message with this code point, the OLT could reauthenticate. The same could be done at the OMCI level, with the notification delivered either via the OMCI response or via a PLOAM message. At the time of writing, no such mechanisms had been defined. In fact, G.987.3 is clear that a MIC error should cause the silent discard of the associated message.

4.6.1.6   Payload Encryption Key Update

Recall from Figure 4.2 that the XGEM frame header contains a 2-bit key index field. The value 00 specifies no encryption, the value 11 is reserved, and the values 01 and 10 select between two possible keys. Whichever code point is in use at a given instant is the current key; the other code point may be either the old (former) key or the new (next) key.

Orthogonal to the two keys of the key index, there are two key domains (we like to call them key rings), the unicast key ring for each ONU, and the broadcast key ring, which is shared by as many ONUs as need multicast encryption. The ONU may have as many as two key rings, one of each type, while the OLT would have a unicast key ring for each ONU that needed encryption, with one more for multicast (broadcast). Each key ring holds up to two keys, one of which is selected dynamically by the key index field of the XGEM frame header. The other key on the key ring is offline so that it can be updated without concern for payload disruption. As the OLT and ONU successfully negotiate each new key, we expect to progress from the current key to the new key by switching key index.

Each GEM port is provisioned through OMCI to use either the broadcast or the unicast key ring (or no key, of course). If the unicast key ring is specified, OMCI also offers the option to provision encryption only downstream or both ways.

Multicast (broadcast) key update is straightforward. If key index 01 is in use, the OLT simply continues to use index 01 while it generates a new key with index 10, distributes it to all ONUs via OMCI, and collects their acknowledgments. At this point, the OLT is free to start using key index 10. Index 01 is no longer in use, so it is available whenever the OLT again wishes to update the key.

If the OLT wished to invalidate the old key, it would be possible do so through an additional OMCI transaction with each ONU, using a state or a code point that would need to be added into the information model. But the OLT could also continue to use the old key as well as the new one, as long as nothing had changed. G.987.3 is silent on this point with regard to the broadcast key.

OMCI was easy because it is an acknowledged channel, and key switching only occurs when the OLT is satisfied that the new key has been distributed correctly to all concerned ONUs. The same simple acknowledged update could have been done with unicast keys, since the PLOAM channel also supports acknowledged message exchanges.

The difference is that, for continuity with G.984 G-PON, it was agreed that the ONU generate the key and send it upstream to the OLT. The acknowledgment thus needed to be from the OLT to the ONU, rather than the other way around. Imagine the following PLOAM semantics:

ONU $\rightarrow$ OLT: New key

OLT $\rightarrow$ ONU: Acknowledge

However, the OLT is responsible for initiating key updates. This implies a third message:

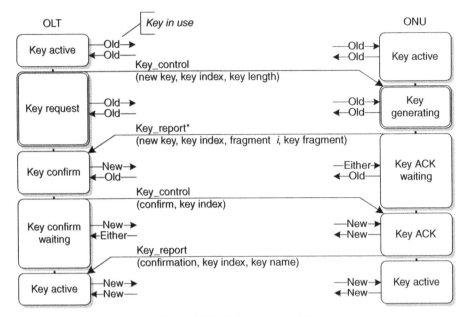

Figure 4.44 Unicast key update.

OLT → ONU: Generate a new key

ONU → OLT: New key

OLT → ONU: Acknowledge

Finally, there were concerns that if the OLT and ONU somehow got out of sync, it might be possible to switch one side to the new key while the other side reverted to an obsolete key that had the same key index. This could have been avoided with three keys on the key ring but was resolved instead by state machines on OLT and ONU, and the message sequence illustrated in Figure 4.44. G.987.3 specifies the state machines precisely and fills in additional detail such as corner case handling.

Let us walk through the sequence of Figure 4.44.

Suppose that encryption has not previously been enabled. Then the OLT's state machine exists (or not) in a state called key inactive, not shown in Figure 4.44. If encryption is then enabled, the state machine is created in (or moves into) the key request state. At the ONU, the state machine is either created in key generating state when the first key_control PLOAM message is received or moves into that state from a key inactive state that is also not shown.

During normal continuing operation, the OLT state machine is in key active state. Suppose that, at the moment, this translates to using key 01 with the given ONU. Suppose that a timer expires, triggering a key update. During the update, the OLT continues to use key 01.

The OLT sends a key_control PLOAM message to the ONU, requesting a new key with index 10, and specifying the length of the desired key. The PLOAM message structure allows for keys of up to 256 bytes, although 16 bytes, 128 bits, suffice for the current need.

As we see in Figure 4.44, the ONU proceeds to generate a new key, which it transmits upstream in one or more key_report PLOAM messages, 32 bytes at a time. The fragments are numbered as a confirmation to the OLT that everything is in order. A single message suffices for the current needs of XG-PON.

Confident that both sides know the new key 10, the OLT starts using it downstream. The ONU accepts traffic with either key during its key ACK waiting transition state, but continues to transmit with the old key 01.

The OLT now sends a key_control PLOAM that requests confirmation of the key, really just a state machine handshake. This informs the ONU that the OLT knows the new key 10. The ONU invalidates the old key 01 and starts transmitting as well as receiving with the new key 10.

Now it is the OLT's turn to accept traffic encrypted with either key.

The ONU generates a 128-bit CMAC of the new key 10, the so-called key name, and returns it in a key_report PLOAM. Given the unicast key *encryptionKey*, the key name is computed as

$$\text{Key Name} = \text{AES-CMAC}(\text{KEK, encryptionKey} \mid \text{constant, } 128) \qquad (4.6)$$

The constant is 0x3331 3431 3539 3236 3533 3538 3937 3933, which just happens to match the ASCII string "3141592653589793".

When it receives the ONU's key_report, the OLT invalidates the old key in the upstream direction. Both sides are now back in key active state, using the new key 10 in both directions, neither OLT nor ONU having disrupted traffic or imposed potentially onerous real-time synchronization constraints upon the other.

4.6.1.7 *Derivation and Use of System Keys*

A few details remain to be filled in. Looking back at Figure 4.43, we see that the 128-bit session key SK is generated by an AES-CMAC function, using the MSK as the key, while the message to be validated comprises the bitwise concatenation of the ONU vendor ID and serial number, the PON tag and a 128-bit constant whose value is 0x5365 7373 696F 6E4B. (Not coincidentally, this happens to be the ASCII string "SessionK".)

$$\text{SK} = \text{AES-CMAC}(\text{MSK, Vendor ID} \mid \text{SN} \mid \text{PONtag} \mid \text{constant, } 128) \qquad (4.7)$$

The session key SK then becomes the key in further AES-CMAC derivations, each with a different constant string. The equation is the same for each:

$$\text{Key} = \text{AES-CMAC (SK, constant, } 128) \qquad (4.8)$$

where the constant strings differ for each key. All of the strings are chosen to be 128 bits long, 16 characters.

Key	Constant (ASCII)
OMCI_IK	"OMCIIntegrityKey"
PLOAM_IK	"PLOAMIntegrtyKey"
KEK	"KeyEncryptionKey"

The MSK and SK are not used for anything else.

PLOAM-IK The PLOAM integrity key validates both upstream and downstream PLOAM messages.

Not to make it too easy, downstream PLOAM messages are prepended with a direction octet whose value is 1, while the direction octet prepended to upstream PLOAM messages has the value 2. The MIC field of the PLOAM message is only 8 bytes, so the CMAC function is only 64 bits long (Eq. 4.9).

$$MIC_{PLOAM} = AES\text{-}CMAC\,(PLOAM\_IK, \text{direction}\,|\,Message, 64) \qquad (4.9)$$

Recall that message validation requires the OLT and ONU to know the common key. To satisfy this requirement for broadcast downstream PLOAM messages, and during ONU activation, the key used in these circumstances is the default, $dPLOAM\_IK = 0x55_{16}$. Appendix II describes the PLOAM messages in detail, including the choice of PLOAM encryption key for each message.

OMCI-IK The OMCI integrity key likewise validates both upstream and downstream OMCI messages. It is used in the same way as is the PLOAM validation key, including prepending a direction octet of value 1 or 2 to the downstream or upstream OMCI message, respectively, before computing the MIC. The OMCI MIC is 4 bytes, so the CMAC function is only 32 bits long.

$$MIC_{OMCI} = AES\text{-}CMAC(OMCI\_IK, \text{direction}\,|\,Message, 32) \qquad (4.10)$$

There is no need for an OMCI default key because there is always a well-defined MSK and SK when OMCI is operable.

KEK The key encryption key is used to conceal the values of the unicast and the broadcast keys as they traverse the PON via the upstream key_report PLOAM message and downstream via OMCI, respectively. Both of the key encryption functions use AES in electronic code book (ECB) mode.

The ONU is expected to maintain state that remembers whether either OMCI or 802.1X strong authentication was performed during its current session on the PON. If not, the ONU does not use the KEK to encrypt new unicast keys upstream, but sends them in the clear. Whether the upstream key is encrypted or not is indicated by a flag in the key_report PLOAM message, so the OLT always knows how to interpret it. It is not specified by the recommendation, but it would be reasonable for the OLT to object if it expected an encrypted key and got a cleartext key.

Regardless of whether strong authentication was used, the OLT encrypts downstream broadcast keys. As we have seen, unless the OLT-ONU association is strongly authenticated, this provides security more apparent than real.

4.6.1.8 Payload Encryption

The careful reader has by now noticed that we have provision to encrypt almost everything, including the encryption keys themselves, as befits the truly paranoid, and we know how to invoke, generate, and update the keys—but as yet we have no provision to encrypt subscriber payload, whose security is the justification for all of this. Yes, we can indeed encrypt payload.

In the ONU2-G managed entity of OMCI, the ONU declares its payload encryption capabilities, and the OLT chooses one of them. At present, AES is the only defined option, with 128-bit keys.

Payload encryption occurs as a process in the step of transferring GEM frames into the XGTC payload (see Section 4.1 and Fig. 4.8, reproduced here as Fig. 4.45). The algorithm is AES in counter mode, which, not surprisingly, relies on counters. Operating at a higher layer, these counters never see the bytes in the underlying server layers, namely the downstream physical frame header, FEC parity bytes, the upstream burst header, upstream PLOAM messages, or upstream DBA reports. To put it a different way, what the counters see is the XGTC header and the GEM frame headers, neither of which is to be encrypted, and the GEM frame payload, which *is* encrypted.

Each downstream GEM port is provisioned through OMCI to be plaintext or to be encrypted with the ONU's unique unicast key or with the common broadcast key. Upstream, the GEM port may be either plaintext or encrypted with the ONU's unicast key. At run time, either unicast or broadcast key 01 or 10 may be specified. Both ONU and OLT must be agile in bringing the correct key into play at a moment's notice.

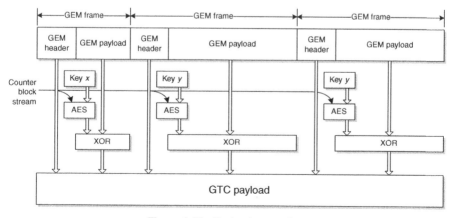

Figure 4.45 Payload encryption.

If AES counter mode is to be secure, its counter values should never repeat—never, at least within some reasonable time frame. We start by using the (super)frame counter as a seed, but we intend to encrypt 128-bit blocks of GEM payload at a time, and there are a lot of 128-bit blocks in an XGTC frame. We cannot just use the frame counter for all of them—that would be a repeated value—so we combine the frame count with an intraframe counter.

In the downstream direction, the intraframe counter initializes to zero with the first bit of the XGTC frame and increments at 128-bit intervals, 16 bytes, including both GEM headers and GEM payload. The same initialization logic applies to the upstream direction, but given that the upstream burst is offset from the start of the upstream PHY frame time, counter initialization is more directly expressed by stating that, at the start time instant of the upstream XGTC frame, the counter initializes to the value $\lfloor start\_time/4 \rfloor$, that is, to the current word count.

After excluding overhead, a 10-Gb/s downstream frame of $125\,\mu s$ contains 135,432 bytes for XGTC payload. It therefore suffices to have a downstream intraframe counter with range $\lceil 135,432/16 \rceil = 8465$, that is, taking on values from 0 to 8464. In the upstream direction, the latest possible start time is 9719 and the largest possible burst is 9720 words. The maximum value that could ever be required from the intraframe counter is therefore 4858 (an exercise for the reader). Each direction requires a 14-bit counter.

We precatenate the intraframe counter with the 50 least significant bits of the (super)frame counter (Fig. 4.46), and duplicate the whole thing, to derive a 128-bit value—the initial counter block—that ticks along during each XGTC frame. The important facts about this sequence are that it is unique over time spans of practical interest, and that it is readily predictable at both OLT and ONU, so that it can be easily regenerated for decoding. To avoid duplicate values between downstream and upstream, we also invert the less significant 64-bit end in the upstream direction before loading it into the initial counter block.

On the first bit of each GEM frame header, we latch the current value of the initial counter block into a 128-bit payload counter. Stable for 128 bits at a time, the output of the payload counter forms the counter input to the AES-CTR algorithm.

Every 16 bytes of payload, the AES-CTR engine increments the payload counter. It is just an ordinary 128-bit counter, that is, with no duplicated 50-bit, 14-bit structure, so the payload counter's value immediately diverges from that of the initial counter block, even if they both happen to increment at the same instants.

That last phrase deserves amplification. Both the intraframe counter and the AES counter increment at 16-byte intervals, but the phase of the intraframe counter is locked to the XGTC payload, which includes GEM headers, while the phase of the AES counter is determined by the start of each successive GEM frame. Because everything in XG-PON is aligned on word boundaries, there are four possible relative phases between the intraframe counter and the AES counter.

Observe also that, while we preload the upstream AES counter at the beginning of the GEM *header*, we do not increment it until byte 16 (the 17th byte) of GEM *payload*. The additional preliminary delay allows the AES engine more time to prepare the first cipher block.

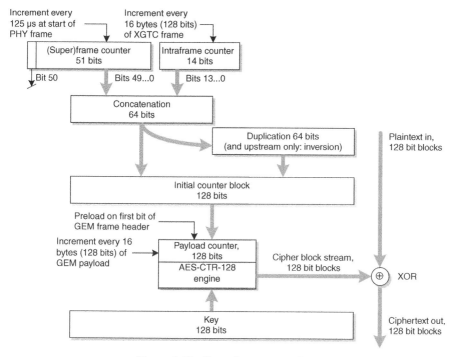

Figure 4.46 Downstream encryption.

Also using the proper key, which may change with each GEM frame, each new value of the AES counter triggers the encryption engine to generate a 128-bit cipher block, which is then exclusive-ORed with 128 bits of cleartext payload to produce ciphertext. At the receiving end, the same cipher block recovers the cleartext.

The upstream direction is basically the same, except for the initialization instants of the counters. We refer the interested reader to G.987.3 for the details.

4.6.1.9 Reduced Encryption Strength

Certain administrations—we refrain from editorial comment—do not permit strong encryption within their jurisdictions. At the time of writing, it appears that AES-128 algorithms are acceptable in some of these jurisdictions, as long as the keying material is restricted to 56 bits. The actual key remains 128 bits long, but 72 of its bits are fixed and well-known values. G.987 XG-PON specifies a leading sequence of 0x55 bytes in this application.

Since software parameters can easily be changed, the length of the key needs to be a compile time option, not a run-time setting. OMCI includes a read-only attribute in the enhanced security control managed entity, an attribute that allows an ONU to declare the effective length of its keys.

As to jurisdictions that do not even permit AES, there is currently no agreed-upon solution.

4.6.2 Security in G.984 G-PON

If G.987 XG-PON security is arguably overkill, G.984 G-PON security might be called PON security lite—lightweight, simple, and perfectly adequate for the needs of many operators. Downstream unicast is the only payload encryption, and with one exception, keys are sent upstream from the ONU in the clear. With the same exception, authentication is a matter of trusting the ONU's serial number or registration ID, and no one expects a counterfeit OLT to suddenly appear in the network.

The exception is that G.984 supports the option for OMCI-based strong authentication. The result of this process is the generation of a master session key MSK, which is used in AES-ECB-128 mode to encrypt the unicast key transmitted upstream in the encryption_key PLOAM message. In G.987 XG-PON, this is done with the key encryption key KEK, but no KEK is defined in G.984 G-PON.

Turning to payload encryption, we first observe that encrypted GEM ports are turned on or off through the encrypted_port-ID PLOAM message, rather than through an OMCI GEM port attribute, as is done in G.987 XG-PON.

As with G.987 XG-PON, the ONU2-G managed entity of OMCI declares the ONU's payload encryption capabilities, and the OLT chooses one of them. AES is the only option presently defined, with 128-bit keys.

Figure 4.47 shows that G.984 G-PON uses a simple request–response PLOAM exchange to update keys.

Details

Because the AES algorithm is fixed at 128 bits, the key is 128 bits long, 16 bytes. The encryption_key PLOAM message carries 8 bytes of key, so there are always two fragments.

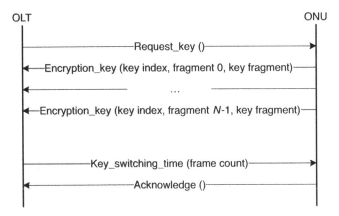

Figure 4.47 Unicast key update, G.984 G-PON.

Each message is repeated three times, including the key_switching_time message and its acknowledgment.

As would be expected, the key switches at the beginning of the frame whose (super)frame count is specified in the key_switching_time PLOAM message. It is good to complete the PLOAM transaction before the specified frame number occurs.

Payload encryption is similar to the G.987 XG-PON algorithm, with a frame counter and an intraframe counter, concatenated and latched into an AES-CTR-128 engine for each encrypted GEM frame.

The (super)frame counter in G.984 G-PON is 30 bits wide. The intraframe counter is 16 bits wide, of which only 14 can vary. The concatenated 46-bit pair is replicated three times to produce a field 138 bits wide. The 10 most significant bits are ignored, yielding a 128-bit so-called cryptocounter.

The G.984 G-PON intraframe counter starts with the value 0 on the first byte of the downstream PCBd, that is to say, on the first byte of the physical synchronization pattern. The counter runs continuously throughout the frame, including during FEC bytes. The counter increments every 4 bytes, so that its maximum value is 9719 at the end of the frame.

As with G.987 XG-PON, G.984 G-PON latches the value of the cryptocounter on the first bit of the GEM frame header. Subsequent increments to the AES counter occur at 128-bit intervals in the payload section of the GEM frame.

G.984 G-PON has no key index; it has an active key register and a shadow key register for new and old versions of the unicast key. The active key is the one in use at any given time, while the shadow register is updated by the PLOAM exchange of Figure 4.47. The key index field in the encryption_key PLOAM message (Appendix II) is merely a sequence number that increments with each new key, allowing the OLT to be sure that all (both) upstream encryption_key messages pertain to the same key.

When the specified (super)frame count arrives, both OLT and ONU shift the shadow key into the active key register and begin using it.

In case of confusion for example, intermittent LOS with recovery via the PopUp PLOAM message, the OLT can cancel a pending key switch by sending a new request_key PLOAM message.

Reduced Key Length in G.984 G-PON   Reduced key length is not defined for G.984 G-PON. Presumably vendors who sell into these markets will implement proprietary solutions, probably along the lines of the G.987 XG-PON recommendation. If interoperability or standards compliance is someday required in these markets, reduced key length could be formalized in an amendment to G.984.3, subject to the usual risk that some implementations might have hardware implementations that could not be adapted to the standard.

4.7 UPDATE

As we went to press, G.984.3 amendment 3 and G.987.3 amendment 1 were being consented in ITU.

Changes proposed for G.984.3 include:

- Restructuring and clarifying G-PON security, including some of the features of G.987.3 such as reduced strength keys
- New optional PON-ID PLOAM message
- New PLOAM, swift_POPUP, to avoid the need for re-ranging if the ONU recovers downstream LOS quickly
- New PLOAM, ranging_adjustment, to allow incremental ranging delay to one or all ONUs, designed for rapid recovery after a protection switch

Changes proposed for G.987.3 include:

- Clarifications of the security clause, including registration ID, encryption control and operation, key and MIC generation and failure
- Clarifications of power management details
- New optional PLOAM message, discovery_control, to assist in isolating rogue ONUs prior to their full activation, along with an extra state (O8, Suspension) of the ONU activation cycle
- New optional PLOAM message, PON-ID. Optional redefinition of the PON-ID structure of the framing overhead (Fig. 4.10)
- Previous acknowledge PLOAM message is now called *acknowledgment*, a change that is reflected in this book.

5

MANAGEMENT

In this chapter:

- *The ONU management control interface OMCI*
- *Establishing the OMCI communications channel OMCC*
- *Fundamentals of the OMCI information model, message exchange, and actions*
- *ONU state model*
- *Alarm reporting control, ARC*
- *Equipment management, software management, reach extender management*
- *PON maintenance*

5.1 THE TOOLKIT

The G-PON management model assumes that the ONU is a logical extension of the OLT, not a separate network element to be managed on its own. This is most clearly evident in the naming convention. A stand-alone network element has its own name, whereas an ONU is always named as a component under the hierarchy of its parent OLT, for example:

Gigabit-capable Passive Optical Networks, First Edition. Dave Hood and Elmar Trojer.
© 2012 John Wiley & Sons, Inc. Published 2012 by John Wiley & Sons, Inc.

```
OLT name -
     Blade number in the OLT -
          PON port number on the blade -
               ONU number on the PON.
```

When the ONU includes an RG function that the operator wishes to manage separately, there may be a logical separation inside the ONU. The ONU's PON identity is as described above, but the CPE identity may be completely independent, just as if the RG were a separate element in its own right. Coordination between identifiers occurs in network management databases, not within or between the equipments themselves. It would not be expected that the OLT or the OMCI information model would even know the name of the associated CPE.

The standard management interface for G-PON systems is called the OMCI. OMCI allows the OLT to perform full fault, configuration, performance, and security functions (FCAPS) at the ONU. OMCI is defined in ITU-T recommendation G.988, as amended. Historical versions exist in the form of G.984.4 and G.983.2.

The term OMCI customarily refers both to the information model and to the set of messages exchanged across the OMCC. The OMCC is set up as part of the ONU discovery process. The information model comprises a set of so-called managed entities (MEs), each of which includes a list of attributes. An attribute may be mandatory or optional, and may have any meaningful combination of read, write, and set-by-create properties. Specific managed entity types support various actions and notifications. There may exist zero, one, or many instances of a given managed entity type, depending on its definition. An instance may be created by the ONU, the OLT, or in some cases by either.

As with all living technologies, new applications arise continuously, and OMCI is extended continuously to serve the need. The G-PON community makes every effort to extend OMCI in backward compatible ways, such that existing implementations never become nonstandard.

5.1.1 OMCI Communications Channel

When the ONU is discovered on the PON, a process described in detail in Chapter 4, one of the first orders of business is to establish the OMCC, which is necessarily a function of the TC layer. Figure 5.1 shows how it happens.

Messages exchanged across the G-PON OMCC are encapsulated in GEM frames, just like any other G-PON traffic. In G.984 G-PON, the GEM port that carries OMCI must be assigned through a TC-layer PLOAM message, but in G.987 XG-PON, the GEM port is implicit and is equal to the TC-layer ONU-ID.

The alloc-ID of the upstream OMCC is implicit in both G-PON and XG-PON and is also the same as the TC-layer ONU-ID.

5.1.1.1 Messages
Table 5.1 lists the messages defined in OMCI. Most messages are initiated by the OLT and result in a simple confirmation from the ONU or an error response that

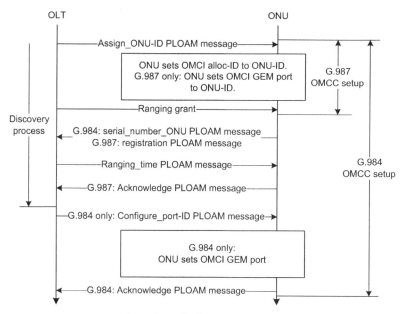

Figure 5.1   Setting up the OMCC.

indicates why the command failed. Other OLT messages serve the purpose of retrieving information from the ONU. Notification messages are initiated by the ONU itself to signal significant events, such as alarms, attribute value changes, or test completion.

OMCI does not include a generalized action message. When an action is needed beyond those defined in Table 5.1, the function is realized through a set command. The parameters and actions of a specific set command directed to a specific managed entity can be tailored to whatever specialized action is needed.

5.1.1.2 Message Structure

The default or baseline OMCI message has a fixed length of 48 bytes. Because a message may contain only an integer number of attributes, this imposes some limitations on the size of information model fragments. Table 5.2 shows the format of a baseline message.

An extended message set is also defined, with variable-length messages up to 1980 bytes. All ONUs and OLTs are required to support the baseline message set, which allows them to initialize in a backward-compatible way. They are then free to negotiate the use of the extended message set.

Table 5.3 shows the extended message format.

The motivation for the extended message set was improved efficiency, primarily to speed up software download. The message types and functions are the same, except that the extended message set also includes a set table command that exploits its longer structure to allow several table rows to be written with a single command.

TABLE 5.1   OMCI Message Set

Message Type	Purpose	Downstream	Upstream
Create	Create a managed entity (ME) instance with its attributes.	Message	Confirmation
Delete	Delete a managed entity instance.	Message	Confirmation
Set	Set one or more attributes of a managed entity.	Message	Confirmation
Set table	Set one or more rows of a table. This message is defined only in the extended message set, explained below.	Message	Confirmation
Get	Get one or more attributes of a managed entity. When directed to a table attribute, the get action causes the ONU to latch a copy of the table for retrieval with a sequence of get next commands.	Message	Response
Get next	Get the latched values of a table managed entity within the current snapshot, as much information as fits into the response body. The OLT issues a series of get next commands to walk through the latched copy of the table and thereby retrieve all of it.	Message	Response
Get all alarms	Latch the alarm statuses of all managed entities and reset the alarm message counter.	Message	Confirmation
Get all alarms next	Get the active alarm status of the next managed entity or entities from the latched alarm status copy. The OLT issues a series of get all alarms next commands to walk through the latched copy of the alarms.	Message	Response
Get current data	Get the current counter value associated with a performance monitoring managed entity. It is always optional whether a PM ME supports this message or not. The alternative is a standard get command, which retrieves PM values from the most recently concluded measurement interval.	Message	Response
Alarm	Notification of an alarm or the cessation of an alarm condition.	—	Notification
Attribute value change	Notification of an autonomous change to the value of a tracked attribute. Not all attribute changes generate AVCs, only those that are expected to be of interest to the OLT or element management system EMS. Changes caused by OLT messages do not generate AVCs.	—	Notification

TABLE 5.1 (*Continued*)

Message Type	Purpose	Downstream	Upstream
Test	Request a test on a specific managed entity.	Message	Confirmation
Test result	Notification of result initiated by a test command.	—	Notification
Reboot	Reboot ONU or circuit pack.	Message	Confirmation
Synchronize time	Coordinate 15-min PM interval boundary between OLT and ONU. Optionally establish a rough time of day reference (accuracy in fractional seconds, not in nanoseconds).	Message	Confirmation
MIB reset	Clear the management information base (MIB), reinitialize it to its default, and reset the MIB data sync counter to 0.	Message	Confirmation
MIB upload	Latch a copy of the MIB. G.988 specifies that some MEs and some attributes are not included in a MIB upload, generally those that are not needed to reconstruct the ONU's provisioning.	Message	Confirmation
MIB upload next	Get the next set of MEs and attributes from the latched MIB copy. The OLT walks the latched MIB image with a series of these messages.	Message	Response
Start software download	Start a software download action. The ONU prepares to receive a new software image.	Message	Confirmation
Download section	Download a section of a software image. The process comprises multiple unacknowledged transfers, followed by a group confirmation when requested by the OLT.	Message	Confirmation
End software download	Indicates that a software image download is complete. The ONU verifies the checksum across the image and stores it persistently.	Message	Confirmation
Activate image	Start executing the downloaded software image.	Message	Confirmation
Commit image	Set the downloaded software image to be executed when the ONU reboots.	Message	Confirmation

This is a convenience, but hardly vital. As to software download, there are other ways to expedite it. Migrating an existing OMCI stack to the extended message set is an investment with little immediate payback. At the time of writing, few vendors had implemented the extended message set.

TABLE 5.2 Baseline OMCI Message Format

Byte Number	Size	Use
1..2	2	Transaction correlation identifier
3	1	Message type
4	1	Device identifier
5..8	4	Managed entity identifier
9..40	32	Message contents
41..48	8	OMCI trailer

Every downstream message is directed to an explicit managed entity instance; every upstream message identifies its originator. This information appears in bytes 5–8 of the OMCI message header, which contain the type of the ME on the ONU, and its instance number.

OMCI messages expose one additional degree of granularity in the information model, namely individual attributes within the ME instance. It is possible for a set or get a message to write or read one or more of the attributes, without touching the other attributes of the same ME instance.

Attribute-level granularity allows for the possibility that a single attribute can be large enough to only just fit into a (baseline) message. Thirty-two bytes are available for message contents, but some messages include per-attribute status flags that indicate whether a specific attribute is present or not, or whether the action on the specific attribute succeeded. This additional overhead reduces the maximum size of a scalar attribute to 25 bytes. A structured table row may be as long as 30 bytes. These constraints are not present in the extended message set, but backward compatibility strongly encourages OMCI evolution to remain within these bounds.

Further Reading The interested reader will find the message definitions in G.988 clause 11, and the messages themselves are defined in detail in G.988 annex A.

TABLE 5.3 Extended OMCI Message Format

Byte Number	Size	Use
1..2	2	Transaction correlation identifier
3	1	Message type
4	1	Device identifier
5..8	4	Managed entity identifier
9..10	2	Message contents length
11..$(N-4)$	–	Message contents
$(N-3)..N$	4	Message integrity check (MIC)

5.1.2 Management Information Base (MIB)

The ONU is a slave to the OLT and as such is not necessarily expected to retain provisioned data across initialization cycles. When it starts up, the ONU contains a so-called default MIB. It contains a few fundamental MEs that convey information such as the ONU's serial number, its capabilities, its port configuration, and parameters of its resident software images. Depending on its capabilities, and especially on its hardware architecture, the ONU will also have automatically instantiated and populated a number of additional managed entities. G.988 defines many of these with language such as, "if the ONU supports this particular feature, it automatically instantiates this particular managed entity." At ONU initialization, the OLT is expected to confirm the ONU's baseline configuration and to audit and if necessary to provision its services.

Once provisioned, and throughout its normal operation, the ONU's MIB represents all of its configured information, along with nonconfigured information resulting, for example, from local events such as state changes or background tests. At any time, the OLT can upload the ONU's MIB to confirm that the ONU is in sync. The OLT can also query individual managed entities to answer more focused questions.

There is no standard file format to capture an ONU's MIB, nor is there a bulk backup and restoration mechanism. This reflects the absence of community demand, rather than any intrinsic limitation. A vendor may implement such functions in a proprietary way, recognizing that they would not be interoperable across vendors.

5.1.3 ONU Information Model

OMCI is a tool for managing ONUs, and as such, it includes a model of the ONU. OMCI is built on the concept of MEs, each of which is identified by a unique ME type or class enumeration (2 bytes). In a given ONU at a given time, each ME type may exist in zero, one, or many instances. Each specific instance is identified by another 2-byte value, the ME ID. Thus, 4 bytes suffice to distinguish any ME in an ONU.

The managed entities of OMCI represent several aspects of a real ONU. Here and in Chapter 6, we examine them in detail.

- Some MEs focus primarily on the ONU as equipment, representing, for example, serial numbers, slots, circuit packs, power, and software images. This chapter is built around these MEs. It explains ONU equipment management functions, at the same time helping to develop an understanding of the model.

- Some MEs represent the access network interface (ANI) side of the ONU, that is, the PON interface itself: traffic containers (T-CONTs), GEM port network connection termination points (CTPs), and upstream priority queues. The functionality of some of these MEs was discussed in Chapter 4; we get further into their details in Chapter 6.

- Some MEs represent the various kinds of user-network interface (UNI) ports available on ONUs: POTS, Ethernet, xDSL (xDSL is shorthand for VDSL2 and all of the varied flavors of ADSL). Business ONUs may have additional types of ports, for example, whole or fractional DS1 or E1. Chapter 6 goes into considerable detail on these MEs and how they are used.
- The previous two categories represent ONU interfaces. Signal processing and flow is controlled by other MEs. Some of them model VLAN manipulation and MAC bridging. Other MEs model POTS services, pseudowires, and IP host functions. Still other MEs collect PM data of all varieties. Chapter 6 describes them.

In total, there are almost 300 ME classes. Many of them are obsolete—some date from the B-PON days and are no longer meaningful; others simply found no market. No matter why these MEs are not used, their class values remain assigned indefinitely. There is no possibility that a class value, once defined, might become undefined or ambiguous. To illustrate the scope of ME definitions, Table 5.4 lists a handful of ME types, together with their class identifiers.

Class values less than 256 were originally assigned in G.983.2 (B-PON), when the class identifier was a single byte. When G-PON standards were developed, it was clear that 2 bytes would be needed. All G-PON MEs, then and now, are assigned class

TABLE 5.4   Exemplary ME Classes

Managed Entity Class	ME Class Value
ONU-G	256
Cardholder	5
Circuit pack	6
Software image	7
T-CONT	262
ANI-G	263
Priority queue	277
Traffic descriptor	280
Multicast operations profile	309
Multicast subscriber config info	310
Physical path termination point Ethernet UNI	11
Physical path termination point POTS UNI	53
Physical path termination point xDSL UNI part 1	98
Pseudowire termination point	282
MAC bridge service profile	45
MAC bridge port configuration data	47
Extended VLAN tagging operation configuration data	171
IP (v4) host config data	134
TCP/UDP config data	136
VoIP config data	138
SIP agent config data	150

identifiers greater than 255; a class value less than 256 identifies an ME that survived from the days of B-PON. G.988 also reserves several regions of the address space for vendor-specific MEs.

The bulk of G.988 comprises the detailed definitions of ME classes. The definition always includes a set of so-called attributes, the first of which is always the ME ID. G.988 defines each attribute in detail, sometimes with a supplementary reference to some other standard that contains a more extensive definition. The ordering of the attributes is crucial, as are the sizes of each attribute and their {read, write, create} properties, because several OMCI messages represent attributes with bit maps and fixed-size fields. An attribute may be mandatory or optional.

Especially for optional attributes that have the set-by-create property, G.988 defines the default behavior if the ONU supports the attribute but the OLT does not. The default value in such a case is always the standard OMCI background of 0 bytes, so the default specification means definition of the meaning of the value 0. Often, G.988 just says that the ONU uses its own internal default, whatever that might be.

This is perhaps the right place to mention that OMCI endeavors to use other standards as much as possible, rather than reinventing wheels. For example, OMCI borrows many aspects of MAC bridging[*] from the IEEE 802 series of standards, while DSL management is based on ITU-T G.997.1, and layer 3 functions such as DHCP and IGMP-MLD use RFCs from the IETF.

5.1.3.1 ME-Type Definition Template

Table 5.5 shows the standard template for the G.988 definition of an ME class.

Managed Entity Name   This string is the formal designator of the ME and is spelled out in full wherever formality is needed. G.988 sometimes uses abbreviations when there is little chance of confusion, most notably the acronyms PPTP (physical path termination point) and PM (performance monitoring).

Introduction   A brief description of the ME's function. This paragraph usually includes a statement about whether ME instances are created automatically by the ONU itself or through the OMCI create command by the OLT.

Relationships   An informal description of the ME's relationship to other parts of the information model. The rigorous relationship is defined by the attribute definitions of the MEs themselves.

Attributes   An ordered list of the properties of the ME. The first attribute is always the managed entity instance identifier, commonly known as the ME ID. As many as 16 additional attributes then form the body of the managed entity. In the next section, we expand the discussion of attributes.

Actions   A list of the actions that can be invoked by the OLT against this ME type. We consider the range of possible actions in a separate section below.

---

[*]All of these terms are explained further in Chapter 6, but not from the assumption of zero prior knowledge.

TABLE 5.5 ME Definition Template

Managed entity name
Introductory and explanatory text
Relationships
Attributes
Actions
Notifications
Supplemental material, if needed

Notifications This section of the G.988 ME definition comprises zero or more tables that define autonomous messages, the messages that can be emitted by an instance of this ME type. Notifications also receive their own subsection below.

Supplemental Material If a moderate amount of explanatory material is needed, it may appear along with the ME definition. Extended discussions generally appear in annexes* or appendices.

5.1.3.2 OMCI Attributes

To recapitulate from above: the G.988 ME type definition template includes a list of attributes, the first of which is always the ME ID, followed by not more than 16 additional attributes.

Managed Entity Identifier The ME ID is always 2 bytes. Its G.988 description always specifies any special rules for identifying instances of the ME. Common patterns include:

- Only one instance, autocreated by the ONU, numbered 0.
- Instances identified by an equipment slot number in one byte and a port number in the other byte. These MEs are usually autocreated by the ONU on the basis of its hardware architecture.
- Instances whose ME ID is identical to the ME ID of some other ME, naturally of a different class. The identity creates an implicit linkage between the two. Performance monitoring MEs are a common example of this pattern, where the implicitly linked ME might be a physical port whose performance was to be monitored.

  In other cases, an implicitly linked ME may exist as one member in a family of related MEs. Each member is distinguished by some characteristic: The patriarch sits at the head of the table, implicitly linked to any number of additional MEs that capture, for example, read-only status information, or upstream versus downstream configuration, or that control optional or

*The difference between an annex and an appendix is that an annex is normative; an appendix is informative.

conditional features. In other cases, there are simply too many attributes to fit into a single ME class, and the extra attributes overflow into implicitly linked sibling ME classes.

- Instances whose ME ID is free to be chosen by the OLT at the time of creation. Often the values 0 and 0xFFFF are excluded from the range of choice. This is because OMCI frequently uses pointers, which are the ME ID values of linked objects, and 0 and 0xFFFF are convenient values for null pointers, as long as they are unambiguous. To put it another way: we need to be able to distinguish a null (meaningless) pointer of value 0 from a valid pointer to an ME whose ME ID is 0, and the way we do it is to exclude 0 from the range of valid ME IDs for the target class, likewise with 0xFFFF.

The ME ID may be read-only if the ME instance is created by the ONU itself, or it may be read, set-by-create if the instance is created by the OLT. The ME ID of an existing object instance cannot be changed.

Up to 16 Additional Attributes   In Section 5.1.3.3, we explain why OMCI is limited to 16 attributes beyond the ME ID.

The G.988 definition of each attribute states its type, range, rules, restrictions, and interpretation. An attribute may be read-only, read-write, or read-write-set-by-create. An attribute may be mandatory or optional. When an ME is extended with a new attribute in a G.984.4/G.988 amendment, the new attribute is always defined to be optional, to avoid making existing implementations nonstandard.

Except for table attributes, about which we shall learn more below, the size of an attribute is fixed by its G.988 definition. This simplifies the parsing of messages but comes at the expense of flexibility.

Simple Attribute Types   Simple attributes include the following types:

- Boolean, one byte whose value can be either *true* or *false*.
- Integers, usually unsigned, occasionally 2s complement, occasionally with special scaling and offset interpretations, of varying size from 1 to 5 bytes, and with varying range restrictions.
- Bit maps of varying size.
- Character and octet strings—Because of the attribute size limit of the OMCI baseline message set, these are restricted to 25 bytes. A character string shorter than the defined attribute length is always terminated by at least one null (zero) byte.
- Pointers, always 2 bytes—A pointer is the ME ID value of an associated ME instance. The type of the associated ME may be fixed by definition or may be determined by a separate ME-type attribute.
- IP addresses—When the OMCI information model was first developed, IP version 6 was far enough in the future to ignore, and all IP addresses defined in

OMCI were 4-byte IPv4 addresses. All attributes in OMCI, with the exception of tables, must have a fixed size, so it is not possible to simply extend an existing 4-byte attribute to the 16 bytes that would be needed for an IPv6 address. The good news is that the information model contains only a few IP address attributes.

To accommodate IPv6, a few new MEs were defined, most notably the IPv6 host config data ME. A few new attributes were added onto existing MEs, and preexisting IPv4 address attributes were reinterpreted. Reinterpretation takes advantage of the fact that the IPv4 address 0.0.x.y is invalid when x and y are not both zero.[*] So in an IPv4 address attribute with value 0.0.x.y, OMCI interprets the 2 bytes xy as a pointer to a large string-managed entity that contains an IPv6 address in text string form.

- Enumerations—One form of enumeration is the attribute that specifies the type of an associated ME. For example, a MAC bridge port may be associated with several UNI types, with an 802.1p mapper, with an IP host, or with other ME classes. Enumerations are customized for their purposes.

OMCI also defines ad hoc attributes as needed. These may include an integer range with special meanings assigned, for example, to the values 0 or 0xFF. Another specialized form is a record, containing several fields, the total not to exceed the OMCI 25-byte size limit.

Table Attributes　OMCI also supports a complex attribute called a table, whose size is not necessarily predetermined. The rows of a read-only table can be as long as desired; indeed, there need be no row structure at all. An arbitrary octet string can be regarded as a table with but a single row.

A table that is intended to be written must be structured as a list of rows, or records, with each row comprising not more than 30 bytes. Some tables include row identifier fields; some do not. Some include write-delete-clear control bits within the row definition; some do not.

There are two motivations for the existence of table attributes. The first is that tables per se are an important tool in the toolkit, a list in which each entry is a row of the table, and each row is a record that is structured into fields.

The second motivation is that tables can be empty or arbitrarily large and can change over the course of time. Large scalar attributes, or attributes of indeterminate size, are therefore also defined to be tables as a way of working around the limited attribute space in a baseline OMCI message. A table avoids the OMCI message constraint of fixed-size, small attributes.

OMCI does not allow for attributes to be fragmented across several messages. The upload of tables is an exception, but nothing is free. As we mentioned, tables can have arbitrarily long rows and can be as large as we will ever need in practical terms. But a table cannot be read back, one row at a time. To find out what is in a table, the

---

[*]The address 0.0.0.0 is only valid in some IGMP messages.

OLT must upload the entire table, effectively as a single long octet string, then parse it. In general, the OLT or the EMS needs a specialized parser for each table attribute.

Extended Table Rows   The length of a writeable table row is limited by the set command of the OMCI baseline set message to 30 bytes. This became an issue when OMCI was adapted to IPv6, in which each address consumes 16 bytes. Especially for multicast management, a table row may need more than one IPv6 address, along with a number of other fields.

It would have been possible to define longer rows in the extended OMCI message set, at the expense of incompatibility with the baseline message set. Someday, a break with the baseline message set may be unavoidable, but the community is not eager for a forced migration. Could we get longer rows into existing tables? Could we add new IPv6 table attributes to existing MEs?

To the first question: yes; to the second: sometimes but not always.

In the multicast tables of interest, the first 2 bytes of each row had already been claimed for overhead. Fourteen of the 16 bits were used as the table key, that is, as a row identifier. The remaining 2 bits controlled the meaning of a set operation:

01   Write this entry into the table, overwriting any existing entry with the same table key.
10   Delete this entry from the table, remaining fields not meaningful.
11   Clear all entries from the table, remaining fields not meaningful.

Fourteen bits for a row identifier? None of these tables will ever need to contain 2^{14} entries. Ten bits ought to suffice, with room to spare. So G.988 amendment 1 claims 4 more bits from the 2 overhead bytes of these tables for use in table substructuring.

The first 3 of our newly claimed bits simply define a so-called row part. For backward compatibility, row part 0 retains whatever structure already exists in the table, assuming that we are extending an existing table attribute. Row parts 1–7 are available to extend the length of each row in the table. In this way, the content of a row could be as long as 224 bytes,* excluding the row overhead bytes. Figure 5.2 illustrates a table that supports four row parts.

Observe that, depending on the specification of the table, some row parts may be optional.

The other reclaimed bit of the row overhead is available for custom use in each table definition. Its only use to date is in the multicast operations profile, an extension of a preexisting managed entity type into the brave new world of row parts. In its access control list attributes, OMCI uses that extra bit as a way to allow the OLT to determine whether the ONU does or does not support the row part extension feature. The so-called test bit always reads back as 0 in an ONU that supports multipart rows,

---

*All row parts must have the same length, not to exceed 30 bytes. But nothing forces the row parts to be 30 bytes long. They are as long as they need to be.

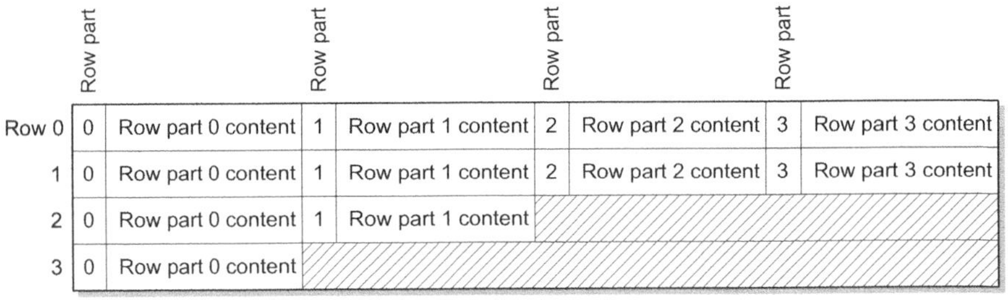

Figure 5.2 Multirow-part table.

even if the OLT sets it to 1. This is an example of the attention to backward compatibility that characterizes OMCI evolution.

Because of its gradual evolution, OMCI does not have a single table architecture, much to the annoyance of the afflicted software developers—every table is a special case. Belatedly, however, there is encouragement that the row part design may prove widely usable in tables that may need to be defined in the future.

5.1.3.3 OMCI Actions

With the definition of each ME class appears a list of the actions that can be invoked by the OLT against instances of the class. The actions map closely onto the OMCI messages of Table 5.1.

- *Get* All MEs support the get action, which allows the OLT to query the ME and its attributes. The get command includes a 16-bit bit map that flags a subset of attributes to be returned in the get response.

  The get response includes its own 16-bit bit map that indicates which attributes are, in fact, present. Normally, the response bit map would be the same as the command bit map, but if the aggregate size of the requested attributes would exceed the capacity of an OMCI message, the ONU may[*] return only a subset of the requested information, whereupon the OLT would be expected to issue another get command for the remainder. In the real world, we expect the OLT to know what is possible and never ask for more than it can get in one response.

  The get response includes two additional 16-bit bit maps, which also consume space in the OMCI message. One of them flags optional attributes that were requested but that are not supported by the ONU; the other identifies attributes whose retrieval failed in some way. These maps are meaningful only if there was an attribute exception; various kinds of exceptions are indicated in a so-called result-reason code that consumes one more byte of the response message.

  Each bit in one of the 16-bit bit maps corresponds to one of the ME's attributes, in order: thus the OMCI limit of 16 attributes per managed entity type

---

[*]The ONU also has the option to reject the get command with an error message.

and the requirement for strict ordering of attributes in messages. Thus the 25-byte size of the baseline gets response message: 32 bytes, reduced by the 1-byte completion code and three 16-bit bit maps.

- *Set* If some of the attributes can be modified by the OLT, the ME supports the set action, which contains a 16-bit bit map specifying the attributes to be modified. The values of the attributes, being of fixed size, are then simply concatenated, one after the other, into as much space as remains in the message. If there is not enough space to fit all of the attribute values, it may be necessary to execute a number of separate set commands with disjoint attribute bit maps.

  Much like the get response, the set response includes a completion code byte and two 16-bit bit maps that indicate whether the ONU supports each of the specified attributes and the success or failure of each attribute modification.

  The extended message set also defines a set table action, which permits a single message to set several rows of a table. This is a way to exploit the longer maximum size of the extended message set to reduce management traffic.

- *Create, Delete* These actions are supported if instances of this ME type can be created and destroyed by the OLT. The create command includes a list of values of the set-by-create (SBC) attributes. No attribute bit map is needed because OMCI requires the create command to contain values for all SBC attributes; the ONU, therefore, knows how to parse the string of octets that accompanies the message. This rule constrains the aggregate size of an ME's SBC attributes to 34 bytes.

  Yes, there are optional SBC attributes, but in this context optional just means that the OLT is permitted to leave the zero default placeholder in the create message for those attributes, and that the ONU is permitted to ignore those fields, even if they are not zero. Because the OLT may leave the bytes at zero and the ONU may interpret them as a valid attribute value, it is important that the semantics of the value zero be both specified and harmless.

  As to deletion, the OLT is warned not to delete an ME instance if the instance is embedded in a relationship with other MEs, unless all of the related MEs are also being deleted. As a general rule, the ONU is not expected to check the integrity of ME relationships. This task falls to the OLT, which is in a better position to know that a missing or meaningless pointer is only a transient step in the middle of a provisioning sequence.

  When models are being constructed, however, the ONU is encouraged to reject messages that attempt to set invalid pointer values. This is a little tricky because although 0 and 0xFFFF are often suggested, there is no well-defined value for a null pointer.

- *Get Next* If an ME supports tables, the OLT reads the table by first performing a get action. This causes the ONU to latch a copy of the table for retrieval. The get response returns, not the table itself but 4 bytes containing the size of the table snapshot. The OLT then issues a series of get next commands to walk through the snapshot. After one minute without a get next against the snapshot, the ONU discards the snapshot.

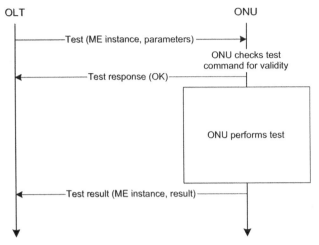

Figure 5.3 Test message sequence.

- *Test* Depending on their definitions, some OMCI MEs support test actions. As would be expected, there is a very wide range of input options and output results, so the related OMCI test and test result messages have a large number of optional formats. Nevertheless, all tests follow a command–response–results pattern, illustrated in Figure 5.3.

    The OLT initiates a test, supplying whatever parameters are appropriate. A test may take considerable time to perform, and OMCI requires a response within 1 s. So the ONU parses the test command and returns a test response message that confirms the validity of the test command but contains no test results. The OMCC and the message stack are then free[*] for whatever else may be needed.

    The ONU performs the test. At some later time, the test is complete, and the ONU originates an autonomous test result message to report the results.

    An ONU may also originate a test result message spontaneously, that is, without a test command from the OLT, if it performs background testing.

A number of additional actions are defined for specialized purposes. Here are a few of them: get all alarms, software download, activate, commit, reboot, and MIB upload. G.988 matches these actions to the specific ME types where they are meaningful. If the OLT issued an action command against an ME that did not support it, the ONU would simply reply with an error code. As for us, we consider these actions below, in the sections where they naturally fit.

---

[*]The baseline message set allows OMCI to have high- and low-priority threads, although implementations typically contain only one. The extended message set recognizes only one thread.

5.1.3.4 OMCI Notifications

The G.988 definition of each ME class includes zero or more tables that define autonomous messages, the messages that can be emitted by an instance of this ME type. We already mentioned test result as one such message.

Attribute Value Change Notifications An AVC notification reports an autonomous change in the value of an attribute that is expected to be of interest to the OLT. Not all autonomous attribute changes meet this criterion; G.988 specifies the attributes whose changes are reported. The AVC message includes the new value of the attribute, except that if it is a table, the message reports only the (new) size of the table.

Alarms An OMCI alarm reports the onset or clearance of an off-normal condition.

The term *alarm* is really a misnomer. When a telecommunications network element reports an alarm, it classifies it according to severity, a value from the set {critical, major, minor, not alarmed}, and usually illuminates a system alarm summary indicator and activates a transducer (e.g., a bell or horn), varying by severity. It expects the alarm to be acknowledged by onsite craft, if any there be, or by the EMS. That is what an alarm is.

An ONU is not a network element: it is a component of a distributed network element. Although they are called alarms, the ONU really reports events. The OLT has the responsibility to assess, filter, and classify event reports and, if appropriate, to convert them into alarms. But the recommendations call an event report an alarm, and so shall we.

Many OMCI MEs are defined with the ability to declare alarms. To avoid excessive traffic in the presence of intermittent faults, it is common to soak defects before declaring them as alarms, typically for 2.5 ± 0.5 s. Before clearing such an alarm, the typical soak interval is 10.5 ± 0.5 s. OMCI does not specify which defects are subject to soaking but does specify that, if soaking is to be done, it is the ONU's responsibility, not the OLT's. This is consistent with the general principle of filtering inputs as early as possible.

Each reportable alarm is represented by a specific bit in a map of 224 bits. The first 208 bits are available for specific alarm types, as assigned by ITU-T; the final 16 bits are available for vendor-specific use. And no, no ME class comes anywhere close to needing 224 different alarm types.

When an alarm is declared or cleared, the ONU notifies the OLT with an autonomous alarm message that includes the 224-bit map of the originating ME instance. The OLT must look for changes, either bits newly set to 1 to represent new alarms or bits newly set to 0 to represent cleared alarms.

ALARM REPORTING CONTROL Alarm reporting control (ARC) is defined in ITU-T M.3100. OMCI supports ARC on selected ME classes: cardholders, the ANI (the PON port), and the physical ports on the subscriber side (UNIs). The idea of ARC is that a manager may wish to suppress alarms when the alarms would merely create clutter in the trouble tracking system.

To illustrate, imagine that we preprovision a backplane slot (cardholder ME) to contain a POTS circuit pack, but we do not plan to physically install the pack until next week. When provisioned to expect a circuit pack that is not present, the normal behavior of the cardholder is to declare a plug-in circuit pack missing alarm. In the operator's NOC, this alarm is not useful. In some sense, it is a false alarm because there is no defect that requires correction. What is worse, the clutter of spurious alarms may distract attention from real alarms that do require action.

By setting ARC, we clear the alarm—assuming it has already been declared—and leave the slot quietly preprovisioned and ready to accept its circuit pack. Next week, we install the physical circuit pack. The ONU checks that it is the correct circuit pack and that it is healthy and autonomously clears ARC. This is sometimes called an automatic in-service feature.

To close the loop with the manager, the ONU sends an AVC notification that ARC has been cleared. Thereafter, the slot and card are fully enabled to report any alarms that may be appropriate.

If we were to install the wrong circuit pack, or if the circuit pack were defective, the ONU would remain in the ARC state. On the one hand, this would give the installer a chance to correct the problem before an alarm was declared; on the other hand, the operator's practices must be designed to ensure that the absence of an alarm is not taken as evidence that everything is correct.

The ARC interval attribute allows a certain amount of fine tuning. If we wish, we can specify that the ME must be free of faults and remain free of faults for some amount of time before the ONU cancels ARC. The ARC interval is more useful for intermittent transmission failures than for equipment diagnosis, but for uniformity, OMCI always provides both attributes as a pair.

The ONU, of course, knows when there is an off-normal condition, whether ARC is set or not; it just does not report it in an alarm message. An option in the alarm retrieval process is to include or exclude alarms that are currently being suppressed by ARC.

ALARM RETRIEVAL Each autonomous alarm message includes an incrementing sequence number, which the OLT can track. If it misses a sequence number, or for any other reason, the OLT may query the ONU to determine the current status of all alarms.

Alarm retrieval bears considerable resemblance to table retrieval. The OLT issues a *get all alarms* message, in which it specifies whether it wishes to see alarms that are under ARC or not. The get all alarms message causes the ONU to reset the alarm sequence number to 0 and to latch a snapshot of the alarm status of all ME instances that are capable of generating alarms. The ONU responds with the number of ME instances in the snapshot. In the baseline message set, this corresponds to the number of upload commands needed, whereas the extended message set can report a number of ME instances in each variable-length message.

The OLT executes a series of *get all alarms next* commands until it exhausts the list; if 1 min elapses without a get all alarms next command, the ONU discards the snapshot.

The OLT then inspects the alarm image, not only to detect which alarms are currently active on which ME instances but also to learn which previous alarms have cleared.

Threshold Crossing Alerts   Threshold crossing alerts (TCAs) report performance monitoring threshold crossing events. The ONU reports a TCA with an alarm message; this is unambiguous because no ME class declares both alarms and TCAs.

PERFORMANCE MONITORING   PM is the process of collecting information about the health and welfare of some aspect of the ONU or its interfaces. This information is defined in specific ME classes, which are usually implicitly linked to parent MEs through common ME IDs. OMCI also defines an extended PM ME format, which can support greater flexibility. At the time of writing, G.988 included only one extended PM class, the Ethernet frame-extended PM ME.

All G-PON PM is collected in 15-min increments. An ONU is not required to contain a ToD clock, but if it does, PM collection is synchronized to quarter-hour boundaries. The OMCI synchronize time message may be used to set the ToD clock.

Each PM attribute is represented by a current value and one historical value. The historical value is that of the attribute at the end of the previous 15-min interval. Each time the 15-min interval expires, the current and historical registers switch places. Another way to think about it is that the newly complete current register is copied into the previous historical register, whose old data is discarded, whereupon the current register is reinitialized. Equivalently, the mechanism can be considered to be a first-in–first-out pipeline of depth 2.

Against a PM ME, the OMCI get message retrieves the historical register. OMCI also defines an optional get current data OMCI message, which retrieves the current values of PM attributes.

THRESHOLDS   Most performance measurement attributes are counters, for example, counters of bytes received, of errored seconds, or of uncorrectable FEC errors. Some attributes support provisionable thresholds; some do not. In our list of examples, there would be no purpose in defining a threshold on bytes received, but if one of the other two counters were to exceed some provisionable value, we would want to know.

At the beginning of a PM accumulation interval, each counter is initialized to zero. If, during the interval, the count reaches the provisioned threshold value, the ONU declares a TCA. OMCI does not have the concept of a transient notification. At the end of each PM accumulation interval, all TCAs are explicitly cleared through follow-on messages.

A few PM attributes are not counters. They may be high-watermark registers, in which case the idea of a TCA still makes sense. Other PM attributes are averages, for example, packet loss rate. These attributes are computed over the duration of the PM interval and are only meaningful in their historical incarnation. If a threshold is defined against an attribute of this type, the TCA can only be declared at the end of the interval, whereupon it is immediately cleared as the next interval begins.

5.1.4 OMCI State Model

The following pair of states appears in quite a number of OMCI-managed entities, and the following discussion pertains to most appearances. Exceptions and extensions are defined in the G.988 documentation of each affected ME type.

Be aware that the ITU-T state model differs significantly from the model common in IEEE or IETF. G-PON follows the ITU-T state model defined in X.731, although only a subset of the full X.731 model is needed. The most visible difference between X.731 and the IEEE/IETF convention is that OMCI does not use the values *up* or *down* to indicate state. Similarly, the values *enabled* and *disabled* refer to the operational state in the ITU-T state model, where IEEE/IETF often use them for administrative control:

1. *Administrative State*   The administrative state reflects the manager's authorization of an entity to perform its service delivery function. If the entity is *locked*, it does not deliver subscriber services, although it remains active for management purposes.

When it is locked, an ME clears alarms that may exist at the moment of lock and declares no new alarms. The ME still recognizes off-normal conditions and may declare them as alarms when it is unlocked. It is encouraged to power down hardware controlled by locked MEs, for example, physical ports. But because a locked ME is still expected to be available for management purposes, there are limits to how far this can go. For example, we would not want ONU or ANI lock to power down the ONU's PON interface.

A certain amount of nested dependency is desirable. That is, if a circuit pack is locked, its ports should cease delivering subscriber service, declare no alarms, and power down as much of their circuitry as is feasible. Likewise, when the entire ONU is locked, it delivers no subscriber services from any port, declares no alarms, and powers down as much of its (non-ANI) circuitry as possible. G.988 suggests that the ONU should be administratively locked as an installation default, to conserve energy if it should happen that the ONU is installed before service is turned up, for example, in a new housing development.

When the manager (OLT or EMS) locks an ME, it immediately ceases to deliver subscriber services. This can be undesirable, especially in voice telephony, where we would almost never wish to disrupt a call in progress. In OMCI, the X.731 *shutting down* state is defined only as a POTS option. When the manager sets the administrative state of a POTS termination to shutting down:

- If there is no call in progress, the ONU sets the state to locked.
- If there is a call in progress, the ONU sets the state to shutting down. As soon as the call terminates, the ONU automatically changes the state to locked.
  In either case, the ONU sends an AVC when it changes the state to locked.

An *unlocked* entity is administratively permitted to deliver subscriber services. Whether the ME is capable of delivering service depends on factors such as its health or its dependency on some other ME that may be locked, missing, or defective. The ability to deliver service is a completely independent matter, signaled through the operational state.

2. *Operational State*   The ONU autonomously sets the value of this attribute to one of the values {enabled, disabled}. Operational state indicates whether the ME is capable of delivering service to subscribers, that is, whether it is defective, perhaps absent, or perhaps out of service because of the failure of some parent entity. The combination of administratively unlocked and operationally disabled states means that the ME is allowed to deliver service but is unable to do so. An operationally enabled ME is fully capable of delivering subscriber service but may be inhibited from doing so by administrative lock. If we observe the existence of subscriber service from some ME, we can deduce both that its administrative state is unlocked and that its operational state is enabled.

X.731 allows distinction between partial and total incapacity; OMCI could also support finer granularity than just enabled and disabled if necessary, but the need has not arisen.

5.1.5   Graphical Conventions in OMCI

As illustrated in Figure 5.4, this book follows the graphical conventions of G.988. Rounded boxes represent managed entity instances that are created and deleted by the OLT, while square boxes are MEs that are created by the ONU itself. Boxes with cutout corners designate performance monitoring MEs, intentionally made less visible because they are rarely crucial to the story being told in a pictorial representation. To facilitate a grasp of the essentials, we frequently omit the PM entirely.

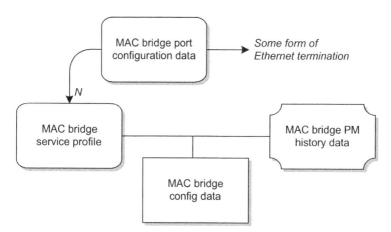

Figure 5.4   Management model notation.

MEs may be related to one another through pointers, shown as arrows, with a quantifier that indicates how many associated instances are permitted. For example, Figure 5.4 indicates that a MAC bridge [service profile] may be modeled with any number N of ports. Each MAC bridge port managed entity instance includes a pointer attribute that refers to its parent bridge. MEs may also be implicitly related through identical ME IDs, shown through lines without arrows.

Because of the undeniable similarity, it is tempting to think of these diagrams as signal-flow representations. All of us, including ourselves in this book, occasionally yield to this temptation. But when a signal-flow interpretation makes no sense, be aware that the diagrams portray only the relationships among managed entities.

Likewise, it is often convenient to speak of an ME as if it performed some function. To be precise, an ME is a collection of information and relationships that causes some engine in the ONU to behave in a certain way. With the exception of the physical equipment and port models, it is not expected that an ONU would contain any discernable module that corresponded to the ME itself.

5.2 EQUIPMENT MANAGEMENT

Before we can ask an ONU to do anything useful for a subscriber, we must install and maintain it as a piece of equipment. Several MEs contribute to this task, as we discuss throughout this section. In addition to explaining equipment management, this section is a vehicle to dig one layer deeper into the OMCI management model.

5.2.1 ONU-G, ONU2-G, and ONU-E Managed Entities

These MEs model the ONU itself. They define some of the ONU's basic characteristics. Two MEs are defined for G-PON, necessary because of the OMCI limit of 16 attributes per ME type. The ONU-E is a subset ME defined for use by G.986 Ethernet point-to-point access networks.

Available attributes include:

Vendor ID, Serial Number   These attributes contain the same information that is supplied by the ONU during the discovery process. They allow verification of the ONU's identity at the OMCI layer.

Version, Equipment ID, Vendor Product Code   These attributes may be used in vendor-specific ways to identify the specific type and hardware release level of ONU.

Battery Backup   If the ONU includes a backup battery, this attribute enables or disables its alarm monitoring feature. Although G.988 does not specify it, it would make sense for this attribute to control the AC power alarm as well.

ONU Survival Time   This attribute represents the manufacturer's commitment of the minimum time between the loss of external power and the silence of the ONU. This time is attributable to energy storage (a capacitor) explicitly provided in the ONU so that the ONU has time to signal a dying gasp before dropping off the PON. The OLT may use this attribute to schedule upstream PLOAM or OMCI opportunities.

The ONU survival time attribute is not intended to include survival time attributable to a backup battery, which is virtually impossible to predict, and with a value of several hours, would not be useful in the OLT's scheduling algorithm anyway.

Traffic Management Attributes One attribute indicates whether the ONU hardware design supports priority-controlled upstream scheduling, rate-controlled scheduling, or both. The QoS discussion in Section 6.3 goes into these matters in considerable detail.

System Uptime This attribute counts 10-ms intervals since the ONU was last initialized (see RFC 1213). Its value is used in some of the other ONU functions, specifically configuration fault management (see IEEE 802.1ag and Section 6.1.3).

States Administrative and operational states are described above.

Registration Credentials Three attributes provide a way for an ONU's registration ID to be conveyed to the OLT through OMCI, rather than through a PLOAM exchange, and for the OLT to signal acceptance of the credentials or otherwise.

Other attributes report the maximum number of GEM ports that can be supported by the ONU, the number of upstream priority queues and the number of traffic schedulers that are available in the ONU's hardware, whether queues and schedulers are fixed in the hardware architecture or whether they can be flexibly associated with arbitrary ports, whether their discipline is fixed or can be provisioned to be either strict priority or weighted round-robin (WRR) queuing, whether their priorities are fixed or provisionable, and more. These attributes report the characteristics of the ONU as a whole; a chassis-based ONU could have additional queues and schedulers as components of individual circuit packs. Section 6.3.2.1 explains these choices.

Actions As well as the usual get and set actions, the ONU-G ME supports some specialized functions:

Reboot Reboot the ONU.

Test Test the ONU as a whole. As outlined above, different ME classes support different test message formats. The format for the ONU ME is completely nonspecific: it invokes whatever tests the vendor deems appropriate for the ONU as a whole. An optional octet string pointer is available in the test command, allowing the ONU vendor to pass vendor-specific arguments into the test function. The test results are equally nonspecific, a single pass, fail or not completed indication, along with an optional general-purpose buffer that contains whatever information the vendor thinks appropriate.

Synchronize Time We briefly mentioned this command in the TCA discussion above. The ONU is generically happy to run PM on arbitrary 15-min boundaries; it includes no time-stamped events.

From the operator's perspective, it is usually desirable for PM intervals to coincide with real-world quarter-hour boundaries. But ONUs are not required to have ToD clocks. The basic synchronize time message synchronizes the start time of all performance monitoring managed entities of the ONU with the time at which the command is sent. All counters of all performance monitoring managed entities are cleared to 0 and restarted. This basic function is intended only to establish rough 15-min boundaries for PM collection and does not establish a clock time.

Various other ONU functions would benefit from a ToD clock, however. Nanosecond-precision ToD synchronization is a separate function, described in Chapter 4.[*] There is a middle ground, however, the desire to have a time of day within a few hundred milliseconds of the true time, suitable for event time stamps. The synchronize time message therefore includes an option to convey time of day, with unspecified but loose accuracy. "At the sound of the synchronize time message, the time will be . . . ."

Alarms The ONU ME class can declare alarms (event reports, really) to signal a variety of problems:

- Equipment failure, ONU self-test failure. It is not forbidden, but it would perhaps not be expected, to declare these alarms as the result of an ONU test command issued from the OLT. The purpose of an alarm is to draw attention to a problem; a test command implies that the ONU is already receiving attention.
- Powering alarm, loss of external power to backup battery charger.
- Battery missing, battery failure, battery low.
- Physical intrusion, reporting an open door on an enclosure.
- DG ONU manual power off. The DG alarm signals that the ONU has lost power and expects to disappear soon. It is not a commitment to drop off the PON, however: power may be restored and the ONU may remain in service. In G.987 XG-PON, DG is also signaled with a bit in the upstream burst, which may be faster than waiting for an OMCI transmit opportunity. In G.984 G-PON, DG is also signaled with an upstream PLOAM message.

  The purpose of these alarms is to assist with possible fault isolation by indicating, first that there is no problem with the ODN or the ONU's PON interface and, second, whether the ONU was intentionally powered down by the subscriber or was the victim of a power failure. These indications can help resolve trouble calls and avoid unnecessary truck rolls.

---

[*]TC-layer ToD synchronization also relies on OMCI, specifically the OLT-G managed entity. For details, consult G.988 clause 9.12.2.

- Temperature yellow, temperature red. The ONU is operating beyond its recommended temperature range and either has (red) or has not (yellow) shut down some services to avoid equipment damage.
- Voltage yellow, voltage red. The line power voltage is below its recommended minimum. Service restrictions are in effect (yellow), such as permitting no more than N lines off-hook or ringing at one time, or (red) some services have been shut down.

5.2.2 Cardholder, Circuit Pack, and Port Mapping Package MEs

This family of MEs models the ONU as a chassis with replaceable circuit packs (cards) that can be plugged into cardholders (slots). An integrated ONU—that is, an ONU without field-replaceable modules—may choose to model itself as a set of virtual slots and circuit packs.

When the ONU has a physical backplane with slots and pluggable circuit packs, the cardholder ME attributes and actions allow for a slot to be empty, to be preprovisioned, to be equipped with the right or the wrong circuit pack, or to accept whatever circuit pack is plugged in (plug and play option). Cardholders support ARC.

Circuit packs may be generically identified through a list of enumerated types (POTS, xDSL, etc.) but usually must be further identified with more specific vendor information: the simple enumeration, for example, cannot distinguish a four-port xDSL card from an eight-port card.

When a card is preprovisioned, or when a plug-and-play slot is equipped with a suitable card, the ONU automatically creates instances of the necessary MEs for that particular card type. These include ANI or UNI ports, along with supporting queues, schedulers, T-CONTs, possibly software images, and anything else unique to that particular circuit pack type. When a circuit pack is deleted from a slot by management action—not the same thing as unplugging a circuit pack for maintenance—the ONU automatically destroys the pack-specific MEs.

Like the ONU in its entirety, circuit packs contain manufacturer information such as equipment identity, version, and serial number, along with capability information such as the number and flexibility of T-CONTs, traffic schedulers, and queues. Circuit packs support administrative and operational states.

Whether the ONU has physical or virtual cardholders, the model expects that at any given time, each cardholder contains ports of a single type, for example, fast Ethernet UNIs. Especially in a chassis-based ONU, this is not necessarily true: for example, the controller card of an MDU might include the PON interface (ANI), a craft port, and an RF video port or a GbE UNI.

The port-mapping package ME provides a way for a given physical or virtual cardholder to include an arbitrary set of port types. It also allows a physical port to be associated with more than one physical path termination point ME, a situation that arises in the reach extender model, where the function of upstream or downstream PON replication can be paired with an optical amplifier on the same physical port.

The ONU vendor has the choice of how to model complex equipment, for example, with virtual cardholders or port-mapping packages. The OLT is expected to conform to the ONU's architecture. This can impose unrealistic expectations on the OLT, or to be precise, on the OLT vendor's expectation of a return on software development investment. The G-PON community strongly advocates interoperability, but its focus is on simple (single-family) ONUs; when more complex ONUs are under discussion, the cost–benefit tradeoff is recognized. If an operator were to require multivendor interoperability of complex MDUs, it would probably be achieved through special contractual and engineering arrangements among the companies directly involved.

5.2.3 Equipment Extension Package ME

The ONU automatically instantiates this managed entity, either for the ONU as a whole or for a particular circuit pack that supports the feature. The equipment extension package allows an ONU to support external sense points that can be wired up to arbitrary sensors. These sensors might include, for example, water level in a handhole application or telemetry points from other equipment such as a rectifier plant or emergency generator that may be installed nearby. The function is very simple: the ONU reports a change in a sense point as an alarm. It is up to the OLT or the EMS to interpret the report.

The same ME allows the manager to control ONU outputs, typically contact closures that might be wired up to actuators of one kind or another. Although the usual list of examples includes sump pumps and environmental conditioners, it would be natural to expect functions such as these to be automatically activated, rather than awaiting external control. Conceivably the sump pump, for example, could be remotely activated as a test, in the expectation that the ONU would signal sump pump active on a related sense point.

5.2.4 Equipment Protection Profile ME

This ME supports equipment protection, for example, a DS1 circuit pack backing up several other DS1 packs in a business application, with revertive behavior or not, and a provisionable soak interval before reversion.

5.2.5 ONU Power-Shedding ME

Depending on its design, the ONU may or may not know whether AC power has failed, or to put it another way, whether it is operating on its backup battery. If it knows, the ONU can shed services selectively to prolong its availability on the backup battery. Each type of service—data, voice, video, and others—can be provisioned in the power-shedding ME with a hold time. If the hold time elapses without recovery of AC power, that service is shut down, thereby extending the availability of higher priority services. If the ONU delivers POTS, POTS is usually regarded as the highest priority.

Individual ports can be declared to be absolutely essential, full stop, and exempted from the timed shutdown that pertains to other ports of the same type. This feature might be used to keep one POTS port alive for emergency use, after having had to abandon all other services on the ONU. As another example, an Ethernet UNI used for mobile backhaul might be kept alive, even after other Ethernet ports on the same ONU were shut down.

Depending on the hardware granularity, it may not be possible to shed power or retain power on individual ports, in which case the ONU is expected to keep service up on the essential port, even if that means some nonessential ports also remain powered up.

5.2.6 ONU Dynamic Power Management Control ME

Energy conservation for XG-PON is defined in G.987.3 and discussed in Section 4.5 of this book. The ONU dynamic power management control managed entity provides the provisionable intervals that support the ONU's doze and sleep functions.

5.2.7 OLT-G ME

This managed entity identifies the type of OLT to which an ONU is connected. It has been regarded askance, since the OLT is supposed to be able to adapt itself to the ONU, rather than the other way around.

However, the OLT-G ME has gained respectability as the parent of the ToD attribute, which permits the OLT to specify a date and time stamp that corresponds to the start of a specified (super)frame count. Section 4.3.2 describes the process of getting time of day across the PON; this ME is where it happens.

5.2.8 ONU Data ME

The ONU data ME has only one attribute, the MIB data sync. The idea is that whenever the OLT makes a change to the ONU's MIB, both parties independently update their copies of the MIB data sync. At any particular time, the OLT can check the ONU's MIB data sync version; if it agrees, the OLT has some level of confidence that it is in sync with the ONU.

Although it has only one attribute, the ONU data ME supports a long list of actions.

MIB Upload, MIB Upload Next   At ONU initialization, and at any time during the ONU's session lifetime, the OLT can audit the MIB. The MIB upload OMCI command causes the ONU to latch a snapshot of its MIB, which the OLT then uploads through a series of MIB upload next actions.

The MIB snapshot does not include ephemeral information such as PM counters,[*] and it does not include tables. A few other MEs and attributes are also excluded from MIB upload, as explicitly defined in G.988.

---

[*]But it does include PM threshold information.

The MIB upload next response identifies an ME instance and sends as many attributes of that instance as will fit in the message body, identified with the usual 16-bit attribute mask. The next MIB upload next response continues the attributes, if necessary. In the baseline OMCI message set, a single message is not allowed to contain attributes from multiple ME instances. In the extended OMCI message set, the next ME instance can begin immediately, if there is enough space in the message envelope.

The sequence of MIB upload next commands and responses eventually walks the entire MIB, in whatever order is determined by the ONU's snapshot. When the operation is complete, the ONU responds with a message filled with null bytes (baseline) or with a zero-size message contents field (extended).

If the OLT issues no MIB upload next command for 1 min, the ONU discards the MIB snapshot.

MIB Reset   Unconditionally at start-up, or subsequently as the result of a MIB audit, the OLT may issue a MIB reset, which destroys the ONU's provisioned information, restores factory defaults, and resets the MIB data sync attribute to 0. The OLT then reprovisions everything.

Get All Alarms, Get All Alarms Next   A secondary use of the ONU data-managed entity is to permit the OLT to upload all alarms on the entire ONU, as described above.

5.2.9 Software Upgrade

The software image ME supports the function of software upgrade. By default, it pertains to the ONU as a whole, but if circuit packs contain independently maintainable software, the same ME is used to manage them. Software can even be downloaded to several such circuit packs in parallel. The possibility of separately manageable circuit pack images has finite value: in a chassis-based ONU, it is likely that the controller would contain an archive that included all possible secondary packs at the correct release level. The circuit packs themselves would not be independently manageable but would instead be slaves of the controller pack. Upgrade of the controller pack would implicitly include upgrade of the other packs.

The model for each software scope, be it the ONU as a whole or a circuit pack, is that there are two images, each represented by an instance of the software image ME. The attributes of the software image include product code and version fields, for vendor-specific use. The current status of the image is available through three additional attributes:

- *Is Valid*   An image may be valid or not, according to criteria that include at least CRC validation of the image but may also encompass vendor-specific checks, for example, of embedded information within the image file itself.
- *Is Active*   At any given time, one of the images is expected to be active (executing)—the exception would be if both were invalid, in which case the ONU would presumably be running from some very minimal PON bootloader.

- *Is Committed* Normally, one and only one image is committed, which means that it is the image to be loaded and executed when the ONU next reboots. The committed image must be valid but not necessarily active. Under exceptional conditions, for example, both images invalid, it would be possible for neither image to be committed. It is never allowed for both to be committed.

Actions The software image ME supports a number of specialized actions related to software upgrade. To briefly summarize the process:

- When an ONU is to be upgraded, the OLT first ensures that the ONU is active on the committed image and addresses itself to the other image.
- The OLT downloads the new image.
- The OLT asks the ONU to validate the new image and store it in nonvolatile memory.
- The OLT activates the new image (reboots the ONU), without having committed it.
- Assuming that all goes well, the OLT then commits the new image, a command that automatically decommits the previous image.

Start Download, Download Section Image download involves a bit more detail, primarily because the baseline OMCI message set can only download 31 bytes at a time, a block called a section (Fig. 5.5). To expedite the process, up to 256 sections are grouped into a so-called window; the size of the window is negotiated between ONU and OLT at the beginning of the process.

Each section includes a sequence number, starting at zero with each window. In each window, the ONU, therefore, expects to receive sections sequentially numbered from zero. If a download error occurs, the most likely symptom is failure to receive one or more sections.

Most OMCI messages from the OLT are acknowledged by the ONU, as indicated by an acknowledgment request (AR) bit in the message header. However, the OLT only

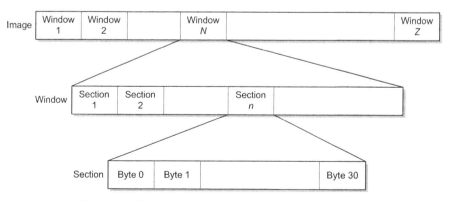

Figure 5.5 Software image structure for (baseline) download.

requests download acknowledgment from the last section in a window; and in fact the AR flag signals completion of the window. The OLT is free to send fewer sections in a window than were negotiated, but it is not permitted to send more.

Based on the acknowledgment at the end of the window, the OLT learns whether something has gone awry in one of the preceding sections. If so, the OLT retransmits the entire window. We have a choice between fewer acknowledgment handshakes when things go well, versus larger retransmission blocks when there is an error.

End Download   After all windows of the image have been downloaded (the final window possibly being abbreviated), the OLT sends the end download command, which causes the ONU to validate the image and store it in nonvolatile memory. When the ONU acknowledges that the image is valid and persistent, the OLT can go ahead and activate, then commit the image.

The ONU is always required to respond to commands within one second, and it may take more than one second to store the image in persistent memory. If the storage operation is not complete, the ONU responds with a device busy code, whereupon the OLT retries the end download command, preferably not instantly.

The process is the same in the extended OMCI message set, although with the longer messages, it is considerably less painful. A section can be as large as will fit the extended message set envelope. Downloading images 31 bytes at a time was in fact a primary motivator for the development of the extended message set.

OMCI also offers an alternative tool for download, a tool that allows the ONU to contact a server and download an image over the data plane, using OMCI only to set up the transfer. The server could be the OLT itself or could reside anywhere in the accessible network. This is one application of the file transfer controller ME.

Once the image is downloaded, it follows the same sequence of activation and commit steps as if it had been downloaded via OMCI.

Activate Image, Commit Image   These commands have already been explained.

Software Upgrade, Vendor-Specific Use   Managed entity IDs that match actual or virtual slots (cardholders) are reserved for software management of the ONU or its circuit packs, along with instance numbers 0 and 1. This is a small part of the 2-byte ME ID space.

Because it is built into vendor software and operator practices, there are cases when the download–activate–commit process may be useful for other purposes. Two examples illustrate the possibilities.

- Configuring VoIP service parameters. It is common for each country or, in some cases, each operator to have its own unique POTS specifications and defaults. Without the ability to customize POTS parameters, the vendor must build a complete set of all possibilities into the software release. Each new market requires a new software release, and over the course of time, the size of the release image grows. The use case for the vendor-specific feature is that, when an ONU is sold into a new market, its configuration can be defined for that particular operator and downloaded onto an existing software release. The release itself need contain nothing more than some minimal default configuration.

- Configuring xDSL equipment in which settings or even entire firmware algorithms may be unique to a given chip set, and well ahead of standardization in OMCI, or possibly beyond the ambition of OMCI.

OMCI allows the nonreserved ME ID address space of the software image ME class to be used for vendor-specific functions such as these. Although they were not part of the initial list of use cases, further possibilities also exist. One that may be particularly useful is the bulk upload or restoration of ONU MIBs as files. If a MIB were contained in a file, it could be archived, analyzed, even modified or constructed from scratch in some remote location such as an EMS.

5.3 REACH EXTENDER MANAGEMENT

Reach extension for G-PON is defined in G.984.6. XG-PON reach extension is defined in G.987.4. At the time of writing, G.987.4 existed only in draft form but contained no surprises.

Much of what has been said about ONU equipment management also pertains to reach extender equipment management. However, an RE is sufficiently unique to warrant additional discussion that focuses on the differences.

Reach extenders are outlined in Section 2.4 and expanded in detail in Section 3.12. A reach extender is remote equipment, usually but not necessarily located at the remote power splitter, whose purpose is to increase the effective optical budget of the PON. The extra budget may be used to extend the reach of the ODN, to increase the split ratio, or both.

There are two simple reach extender architectures, explained in considerable detail in Chapter 3: OEO and OA. Figure 5.6 illustrates the case when both directions of the RE employ OEO regeneration.

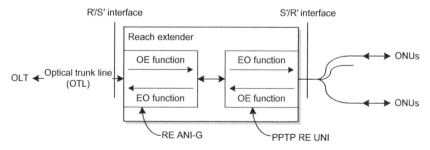

Figure 5.6 OEO reach extender.

The pertinent OMCI managed entities are the RE ANI-G and the physical path termination point RE UNI. Although we show only a single PPTP RE UNI, it is also possible to have an internal electrical split, with several RE UNIs, each serving its own optical splitter and set of ONUs.

The RE ANI-G models the transceiver at the R'/S' interface. As well as the expected administrative and operational states and ARC attributes, we find attributes to measure optical transmit and receive levels and to set optical alarm declaration thresholds. It is expected that a composite RE may extend several PONs, each on a separate wavelength, so the RE ANI-G also includes attributes for downstream and upstream frequency (wavelength).

Finally, one of the PONs extended by an RE is used for the RE's own management channel, and we find a usage mode attribute to declare whether this is the one.

On the drop side of the RE, the PPTP RE UNI is also basically a model of a transceiver. One or more PPTP RE UNIs points to an RE ANI-G. We find the expected set of states and optical layer attributes and a pair of read-only attributes that represent the ONU's need for extra interburst guard time or extended burst preamble.

The other basic technology possible in a reach extender is optical amplification (OA) in both directions. Figure 5.7 allows for the possibility of one or more optical amplifiers on the ONU side, much as we saw in the OEO case. The idea is that a single amplifier would suffice in the downstream direction, but each of several splitters may have its own upstream amplifier.

The pertinent OMCI managed entities in this case are the RE upstream and downstream amplifiers. The RE downstream amplifier contains the familiar state, ARC and optical layer attributes, along with optical power measurement values and alarm thresholds. The RE upstream amplifier contains these attributes as well, a pointer to an RE downstream amplifier, and a control for constant gain, constant power, or autonomous operation.

Hybrid RE architectures are possible in which one direction is electrically regenerated while the other direction is optically amplified. Because of the attributes they carry, G.988 insists that a complete set of RE ANI-G and PPTP

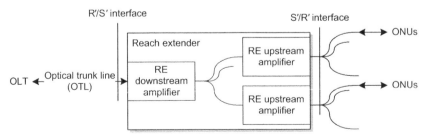

Figure 5.7 OA reach extender.

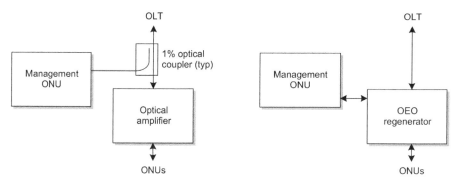

Figure 5.8 Reach extender management ONU.

RE UNIs exist if OEO regeneration is supported in *either* direction. This creates a bit of a problem in the OMCI model because a physical port is normally associated with only one ME. Now we find an OLT-side physical port associated with both an RE ANI-G and an RE downstream amplifier or an ONU-side port associated with both a PPTP RE UNI and an RE upstream amplifier.

This required an extension to the OMCI port-mapping package ME definition, to allow a given port to be associated with more than one termination type. The primary termination type is recommended to be the OEO ME, whose attributes such as administrative state and optical alarm thresholds take precedence over the corresponding attributes of the OA type.

As shown in Figure 5.8, RE management is performed by a logical ONU function contained within the RE (called an EONT or EONU in the recommendations). If the RE has an optical amplifier on the OLT side, the management ONU function is fed from the trunk fiber through a tap that drains off only a very small amount of light, typically 1%; conversely in the upstream direction.

The irony of this model is that, while the RE physically contains the management ONU, the management model deems the ONU to be the fundamental entity, so that logically, the management ONU contains the RE. All equipment aspects of the RE are managed as if they were contained within the management ONU. By the way, it is perfectly legal for the management EONU to have a few subscriber ports delivering end-user services directly from the RE. If the RE terminates multiple fibers or multiple wavelengths, one fiber or wavelength is deemed to be primary, the one that supports the EONU. Logically, it is just one of the ONUs on its PON.

Managed Entity Details   A few further details about the RE MEs:

- *Administrative State*   As would be expected, administrative lock causes the blocking of subscriber traffic and the preservation of management communications. Administrative lock on a PPTP RE UNI causes a loss of signal to (and from) downstream ONUs, which are regarded as clients of the RE, rather than as targets for management communications that ought to be retained. Locking a primary ANI, or the only ANI, does not cause upstream loss of signal at the

OLT because the management channel needs to be kept alive. However, locking a secondary RE ANI-G in a multi-PON or a protected RE, that is, an ANI that is not carrying the RE OMCC, would disable and power down its PON interface.

- *Received Optical Signal Levels* Together with provisionable alarm thresholds, these are useful in diagnosing incipient problems. The model in the upstream direction includes the ability to record a power measurement from each ONU individually, with the idea that updates occur continuously at whatever rate is possible in the RE's architecture. During the burst from a given ONU *i*, the RE triggers a power measurement, then stores it at row *i* in a received power attribute table.

For this to be possible, the RE needs to understand which ONU is generating a burst at what time. The RE, therefore, needs to parse the bandwidth map and compensate for the round-trip delay. Because the RE is located closer to the ONUs than the OLT, the delay compensation performed by the OLT is not correct from the RE's perspective. The RE may need to snoop downstream ranging_time PLOAM messages and offset them with its own measurements, derived by monitoring upstream serial_number_ONU or registration PLOAM messages. This complexity is, of course, the reason that, as described in Chapter 3, RE receivers are resetless.

5.4 PON MAINTENANCE

Let us return from the reach extender to the wider scope of the entire G-PON, where we observe that G-PON faults can be hard to diagnose.

The point-to-multipoint nature of the PON is part of the reason, not only because it is difficult to isolate failures but because it is undesirable to take the entire PON out of service for diagnosis, as long as some of the subscribers still have service.

Another contributing factor is that, for economic reasons, a G-PON is often engineered very near the limits of the optical budget, either in terms of distance or split ratio or both. Relatively minor changes in the ODN or in the devices at one end or the other can consume the limited margin and cause failures on a PON that formerly worked well.

G-PON equipment is never pampered. It may be deployed in severe environments, it may be powered remotely, it may be vulnerable to tampering or vandalism. Diagnosis is complicated because the failure modes include pretty much anything imaginable.

Finally, a G-PON may support service interfaces to almost every known form of wireline telecommunications interface: Ethernet, DSL, POTS, DS1, E1, and so forth. Each service interface has its own performance and test requirements.

The G-PON standards provide for monitoring, testing, fault identification, and diagnosis at every layer from physical through the ultimate telecommunications service delivered to the subscriber. We outline performance monitoring and alarm declaration in Section 5.1.3.

At the physical layer, OLTs, ONUs, and REs are encouraged to monitor, to collect, to make available for testing, and to report thresholded violations of optical parameters. The pertinent OMCI managed entities include the ANI-G, PPTP video ANI, RE ANI-G, PPTP RE UNI, RE upstream amplifier, RE downstream amplifier, and RE common amplifier parameters. Depending on the ME and the particular parameter, the information may be available as a read-only attribute, as the result of a test action, or as an alarm with or without provisionable thresholds. The same information and behavior is intended to apply at the OLT, but the OLT's information model is beyond the scope of the standards. Physical layer measurements include:

- Transmit optical power.
- Received optical average power.
- Received optical burst power, per-ONU, upstream only. We describe this in the context of REs above. The OLT is encouraged to make the same measurements; it has a considerable advantage over the RE in knowing which ONU's burst to expect when.
- Laser bias current or end-of-life indication.
- Optical module temperature.
- Optical module power supply voltage.

At the bit transport layer, performance monitoring is based on counters:

- BIP error counts in both directions, except that G.987 XG-PON does not have BIP in the downstream direction because FEC is always enabled. G.984 G-PON returns BIP counts from the ONU in periodic remote_error_indication PLOAM messages.
- FEC statistics, when FEC is enabled. Counts of total code words, corrected code words, corrected bytes, uncorrectable code words and seconds during which at least one FEC error was observed. The OMCI FEC PM history data ME is the vehicle to report ONU statistics to the OLT.

G.987 XG-PON includes TC-layer PM. The XG-PON TC performance monitoring history data ME counts a variety of anomalies in the downstream flow, including HEC errors everywhere HEC is used (Chapter 4), grants that specify unknown profiles, or invalid encryption keys. It also counts transmitted GEM frames and fragmented SDUs.

At the PLOAM layer, the G.987 XG-PON ONU collects total PLOAM message count, along with counts of each type of PLOAM message received and MIC errors, using the XG-PON downstream management performance monitoring history data ME. The XG-PON upstream management performance monitoring history data ME counts PLOAM messages transmitted upstream, in toto and by message type.

OMCI can be tracked with counts of baseline and extended messages received, and the number of messages discarded because of MIC errors.

G.987 XG-PON ONU energy conservation can be monitored with the energy consumption performance monitoring history data, which accumulates doze and sleep time and optionally the actual amount of energy consumed.

Chapter 6 describes some of the PM and test operations defined for the various services supported by a G-PON.

5.5 OBSOLETE FRAGMENTS OF INFORMATION MODEL

Technology and markets move forward, and what seemed useful at one time may turn out to be of little interest later on. Here, we mention a few features of the OMCI model that faded away.

Going back to the beginning, recall that the first of the modern PONs was A-PON, ATM PON, which evolved into G.983 B-PON. In these technologies, the PON itself carried ATM, and all payloads were mapped into ATM cells. When G-PON first began to be standardized, it was proposed to include three PON partitions: an ATM partition, a GEM partition, and a TDM partition. Neither ATM nor direct TDM ever progressed into the marketplace. Both were deprecated out of G.984, and were not even considered in G.987 XG-PON. The access network of interest is packet based, with Ethernet frames as GEM payload.

Nor are GEM mappings immune to the forces of history. G.984 defines mappings of IP directly into GEM, and an SDH mapping into GEM. Neither survived into G.987. An MPLS mapping into GEM is defined in the current recommendations; it remains to be seen whether it will prove to be useful.

An OMCI model was defined with the goal of provisioning a home gateway router and an IEEE 802.11 wireless local area network (LAN) at the customer premises. Both of these features are very important in residential access, but as it turned out, no one wanted to manage them with OMCI—they were managed by subscribers directly, or if the operator intended to manage them, BBF TR-69 was a better fit for operators' business and operations models. Both of these features were removed from G.988.

Removing a feature such as the ATM partition from the TC layer means that it really is gone—it is not even supported by the silicon, much less the software. Discontinuing the documentation of OMCI management capabilities is primarily a way to remove clutter from the documentation. The definitions and code points are permanently reserved, however, so that if a new application has reason to resurrect some legacy MEs, it is free to do so. A case in point arose in the m of ATM into pseudowires, where some of the G.983.2 ATM MEs from B-PON turned out to have second lives.

UPDATE

As we went to press, G.988 amendment 2 was being consented in ITU. Changes within the scope of this chapter include:

- Model for PoE (power over Ethernet) control
- Ability to add vendor-specific command result information into padding bytes of OMCI response messages.

6

SERVICES

In this chapter:

- *Ethernet management*
  - o *Bridge model*
  - o *Traffic classification*
  - o *VLAN tagging*
  - o *Forwarding*
  - o *Connectivity fault management*
- *Multicast*
  - o *IGMP/MLD snooping and proxy*
- *Quality of service, within and among ONUs; DBA*
- *IP services*
  - o *IPv4 and IPv6 stack models*
- *POTS*
  - o *Analog telephone adaptor ATA in the ONU*
  - o *SIP and H.248 signaling*
  - o *ATA external to the ONU*
  - o *TR-69 management*
- *Pseudowires*
  - o *Circuit emulation*
  - o *Ethernet and ATM PWs*

Gigabit-capable Passive Optical Networks, First Edition. Dave Hood and Elmar Trojer.
© 2012 John Wiley & Sons, Inc. Published 2012 by John Wiley & Sons, Inc.

The preceding chapters explain the G-PON infrastructure and the basic tools. In this chapter, we finally get to the question of delivering telecommunications services to end users, subscribers. This chapter shows how G-PON and OMCI model these services and manage them. To a considerable extent, service definitions are independent of access network technology, but some aspects are affected by PON as the server layer, and the specifics of the OMCI management model tie this chapter directly to G-PON.[*]

Each telecommunications operator has its own service model, so the services described in this chapter are necessarily generalizations. Common aspects of the system requirements are defined in two standards streams, both of which are recommended for further study.

ITU-T G.987.1 defines the FSAN operators' requirements for XG-PON. G.987.1 assumes G.984 G-PON as a baseline and is to a considerable extent an explicit restatement and extension of G.984 G-PON requirements, some of which are implicit. It is a view that emphasizes the desired continuity with G.984, along with the operators' perceived need for enhancements.

Broadband Forum BBF has been considering the access network and customer premises devices for many years, but only recently in the specific context of G-PON. BBF TR-101 is widely accepted as the baseline set of requirements for migration from ATM-based access to an Ethernet access network. More recently, TR-156 and TR-167 have extended TR-101 from a DSL mind-set into specific requirements on G-PON access networks. If a single model for G-PON Ethernet services can be said to be common across the worldwide operator community, it would be that of TR-156. This is not to say that all operators are 100% aligned with TR-156, but just that TR-156 establishes a very substantial common ground with most operators.

6.1 BASIC ETHERNET MANAGEMENT

If a single technology can be said to pervade all modern networks, it would surely be Ethernet. Ethernet frames can be mapped into any number of server layer protocols, including physical layers defined in IEEE 802.3 itself. Any number of client protocols are able to map their protocol data units (PDUs) into Ethernet frames. Ethernet is defined in IEEE 802.3, which specifies several physical layer options and a frame format. Except when dealing directly with external interfaces, we almost always refer to the frame format meaning of the term, along with layer 2 (L2) functions such as VLAN tagging and MAC bridging.

The MAC bridge is arguably the heart of Ethernet and arguably the heart of the OMCI information model.

---

[*] At the time of writing, one part of the EPON community had also chosen OMCI as its management model.

What Is a MAC Bridge?

The basic idea of a bridge is simple: in the absence of any other information, a MAC bridge accepts Ethernet frames on one port and replicates them to all other ports. A MAC bridge learns the source MAC addresses of incoming frames and thereafter forwards outgoing frames only to ports where the pertinent destination addresses are connected. Only when it does not recognize a destination address does the bridge flood a frame to all ports. Learned address sedimentation is limited by timing out and removing MAC addresses that have not recently been used.

If two bridges are connected together on two or more ports, directly or indirectly, there is a risk that they could forward frames to each other forever. To avoid this, bridges can implement a spanning tree protocol (STP), usually either rapid STP (RSTP) or multiple STP (MSTP). Through STP, bridges recognize and break forwarding loops by disabling one of the links that would otherwise close the loop. If it is known that a bridge cannot be part of a loop, for example, in most—but not all—ONU applications, it is not necessary for it to run an STP.

Then there is multicast, in which we often wish to replicate multicast frames onto several ports based on control actions, and certainly not based on MAC learning. Multicast replication is theoretically based on IP address, but can also be performed at layer 2 with MAC addresses.

Finally, the MAC bridges of interest are aware of VLANs. A VLAN is identified by tags in the Ethernet frame headers. Ports of a bridge may be configured to be part of a particular VLAN. Among these ports, learning, forwarding, and flooding occur as expected for frames that are tagged with that particular VLAN ID (VID). Ports that are not members of the specified VLAN see none of its traffic.

A VID is 12 bits long. Of the 4 additional bits that remain in the 2-byte VLAN tag, 3 are used for priority; we often refer to them as P bits. A tagged frame with VID = 0 is referred to as a priority-tagged frame.

The fourth bit is a canonical format indicator (CFI) bit, whose use in our context is as a drop eligibility indicator (DEI) bit. In case of congestion, a bridge preferentially discards frames with the DEI bit set.

There is more (there is always more!), but this overview may be helpful as a brief review or introduction.

Figure 6.1 illustrates the basic MAC bridge configuration model in OMCI. From the rounded-block notation, explained in Chapter 5, we see that the MAC bridge service profile is created by the OLT, as are the MAC bridge ports. The model allows an ONU to contain any number of MAC bridges, bounded only by the ONU's capabilities.

The MAC bridge service profile managed entity represents the core functionality of the bridge. A bridge can have any number of ports, where each port connects to an Ethernet termination point of some type. Let us examine the MEs of Figure 6.1.

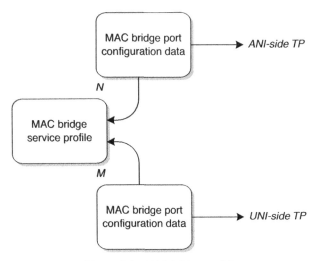

Figure 6.1 MAC bridge model.

MAC Bridge Service Profile This ME models the bridge itself. Its attributes include:

- Spanning tree control, along with values to be used in spanning tree negotiation.
- MAC address learning, enable or disable, with control of MAC learning table depth and MAC address timeout. Many, if not most G-PON applications require only provisioned VLAN forwarding, not MAC learning.
- Bridging between UNI-side ports, on or off. In a single-family ONU, the local bridging feature permits communication among several PCs connected to the ONU's several Ethernet ports. In an MDU, subscribers would emphatically *not* want their home networks to be promiscuously bridged together. But in an MDU, we would probably create a separate bridge for each separate UNI anyway.
- Flooding or discard of frames with unrecognized destination MAC addresses. Although a generic bridge floods unrecognized frames, the context of an ONU MAC bridge model—provisioned VLAN forwarding—is such that flooding may be unnecessary and undesirable.

MAC Bridge Port Configuration Data Each port of the MAC bridge is modeled by an instance of this ME. It includes a pair of pointers, one to link to the MAC bridge service profile, the other to connect with a subscriber-side (UNI) of one kind or another, or to a PON-side (ANI) termination point. Its other attributes include:

- Bridge port MAC address, in case the port has its own. All ports may also share a single MAC address that belongs to the bridge, at the vendor's option.

- The depth of the MAC address learning table on this port. If it is nonzero, it overrides the setting of the same attribute in the parent MAC bridge service profile. Again, an attribute that may pertain to the bridge as a whole, or to the ports individually.
- Pointers to inbound and outbound traffic descriptors, which may be used to mark traffic according to a service contract, based on its arrival rate. We discuss QoS in more detail in Section 6.3.
- Port parameters for use in spanning tree negotiations.

Nothing is ever as simple as it seems. Figure 6.2 shows several additional MEs that can be associated with a MAC bridge.

MAC Bridge Configuration Data   The MAC bridge config data ME is automatically instantiated by the ONU when the MAC bridge service profile is created. It is a read-only ME that contains the bridge's MAC address, along with several attributes that report values from the bridge's spanning tree protocol (RSTP or MSTP) negotiations.

Dot1 Rate Limiter   The OLT may choose to create a dot1 rate limiter (dot1 in the name is an acknowledgment of its parentage: IEEE 802.1), one of whose purposes is to help protect the network from some forms of DoS attacks. The rate limiter

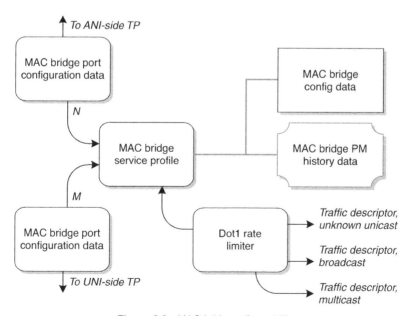

Figure 6.2   MAC bridge adjunct MEs.

provides a way to restrict the amount of traffic allowed in the upstream direction, traffic classified by:

- Unicast traffic with an unknown destination address (traffic that may be flooded within its VLAN)
- Broadcast traffic
- Multicast traffic

Rather than directly specifying parameters for each class of traffic, the dot1 rate limiter contains three pointers to traffic descriptor MEs. The numbers appear in the traffic descriptors, values for committed and peak information rates, committed and peak burst size, and attributes to recognize and mark drop-eligible traffic for possible discard during periods of congestion. We discuss the operation of traffic descriptors in Section 6.3.

MAC Bridge Performance Monitoring History Data   The OLT may choose to create a MAC bridge PM ME. If so, its ME ID must be the same as that of the MAC bridge service profile. This particular ME counts discarded (timed-out) MAC address learning entries.

Many more MEs come along with each port on the bridge (Fig. 6.3), some of them autonomous, some created at the behest of the OLT.

No actual implementation is likely to be (quite) as complex as suggested by Figure 6.3. VLAN tagging is done by one, not both, of the VLAN tagging operation MEs (extended version preferred; older one hallowed by history). Likewise, two forms

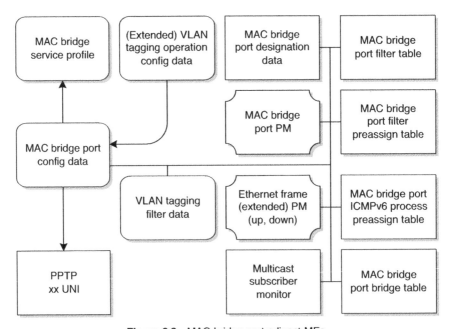

Figure 6.3   MAC bridge port adjunct MEs.

of Ethernet PM are defined; the extended PM is more flexible—it can filter by individual VLAN ID or priority, either upstream or down—but is more complex to provision.

The VLAN MEs in Figure 6.3 receive considerably more discussion later, as does the multicast subscriber monitor. Here, we summarize the key features of the others:

MAC Bridge Port Filter Table Data   This ME is a table of MAC addresses. An action is associated with each entry: provisionably forward or filter (discard) frames with the specified address, as provisionably either source or destination. If the filter table is associated with an ANI-side bridge port, it filters upstream traffic, conversely if it is associated with a UNI-side port.

If a forwarding action is provisioned, all other MAC addresses are filtered; if a filter action is provisioned, all other MAC addresses are forwarded. The semantics of a mix of forwarding and filtering rules are undefined.

MAC Bridge Port Filter Preassign Table   This ME is a simple way to provision the filtering (discard) of well-known MAC address ranges and Ethertypes, for example, IPv4 multicast or broadcast, IPv6 multicast, address resolution protocol (ARP), point-to-point protocol over Ethernet (PPPoE) broadcast, NetBEUI, and Appletalk.

MAC Bridge Port ICMPv6 Process Preassign Table   New in G.988 amendment 1, this ME allows the provisioning of transparent forwarding, snooping, or discarding of various forms of IPv6 Internet control message protocol (ICMP) traffic. Multicast listener discovery (MLD) is an ICMP feature; it is advisable to avoid provisioning MLD rules in this ME if the ONU is intended to participate in IPv6 multicast, and instead provision the multicast-specific MEs, about which we learn in the next section.

MAC Bridge Port Bridge Table Data   This read-only ME reports information about the MAC addresses known to the bridge port, whether each address was statically or dynamically learned, how long it has been known, and whether it is used to forward or filter traffic.

MAC Bridge Port Designation Data   This read-only ME reports information that reflects the bridge port's STP negotiations. If the ONU is not capable of STP, it does not instantiate this ME.

MAC Bridge Port PM History Data   PM collected here includes counts of forwarded frames and frames discarded for various reasons. It can declare threshold crossing alerts against counters that indicate exceptions.

Ethernet Frame PM   There are three MEs in this family. Two of them can only be attached to a bridge port, one for upstream traffic, the other for downstream. The third member, extended PM, is more flexible:

- It can be associated with any of a number of Ethernet-related managed entity types.
- It can look at a specific GEM port.

- It can filter by VLAN ID and priority fields.
- It can be configured to monitor either upstream or downstream traffic.
- It can be disabled on demand. Baseline PM managed entities can only be disabled by deleting the managed entities themselves.

All three collect the same information: total packets, total octets, broadcast and multicast packets, packets with various kinds of errors, and a histogram of packets by size. All three can declare TCAs against counters that indicate exceptions.

Now that we have seen some detail of the MAC bridge and its ports, let us step back. Omitting the fine-grained details of the previous configurations, Figure 6.4 exemplifies a MAC bridge in a wider context.

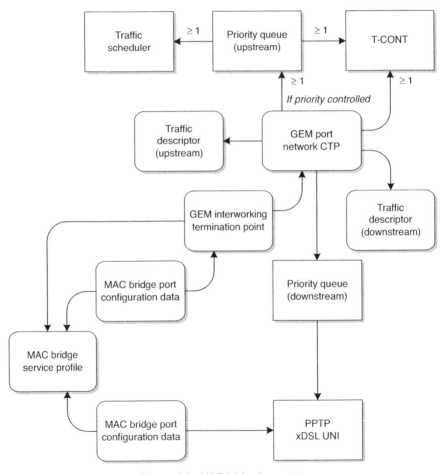

Figure 6.4 MAC bridge in context.

This example shows a single xDSL PPTP on the subscriber (UNI) side, which is understood to be transporting Ethernet payload. On the PON side—the ANI—we show one of the small number of standard configurations, namely association with a GEM interworking termination point. The variations on the standard ANI-side configuration include priority mapping and a multicast association, which we consider in subsequent sections.

The GEM interworking termination point (TP) associates the bridge port with a GEM port network CTP, thence to a T-CONT, either directly or through a priority queue, whose QoS is managed with a traffic scheduler.

GEM Interworking Termination Point   The GEM interworking TP is responsible for mapping client payloads, in this case Ethernet into GEM. It appears in other models as well, interworking GEM to functions such as TDM circuit emulation. What this ME contributes to the management model is the definition of the non-GEM payload type, in most cases by way of a pointer to a non-GEM service profile, in this case the MAC bridge service profile.

GEM Port Network CTP   The most important attribute of the GEM port network CTP is the GEM port ID. Other than that, the purpose of this ME is to carry pointers that glue the various other MEs together, including upstream and downstream QoS MEs (Section 6.3). It defines whether the flow is bidirectional, or if unidirectional, in which direction. In G.987 XG-PON, it also determines whether traffic for the GEM port is encrypted and, if so, in which direction and with which key.

What Is a GEM Port, Anyway?

A GEM port is nothing more nor less than a flow identifier, whose scope is limited to the G-PON itself. In the context of Ethernet traffic, a GEM port might be thought of as another VLAN tag, added at the transmitting end of a G-PON connection and stripped at the receiving end.

Okay, then, why does G-PON not just use another VLAN tag? For several reasons, probably the most important of which is that, at least conceptually, Ethernet is not the only kind of traffic that maps to GEM.

As well as client protocol agnosticism, a uniform GEM encapsulation scheme is useful for purposes such as fragmentation and dynamic encryption control, as discussed in Chapter 4. It would have been perverse to define a GEM layer and not to include what is essentially a routing tag in a layer where it is easy to parse and where it does not affect the client signal in any way.

The GEM port is often spoken of as if it were a managed object in its own right. In fact, the GEM port identifier is an attribute of the GEM port network CTP.

Figure 6.4 shows a single GEM port network CTP directing its downstream traffic to a priority queue associated with a specific UNI. In the general case, the bridge serves more than one UNI, and the downstream egress UNI is determined by the bridge itself.

If the logic of provisioning or bridging directs downstream traffic to different UNIs, the downstream queue specified by the GEM port network CTP is taken as a template. That is, if we designate queue number 2 for one port, we implicitly ask for queue number 2 on all ports that may receive traffic from that GEM port. Of course, if a single UNI contributes all of the upstream traffic for a given GEM port, it only makes sense to provision the downstream queue pointer to that same UNI.

We defer additional exploration of the priority queue, traffic descriptor, and traffic scheduler to Section 6.3, while we turn our attention to tagged frames.

6.1.1 Priorities and VLANs

Ethernet frames may be untagged, priority tagged, or VLAN tagged. As a convenient reference, Figure 6.5 illustrates the format of an Ethernet frame (see IEEE 802.3, 802.1Q, 802.1ad).

More than one tag may be inserted between the source address (SA) and the length/type fields. The leftmost tag is called the outer tag, while additional tags would be designated inner tags. OMCI allows for two tags, although some applications may stack more than two.

How would we know whether that next byte is a tag? By the value of the TPID octets. IEEE 802.1Q defines 0x8100 to indicate the presence of a VLAN tag. In 802.1ad, this is further refined to mean a C-VLAN, customer VLAN, and the value 0x88A8 is added to designate a service VLAN, S-VLAN.

The CFI bit is of little interest in its own right, but when the outer tag is an S tag, that is, a service provider tag, the same bit is called DEI, the drop eligibility

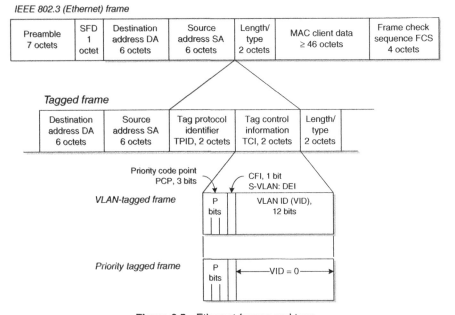

Figure 6.5 Ethernet frames and tags.

indicator—and is sometimes regarded as such even in C-tagged traffic. A traffic policer may set DEI = 1 on frames that exceed their contractually guaranteed traffic rate, thereby flagging them as yellow traffic; if the next bridge experiences traffic congestion, these frames are preferentially discarded.

Drop eligibility can also be conveyed implicitly in the priority code point (PCP) bits (commonly called P bits) of C tags, according to categories defined in 802.1ad and provisioned in the traffic descriptor ME. The OMCI priority queue can then be provisioned to regard certain priority classes as eligible for early discard. Section 6.3.2.1 explains these points in additional detail.

As well as at layer 2, priority can be conveyed at layer 3, through an IP DSCP value. DSCP is a 6-bit field in the IP packet header. In IPv4, it is part of an octet called *differentiated services*; in IPv6, the octet is called *traffic class*. A 3-bit subset of this field represents the older IPv4 type-of-service (TOS) field.

6.1.1.1 Priority Mapping in OMCI

The OMCI model concerns itself primarily with upstream traffic management. Traffic should be policed as near as possible to its network ingress point; by the time we have incurred the resource cost of forwarding a packet all the way through the network toward an ONU's UNI, we should make every effort to deliver that packet.

Considering the upstream direction, we now turn to traffic priorities, which may be determined based on any combination of the following factors:

- Some traffic may have a predetermined contractual priority simply because it appears on a particular physical port. This could be appropriate for transparent business services.

- Some traffic arriving upstream at a UNI may already contain explicit priority indications that were added by the subscriber's equipment, either IP DSCP or Ethernet P bits. If the subscriber device is trusted,[*] the ONU may simply accept the incoming VID and priority. Or, instead of passing preexisting tags transparently, the operator may add a new tag, effectively tunneling the original flow, or may modify the existing tags. The ONU may be expected to translate or copy VID or priority fields from an existing tag to the new one.

- Some traffic may require further inspection of subscriber packets, with service levels determined by Ethertype, MAC address, IP address, or higher layer protocol or port assignments. OMCI currently has limited ability to model deep packet inspection but could be extended based on community needs.

6.1.1.2 The 802.1p[†] Mapper

The ONU classifies upstream traffic when it enters the UNI. VLAN tags and P bits may be added, overwritten, copied, or translated from their original subscriber-

---

[*] One example of a trusted device is an RG associated with the ONU, authenticated and managed by the operator through BBF TR-69.

[†] 802.1p is a common notation for the Ethernet priority field (PCP bits, often just called P bits). The field was first specified in IEEE 802.1p, which was merged into 802.1D in 1998.

assigned values (if any), based on the factors listed above. By the time upstream traffic reaches the ANI side of a MAC bridge, it is usually tagged with its ultimate VID and P bits—ultimate as far as the ONU is concerned; the OLT may modify it further. In the OMCI model, the 802.1p mapper service profile ME separates this traffic on the basis of its P bits. It can direct this traffic into as many as eight GEM ports, each of which can then be connected into shared or dedicated QoS engines.

Figure 6.6 exemplifies the use of the 802.1p mapper service profile. The mapper may be thought of as a tree (an octopus?), whose root collects all traffic flowing upstream from a MAC bridge port. The octopus distributes traffic, one P-bit value per tentacle. In Figure 6.6, we assume that traffic with P-bit values 0 and 1 is not to be distinguished, so we map both to GEM interworking TP A. Traffic with other P-bit priorities is mapped to other GEM interworking TPs, which are themselves mapped to distinct GEM port network CTPs. Because P-bit values 0 and 1 were indistinguishable at this level, they both map to GEM port network CTP A.

For the sake of illustration, suppose that P-bit values {0, 1} and 2, although needing separate GEM ports, nevertheless share a common class of service. So we map both GEM ports A and B into the same priority queue K, while traffic from GEM port C is mapped to priority queue L.

The ME relationships are provisioned and therefore flexible, so it is fair to ask what happens if a tentacle of the mapper is not bound to a GEM interworking TP.

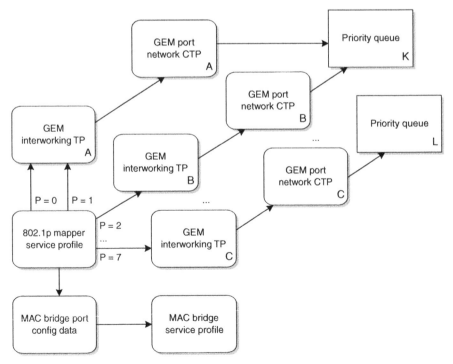

Figure 6.6   Bridge with P bit mapper.

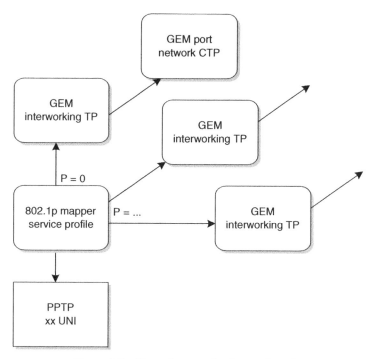

Figure 6.7   Alternative mapping (not preferred).

Having nowhere to go, traffic with the corresponding P-bit value would simply be discarded.[*]

Another feature of the 802.1p mapper ME is its ability to map DSCP prioritized frames. The mapper includes a table that provisionably associates the 64 possible values of the 6-bit DSCP field with the 8 possible P-bit values. The mapper directs the frame to one of the GEM interworking TPs according to the P-bit category provisioned in this table. The client Ethernet frame is not modified in any way.

The final mapper option is to direct an untagged frame to a given GEM interworking TP *as if* it were prioritized with a provisioned P-bit value, again without altering the frame itself in any way.

Figure 6.7 illustrates another possibility, in which the mapper is connected directly to the UNI, without benefit of an intervening bridge. This is legal in OMCI, but in the interest of interoperability, the industry converged on the bridge-based model shown in Figure 6.6. The model of Figure 6.7 is discouraged.

6.1.1.3 Filtering
We now know how to get upstream traffic into the correct GEM port and queue, but what if we have two VLANs that require separate treatment? Maybe they have the

---

[*] It really *is* irresistible to think of these diagrams as signal flows!

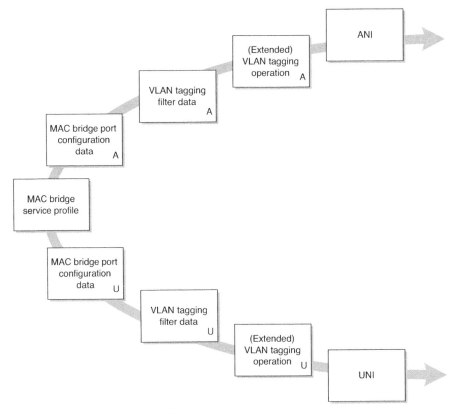

Figure 6.8 VLAN filtering.

same P bits but need to be mapped to separate GEM ports upstream. Or maybe we want to impose explicit security on a port by only allowing certain specific VIDs in one or both directions.

OMCI offers an ME called the VLAN tagging filter data to help with these and other situations. Figure 6.8 illustrates how the VLAN tagging filter fits into the traffic flow.

VLAN tags and priorities on upstream traffic are classified and modified by (extended) VLAN tagging operation configuration data ME U—associated with the UNI—whose moniker is close enough to that of the VLAN tagging filter to be confusing—the word *filter* helps to keep them distinct. The result of the tagging operation is then processed by VLAN tagging filter U and, having been forwarded by the bridge, by VLAN tagging filter A (we probably only have one, not both of these). Whatever gets through filter A could have its tags modified yet once again by extended VLAN tagging ME A. Downstream traffic follows the same sequence in reverse.

The VLAN tagging filter data ME includes a list of up to 12 TCI values. Recall from Figure 6.5 that a TCI comprises:

• A 3-bit priority field, the P bits

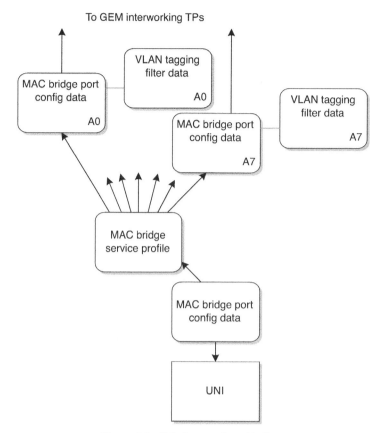

Figure 6.9 Priority mapping with filter.

- A single bit, the DEI, defined for S tags (or generically, the CFI)
- A 12-bit VLAN identifier, VID

For all TCI values in the list, the filter can be provisioned to look only at VID, at priority bits only, or at the entire TCI. When it finds a match, it can be provisioned either to pass the frame or discard the frame. Finally, it can be provisioned to consider bridge ingress traffic,* egress traffic or both.

Quite a flexible ME! Consider another possible application, to prioritize traffic. See Figure 6.9. Suppose we do not care about VID, but we want to distinguish traffic by P bit. Earlier, we saw how to route this traffic with the 802.1p mapper. To do the same thing with the VLAN tagging filter data ME, we create eight ANI-side bridge

---

*The filter does not think in terms of upstream or downstream flows; it sees traffic either entering the bridge or leaving the bridge.

ports, A0 through A7, with a VLAN tagging filter on each. We provision each filter to ignore VID, but to forward only frames that match its own particular value of P bits. Then we connect the bridge ports to GEM interworking TPs, in the same way we did with the mapper.

Going beyond the capabilities of the dot1p mapper, we can even provision two filters to the same priority, but with different VIDs. This is a powerful ME!

In the last two or three pages, we have outlined prioritization with and without a MAC bridge, with and without a P-bit mapper, with and without VLAN tagging filters. These are examples of the flexibility in OMCI that led to interoperability issues, and the formation of the OISG mentioned in Chapter 1, whose purpose was to agree on industry best practices.

6.1.1.4 VLAN Tag Management

Having learned how to classify and prioritize traffic, our remaining task is to modify its VLAN tags. P bits are part of the VLAN tags, so tag management is also a tool for reprioritization as well as for VLAN encapsulation or interchange.

There are two MEs dedicated to tag management—yes, more options—the VLAN tagging operation configuration data and the *extended* VLAN tagging operation configuration data. The original version can be provisioned to pass upstream traffic transparently or to add a tag, either overwriting the original tag if the traffic was already tagged or preserving the original tag, if any, and prepending a new outer tag. It either passes downstream traffic transparently or strips the outer tag. This is perfectly adequate for many real-world purposes,[*] but in the interest of a common provisioning model for all cases, it is fading away in favor of the extended tagging ME.

As with the original ME, the extended VLAN tagging ME can be attached to various flow point MEs on either the UNI or the ANI side of a MAC bridge. As with the original ME, its primary function is on upstream traffic. Downstream flows can either be provisioned to be transparent or to be the inverse of the upstream operation. The meaning of *inverse* is not always immediately apparent.

The core of the extended tagging ME is the received frame VLAN tagging operation table. Each row of the table contains a rule and an action. Incoming upstream frames are matched against each rule; when a frame matches, the specified action is applied to it.

A frame could match more than one rule; the first match is the one that matters. *First* means *in table sort order*, that is, the order in which the rules would sort.

---

[*] Double-tagging operations can be divided, with part done in the ONU and part done in the OLT. This option has spawned something of a religious war, raising the issue whether the OLT is even allowed to do some of the heavy lifting. (Of course it is; the OLT has to have such capabilities anyway.)

The extended VLAN tagging ME an input tag protocol ID (TPID) attribute, which we can use in the filter rules. The rules table offers seven filter (classification) fields, which we can specify or ignore independently.

- *Outer tag P-bit priority*—One of the code points is to ignore all (other) outer tag fields.
- *Outer tag VID*
- *Outer tag TPID/DEI*—Choices include 0x8100, which designates a VLAN-tagged frame, and a match against the input TPID attribute. Given a TPID match, the DEI bit can be specified to match either 0, or 1, or to be ignored.
- *Inner tag priority*
- *Inner tag VID*
- *Inner tag TPID/DEI*, the same choices as for the outer tag. Since the inner tag should never be an S tag, and strictly speaking, the DEI bit is defined only for S tags, this is an example of stretching the formal standard.
- *Ethertype*—Choices include IPv4oE, IPv6oE, PPPoE, ARP, or no filtering.

Once a frame matches a rule, what can we do with it? Here also we have seven possibilities, which can be combined arbitrarily.

- *Tags to Remove*   Strip 0, 1, or 2 tags; one of the code points in this field also discards the frame—so we have yet another way to perform filtering!
- *Outer Priority*   Add an outer tag, or not, with P bits provisioned, or copied from the outer or inner P bits of the incoming frame, or derived through a provisionable mapping table from the DSCP bits of the incoming frame.
- *Outer VID*   Insert a specified value into the VID field or copy VID from the outer or inner tag of the incoming frame.
- *Outer TPID/DEI*   Copy the TPID field from the inner or outer received tag, or set it to the value of the separate output TPID attribute, or to 0x8100, which just indicates a standard 802.1Q VLAN. Copy the DEI bit, if TPID indicates an S tag, from the inner or outer received tag or set it to 0 or 1.[*]
- *Inner Priority, Inner VID, Inner TPID/DEI*   The choices are the same as for the outer tag.

It should be apparent that we can do pretty much anything we want to with an Ethernet frame, as long as we only want to look at layer 2—with a bit of leakage from layer 3 (DSCP)—and as long as we are satisfied to make changes only at layer 2. Since G-PON is intended to function at layer 2, this fits well. A G-PON ONU can

---

[*] DEI management is a way to unconditionally declare all—or none—of a given stream to be yellow traffic.

look at layer 3, for example, IGMP or DHCP messages, but such messages are trapped to the ONU's processor and handled in specialized ways by software, rather than treated as just part of the ordinary bridged traffic flow. It is also true that operators sometimes see a need to classify ordinary subscriber traffic based on layer 3 and even layer 4 information. To date, this has not arisen as an ONU requirement. If it were to become necessary, OMCI could be extended accordingly.

6.1.2 Downstream Forwarding

In the downstream direction, the TR-156 service model requires four queues at the OLT PON port and asks for six, with strict priority as a requirement and weighted fair queuing as a desirement. While these are valid requirements, we do not see them in OMCI. The OMCI management model only specifies the ONU and does not specify matters internal to the OLT.

Once traffic has been transmitted downstream from the OLT, the OMCI model assumes that there are no congestion points prior to the UNI, where queuing and scheduling may yet again be needed. Again, TR-156 requires four strict priority downstream queues on each Ethernet UNI and asks for six.

Recall from Figure 6.4 that the GEM port network CTP includes an attribute that points directly to one of the downstream queues at the UNI. The ONU's downstream queue architecture may be fixed or configurable. Strict priority queueing is the default.

The downstream queue pointer from the GEM port network CTP (Fig. 6.4) forces all traffic into a given queue, or into a template queue if several UNIs are involved. Particularly if several priorities were expected in the downstream traffic flow for a given GEM port, it might be preferable to leave the queue pointer blank and let the bridge determine the proper priority handling. The ONU would be expected to select a queue based on the frame's P bits, with a default queue for untagged frames. Table 8-2 of IEEE 802.1Q recommends how to combine P–bit priority classes into fewer than eight queues.

As to downstream tag manipulation, our tool is the (extended) VLAN tagging ME. The downstream direction can be provisioned to be transparent, but we usually want it to be the inverse of the upstream operation. If we added a tag to upstream traffic, we probably want to strip a tag in the downstream flow. If P-bit values were translated in the upstream direction, well, that can be a problem, since the upstream mapping could be many to one. The rule in the downstream direction is to reverse the mapping of the first matching rule in the list. That is to say, the frame is treated as if it were an upstream frame, matched against rules in the list, in list order, and the first matching rule is applied in reverse to the downstream frame. While strange corner cases could be constructed, this rule has until now proven to cover the real-world cases of interest.

Update   As we went to press, G.988 amendment 2 had just been consented in ITU-T. Among other things, it included substantial extension of the extended VLAN ME's ability to classify and process downstream frames.

6.1.3   Layer 2 Maintenance

IEEE 802.1ag defines a set of maintenance entities and operations for Ethernet, collectively known as configuration fault management (CFM). ITU-T recommendation Y.1731 extends 802.1ag to include other important features such as AIS.

802.1ag provides three basic functions:

- Assurance of correct connectivity between the endpoints of a trail, with an alarm when connectivity is lost or when incorrect connectivity is detected. Using continuity check messages (CCMs), this function may run continuously over the lifetime of a connection. However, in view of the processing load imposed by hundreds or thousands of circuits, it may be desirable to generate CCMs rarely—10 min is the greatest defined interval between CCMs—or to turn them off altogether except when troubleshooting.

- On-demand loopback tests. The query is called a loopback message (LBM); the response is a loopback reply (LBR). The OMCI test message includes an option to invoke 802.1ag loopback tests.

- On-demand path trace tests, allowing the originating endpoint to determine the sequence of nodes through which the Ethernet path passes. The query is a link trace message (LTM); a link trace reply (LTR) returns from each point along the link at the same maintenance level. The OMCI test message includes an option to invoke 802.1ag link trace tests.

802.1ag assumes a multilevel network architecture, eight possible levels, in which the ultimate end-user subscribers occupy higher numbered levels. The network operator occupies the lower levels, and intermediate service providers occupy intermediate levels. The concept is to permit separate maintenance across the domain of interest of any of the participants, while hiding the detail of server layers.

Figure 6.10 illustrates how this might work. Suppose we have subscribers a and b, buying service from provider P over a network that spans two operators A and B. We model the network elements along the path as bridges, labeled A1 through B4. The example illustrates that maintenance levels are nested, that they become more granular as the level decreases, and that they never overlap at a given level.

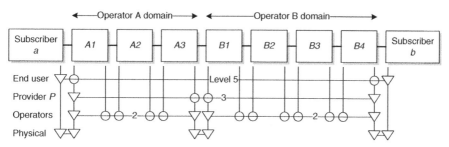

Figure 6.10   Maintenance levels.

Subscribers a and b have end-to-end verification of their connectivity at level 5. They have maintenance endpoints (MEPs, triangles) in their own outgoing bridges, and maintenance intermediate points (MIPs, circles) at the edges of their server network. They see nothing of the intermediate details.

Provider P buys network capacity wholesale from operators A and B and sells the service to subscribers a and b. P needs to know that everything is okay between bridges A1 and B4, and, if there is a problem, whether the problem is in network A or B. Therefore, P performs maintenance between bridges A1 and B4 at level 3, with MIPs at the A–B boundary.

Each operator is likewise concerned about the service across its own domain and localizing problems to particular bridges within its domain. The operators may share level 2 (recommended level) since their domains do not overlap.

Finally, it may be desirable to set up MEPs at the physical Ethernet level between each pair of bridges (only interdomain links shown).

MEPs are entities that can originate Ethernet CFM messages; MIPs can respond to LBR and LTR messages but not originate them. MEPs are explicitly created; MIPs are implicitly created according to provisioned criteria.

This is a very high-level view of the architecture. OMCI provides half a dozen managed entities that support the necessary domain definitions, layers, functions, testing, monitoring, and reporting. The interested reader is encouraged to plunge into the complexities of 802.1ag and Y.1731, and then to study how they are supported by OMCI.

6.1.4 CPE Authentication

Some operators require the subscriber's terminal to be authenticated as a condition of service. Usually this is an RG provided by the operator. Successful authentication gives the operator some level of assurance that the RG can be trusted, in particular with regard to issues such as setting IP packet or Ethernet frame priority, and perhaps in blocking some forms of security attacks.

We include this function in the Ethernet services section because the only CPE authentication mechanism presently defined is IEEE 802.1X, which pertains to Ethernet interfaces. Higher-layer services are presumed to be carried across the UNI in Ethernet frames and to be bound to an Ethernet MAC address.

The CPE acts as an 802.1X supplicant, while the ONU plays the role of authenticator. To authenticate the CPE, the ONU sets up a session with a RADIUS[*] server somewhere back in the core of the network and transfers EAP[†] messages back and forth. In turn, the RADIUS server may itself invoke the services of an authentication, authorization, accounting (AAA) server even further in the depths of the back office.

The format and content of the EAP messages is opaque to the ONU, determined by negotiation between the RADIUS server and the CPE. The EAP exchange is probably encrypted, with a key that is not known to the ONU. The ONU snoops the

---

[*] RFC 2865.
[†] RFC 3748.

RADIUS messages—the RADIUS message itself is *not* encrypted, merely the EAP that is encapsulated within the RADIUS message—to understand the final accept or deny conclusion of the transaction. As authenticator, the ONU then acts accordingly to allow or block subscriber traffic from the supplicant MAC address.

OMCI includes four MEs to configure the 802.1X CPE authentication function, to connect to redundant RADIUS servers, and to collect performance monitoring statistics.

Dot1x Configuration Profile If it is capable of supporting 802.1X authentication, the ONU creates this ME autonomously. The ME includes access information for redundant RADIUS servers, and a fallback policy in case no server is available. The fallback policy may be to authenticate unconditionally or to deny authentication unconditionally.

Dot1x Port Extension Package This ME is implicitly linked to an Ethernet UNI of some kind. It determines whether 802.1X authentication is required and for which direction of transmission. Its attributes include timers for authentication server failure and scheduled reauthentication.

Dot1x PM History Data This ME counts EAP frames of various types and in various directions. It can declare TCAs if it observes too many defective frames.

RADIUS PM History Data In much the same way, this ME counts different varieties of RADIUS messages transmitted and received and declares TCAs if appropriate.

It is also perfectly feasible for the ONU to tunnel 802.1X frames through to the OLT, such that the OLT acts as authenticator. Since it does not affect the interface between OLT and ONU, this is beyond the scope of the recommendations; but in such an architecture, it would be the OLT's responsibility to block subscriber traffic to and from an RG until 802.1X authentication succeeded. Interoperability between these options has not yet been a market concern.

Some operators also wish to authenticate the ONU itself (Section 4.6), in which case the OLT acts as authenticator, the ONU as supplicant.

A separate use of 802.1X is also imagined, in which a device subtended from the residential gateway is authenticated. This could be applicable if a roaming device wished to attach to a wireless RG, for example. It differs because, in this latter case, the RG is presumed to be a trusted device that itself acts as the authenticator and the RADIUS agent. The G-PON system is transparent to this operation. Nothing prevents the use of 802.1X, first for the OLT to authenticate the ONU, then for the ONU to authenticate the RG, followed by the RG in turn authenticating its peripherals, and so ad infinitum.

6.2 MULTICAST

Multicast is an Ethernet service, but it includes enough complexities to warrant a section in its own right. Multicast—IPTV—is one of the key services that justifies the business case for G-PON. The intrinsic high bandwidth and broadcast nature of the PON physical layer is well suited for multicast.

6.2.1 Overview

No access technology has the capacity to economically carry all possible multicast streams past all subscribers all the time. From the multicast router along the tree all the way to the subscriber, it is important to forward traffic onto a given branch only if there is an actual listener at any given time, and to prune the tree when the last listener leaves. At each point in the flow from router to subscriber, multicast-aware nodes in the network have the ability to replicate incoming multicast groups onto any number of outgoing downstream ports, selecting only ports on which there is a current listener.

IGMP is the widely deployed protocol that enables networks to conserve bandwidth by propagating multicast groups only on demand. IGMP version 1 is deprecated. IGMP version 2 is defined in RFC 2236; it is widely deployed but officially obsolete. IGMP version 3 is defined in RFC 3376. IGMPv3 is backward compatible with IGMPv2 (and even v1), though with reduced functionality. In IPv6, the equivalent is multicast listener discovery MLD (version 1: RFC 2710, updated by version 2: RFC 3810).

A multicast client is conventionally referred to as an STB. When it wishes to receive a given multicast group (channel), the STB generates a join[*] request upstream. Each IGMP-aware network node along the upstream path recognizes the message. We discuss security aspects later; for now, assume that the STB is authorized to receive the group.

IGMP Snooping   If the network node implements so-called IGMP snooping, it listens to join requests, forwards them further upstream, and starts replicating the requested group immediately. If the requested group is not already present in the downstream flow, then, of course, nothing happens, but if the stream does happen to be present already, this is a way to get a rapid channel change. Having been propagated upward by each snooping node, the join request eventually reaches a node that can replicate the stream, so in any event, the STB does eventually get its channel. All (or almost all) G-PON ONUs are capable of snooping IGMP.

IGMP Proxy   The strategy of sending a join request all the way to the multicast router for every channel change does not scale well, and there is really no point in sending a join further upstream than the node that is already replicating the traffic. If the replicating node functions as an IGMP proxy (RFC 4605), it forwards a join request further upstream only if it is not already replicating that group to existing clients, either on that same port or on some other of its ports. In this way, only one

---

[*] Join and leave terminology is hallowed by usage. The various versions of the protocol have differing designations of the actual message types.

join message flows upstream for a given group, regardless of the number of listeners served by the proxy. The proxy acts as a virtual set-top box in lieu of the many real STBs—or other proxies—further out on the multicast tree.

Well, but what happens when the subscriber stops watching some particular channel? How do we know when to *stop* replicating a given multicast group down the tree? A well-behaved STB will send a so-called leave message when it departs. By counting the number of joins and leaves, nodes further up the tree can determine when no client remains active.

Even with the best of intentions, however, the STB may not be able to signal a leave. The subscriber may simply power it down, for example. So an IGMP router or proxy periodically (default 125 s) queries the STBs to determine whether anyone still cares about each of the multicast groups it is replicating.

In a feature called fast leave, an IGMP router or proxy generates a last member query asynchronously, when it receives what it believes is the last leave message along a branch. Active clients respond with membership reports to confirm their continuing interest. If it receives no response for a given multicast group, the IGMP router prunes that branch of the tree. The router or proxy sends the last member query message more than once (default twice) before deciding.

If a network node is an IGMP snoop, the query messages flow downstream past it, and the interested downstream nodes respond individually. An IGMP proxy, on the other hand, captures the query message and responds on behalf of everything further down the tree. The proxy takes on the responsibility to know whether there are any active clients along its own branch, which it does by sending its own query messages and capturing the replies. A network scales better with proxies; BBF TR-156 mandates a proxy in a G-PON OLT. Depending on its size, an MDU or a G-PON-fed DSLAM may also contain an IGMP proxy.

A full proxy hides the detail of everything downstream from itself. To the upstream multicast router, the proxy looks like a single client. To the various downstream STBs, the proxy appears to be the multicast router. As an alternative, an SPR node (snooping with proxy reporting) passes interesting client messages upstream, having filtered them to remove redundancy. RFCs 4541 and 4605 discuss snooping and proxy operation further.

To peel one more layer off the onion: by Internet Assigned Numbers Authority (IANA) allocation, IPv4 multicast is carried in the destination address space 224.0.0.0 through 239.255.255.255. The low-order 23 bits of these addresses may also be used as partial Ethernet MAC addresses (reserved range 01.00.5E.00.00.00– 01.00.5E.7F.FF.FF). This mapping facilitates some level of equivalent layer 3 multicast functionality from layer 2 devices. Because G-PON systems are primarily layer 2 devices, this is of considerable interest.

However, because there are 28 variable bits in the IP multicast address range, and only 23 free bits in the MAC address mapping, a given multicast MAC address may correspond to any of 32 possibilities. In a limited service environment, it may be possible to assign different multicast destination IP address ranges to different providers. In larger environments, with multiple service providers, each a separate business, each overlapping the others' geographic range, this may not be feasible.

Not that there exist too many multicast groups for the address space, but just that it may be too complex to manage several mappings in overlapping geographic regions that may each have different constraints.

There are three ways to deal with this problem. One is for the multicast replication node—the ONU, in this case—to look at the layer 3 address, the IP destination address itself, rather than the MAC address. This disambiguates the MAC address, but it does not necessarily solve the underlying problem, which is IP address duplication.

The second approach is to also consider the IP source address of the stream, as well as the multicast destination address. This is referred to as source-specific multicast (SSM), which contrasts with any-source multicast (ASM). RFC 4604 explains SSM support in IGMPv3 and MLDv2.

It is even possible that the same source and destination IP address pair may be offered by competing providers. In this case, streams can only be distinguished by VLAN. This feature allows subscribers to purchase programming from competing middlemen, who in turn acquire it from a single originator, and perhaps offer packages with different combinations of multicast groups in accordance with their perception of market opportunity.

In the next few pages, we shall see how OMCI supports layer 2 or layer 3 IGMPv2 and v3 (and v1), MLD v1 and v2, snooping and proxy, ASM and SSM, with VLAN selectivity.

6.2.2 Multicast in G-PON

In most cases, a PON will be set up with only one GEM port for all multicast content, and all ONUs that have multicast subscriptions will be configured to use it. Figure 6.11 shows the ONU management model.

On each ONU, a GEM port network CTP is configured with the PON's common multicast GEM port. Because this flow is downstream only, we need not associate the GEM port network CTP with a T-CONT. As we know, the GEM port value is used by the ONU hardware to filter traffic from the PON, so specifying this GEM port helps offload the ONU hardware.

We tie the GEM port network CTP to a multicast GEM interworking termination point. The primary feature of the MC GEM IW TP is that it includes a multicast address table,[*] each row of which includes not only an IP multicast destination address range but also a GEM port! A GEM port?

Why do we need that GEM port network CTP at all? Only for its secondary attributes, such as the downstream priority queue pointer. Although it is nowhere specified, we expect that the multicast address table in the MC GEM IW TP ought to include the same GEM port mentioned in the GEM port network CTP, possibly among other GEM ports.

Other GEM ports? Yes. We said that the common model was a single GEM port for the whole PON, and it is true. But in a slightly more elaborate service model,

---

[*] One logical table, represented in the OMCI model as two table attributes, one for IPv4 and one for IPv6.

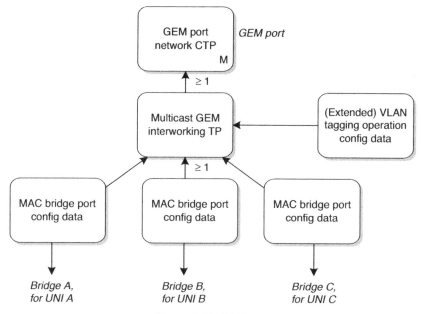

Figure 6.11   Multicast.

the operator may assign multiple multicast GEM ports to distinguish different service packages, presumably destined to different ONUs. If there are several GEM ports, and an ONU cares about more than one, then it suffices to list them in the multicast address table; there need not be an additional GEM port network CTP for each. This is a good time to remind ourselves that the OMCI model is not a signal flow diagram; it is a way to aggregate useful collections of management parameters.

As illustrated in Figure 6.11, any number of subscriber ports can connect through MAC bridges to the multicast GEM interworking TP. VLAN selectivity could also be done here with a VLAN tagging filter, in the very simple service model. In a more complex model, VLAN specification is recommended to be configured in the multicast operations profile instead, an ME we discuss below.

Level 0 Multicast   We might describe the configuration above as a level 0 multicast service. Every subscriber has access to any of the groups that are associated with the multicast GEM interworking TP. The level 0 model includes no explicit provision for IGMP; the ONU might or might not support IGMP with default parameters. Reasonable default behavior would be for the ONU simply to snoop the join, leave, and group query messages and assume that any requested group was authorized. The purpose of IGMP in the ONU is merely to avoid overloading the UNI with downstream traffic that is of no interest to the subscriber. Particularly in single-family ONUs, level 0 suffices for many real-world use cases.

As it traverses the network, multicast traffic is VLAN tagged. The OLT maps the downstream multicast VLAN to a given GEM port, and the ONU uses the GEM port as the key to pull the appropriate content off the PON. Multicast traffic at the subscriber's UNI is typically untagged. To strip the tag, an (extended) VLAN tagging operation config data ME can be associated with the multicast GEM interworking TP, provisioned to add the specified VLAN tag upstream. There is no upstream flow, of course, but the inverse operation in the downstream direction causes the tag to be stripped. (The OLT could also strip the downstream tag, if it is known that it serves no further purpose.)

Where are IGMP[*] messages carried? Downstream query messages reside in the multicast stream itself; upstream messages are usually forwarded in the subscriber's unicast high-speed Internet stream, expecting that they will be trapped upstream. Upstream messages are typically untagged across the UNI, just as is the rest of the subscriber's Ethernet traffic, and the level 0 model treats it the same, adding the same VLAN tag that applies to the subscriber's other Ethernet traffic. If the OLT (as proxy) or the multicast router can see the IGMP traffic in the subscriber's VLAN—and typically they can—we are okay.

Level 1 Multicast Figure 6.12 portrays a more detailed model, which we might designate a level 1 multicast service (levels 0 and 1 are enough; we promise not to add a level 2 version a dozen pages further along). The new MEs at the subscriber level include a multicast subscriber config info ME, a multicast operations profile, and a multicast subscriber monitor. As with level 0 multicast, IGMP traffic is typically directed upstream through the same GEM port used for the subscriber's high-speed Internet access, but not necessarily with the same tagging.

Through the multicast subscriber config info, each subscriber is associated with one or more shared multicast operations profiles. Different UNIs may subscribe to different groups of profiles.

Multicast Operations Profile The multicast operations profile includes a number of useful attributes.

- Here we have the ability to specify whether we wish the ONU to act as an IGMP snoop, snoop with proxy reporting, or full proxy (not all ONUs will support all of these options, of course). When it performs proxy functions, the ONU needs a whole list of parameters from RFC 3376, which are also provisioned here.
- We can specify whether we want IGMPv1, v2, or v3 (only version 3 is recommended). In the IPv6 world, the choice is MLD v1 or v2.
- If upstream IGMP messages are to be tagged, or if a preexisting tag is to be modified, here is where we specify it. Upstream tags are usually managed through the (extended) VLAN tagging ME, but can be overridden specifically

---

[*]For purposes of this discussion, the term *IGMP* includes MLD as well.

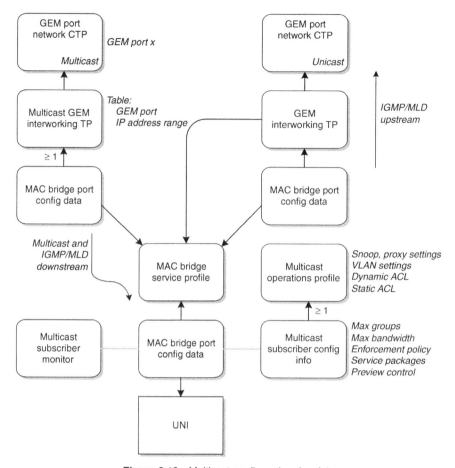

Figure 6.12 Multicast configuration, level 1.

for IGMP messages with this function. As well as specifying the VID, we can set priority and DEI.

- We can impose rate limiting on upstream IGMP traffic. This can mitigate some forms of denial of service attacks.
- The lost groups list is a diagnostic tool that records multicast groups for which there is a join request but no downstream flow. It can help identify network configuration problems or failures further upstream. Because a join request may have to propagate all the way to the multicast router before traffic flow actually begins, it is okay for a group to appear in this list, as long as the appearance is transient.
- We can specify whether the ONU should forward or discard a join request that fails to match any entry in the ACL. While the ONU never replicates unauthorized groups to the subscriber, it is possible that the operator might wish to see

unauthorized requests for market research, advertising, or instant authorization purposes. Or the operator may regard unauthorized join messages as clutter, to be discarded at the ONU. TR-156 does not consider the options; it states a firm requirement to discard join messages that do not match the ACL.

- The most important attributes are the dynamic and static access control lists. The dynamic ACL operates in conjunction with IGMP-MLD. The static ACL represents groups that are unconditionally replicated to the associated UNIs, without IGMP join messages. These groups represent a basic always-on service level, the most common channels, but also represent a continuing load, especially on the capacity of the UNI.

Let us look into the ACLs in detail.

Access Control List   The ACL is a table that specifies the allowed multicast groups. It is therefore sometimes called a white list (as distinguished from a black list, which would list the *dis*allowed groups). Each entry lists a GEM port and a range of multicast destination addresses, IPv4 or IPv6. These entries should be a subset of the multicast address table configured in the multicast GEM interworking TP. In an ONU with several subscribers, it may be easiest to configure the multicast GEM interworking TP with the full range of multicast IP addresses, and let the operations profile do the filtering.

Each ACL entry contains the following additional information:

- Downstream VLAN ID, used to distinguish the stream of interest from other possible streams in other VLANs.
- IPv4 or IPv6 source address, which allows the ONU to support SSM, source-specific multicast. SSM is one of the key features that cannot be supported with level 0 multicast.
- Imputed group bandwidth. Like most data traffic, IPTV flow rates are not perfectly constant. The imputed bandwidth is a provisioned value, used for admission control against physically or contractually limited subscriber bandwidth. We simply assume that each channel listed in this table entry is, for example, a 4-Mb/s flow.
- Preview fields, fields that permit the subscriber to receive the multicast stream for a limited time only. Preview policy pertains independently to each group included in the table entry. Provisionable preview parameters include:
  (a) Preview length, the number of seconds of the group to be allowed during a preview.
  (b) Preview repeat time, the number of seconds after which an additional preview is permitted for a given group.
  (c) Preview repeat count, the maximum number of preview episodes permitted.
  (d) Preview reset time, the clock time at which the preview repeat counter is reset. The repeat counter may also be explicitly reset by OLT action.

Previews

OMCI preview functionality is strictly limited. The subscriber is allowed to receive a given group for a given amount of time, after which the preview is canceled and the screen presumably either goes dark or freezes. More sophisticated preview functions, including advertising banners, coordination of timing, and opportunities to purchase the program, are a function of the set-top box and the middleware, and are intrinsically impossible through OMCI. The limited capability available through OMCI may serve the needs of some operators until they develop more sophisticated marketing tools.

Multicast Subscriber Config Info   The intention of the multicast operations profile is that several subscribers may share the same service parameters. But some multicast information is unique for each subscriber. That information is provisioned in the multicast subscriber config info ME.

- Attributes of the multicast subscriber config info ME limit the maximum number of simultaneous groups and the maximum imputed bandwidth to be allowed, and whether the subscriber is permitted to exceed the maximum bandwidth at his own risk, or whether it is a hard limit.

  Suppose that a given subscriber contracts for 30 Mb/s of capacity and is already using 25 Mb/s at some given time. A new join request for a group whose imputed bandwidth is 6 Mb/s will be ignored, while a new join request for a group whose imputed bandwidth is 4 Mb/s will be honored. In any case, the instantaneous flow across the subscriber's UNI may be more or less than the imputed total.

  If the UNI is permitted to exceed the imputed maximum, there is risk of queue congestion and traffic delay or discard among the real-time video streams. If this were to occur, it would degrade the quality of all displays. In our example, a 100-Mb/s Ethernet UNI would not be affected by traffic peaks exceeding 30 Mb/s, but a DSL subscriber drop might be.

- The multicast subscriber config info ME points to one or several multicast operations profiles for that particular subscriber. A single profile suffices for a simple package offered by a single provider. Multiple profiles permit the subscriber to sign up for a number of service packages that might be offered by a number of independent service providers, through an attribute called the multicast service package table. In addition to a multicast operations profile, each entry in this table specifies:

- (a) A UNI-side VLAN ID. The previously specified VID was on the ANI side, downstream; this one identifies the tag on upstream IGMP messages. Its use here is to classify upstream IGMP messages by VLAN tag to the proper service provider. It presupposes that the STB tags its IGMP messages in a way that permits them to be sorted out into different service packages.

(b) Bounds on the number of simultaneous groups and the maximum imputed bandwidth that are allowed by this particular package. The aggregate subscriber load is also bounded by the global group count and bandwidth limits of the multicast subscriber config info.

- The allowed preview groups table might better be called a pay-per-view table. It is populated when a subscriber purchases a multicast program, possibly after having previewed it. Each entry in the table specifies a source and destination IP address, with appropriate ANI- and UNI-side VLAN IDs. Each row also includes a field that controls the total duration of the paid view and a timer field that indicates the remaining duration at any given time. The table entry can also allow unlimited viewing time, which could be useful if the subscriber were to sign up for a few specific multicast groups a la carte, for which no explicit service package had been defined.

Upstream IGMP-MLD Flow As well as the downstream multicast structure, Figure 6.12 shows a unicast GEM port, along with its usual associated upstream MEs. Upstream IGMP messages are sent over the subscriber's normal Internet connection, suitably encapsulated in a VLAN as specified in the multicast operations profile.

For the usual high-speed Internet access traffic, VLAN tag management is performed under the control of the extended VLAN tagging configuration data managed entity. But multicast is different: it may require customized downstream tag management, and upstream flows may need to be treated differently based on the fact that they are IGMP or MLD. Recall that the extended VLAN tagging ME cannot perform deep packet inspection. More flexible and granular control is needed for multicast and IGMP.

The multicast MEs allow tags to be added, translated, or stripped in the downstream direction. In the upstream direction, tags can be added to IGMP/MLD traffic, or preexisting tags from the STB can be translated.

Multicast Subscriber Monitor The multicast subscriber monitor is a status, accounting, and diagnostic ME. At any given time, it contains a list of the currently active groups (separate attributes for IPv4 and IPv6), the current total imputed multicast bandwidth in use, and counters for IGMP join messages and attempts to exceed the provisioned maximum bandwidth.

This is as good a place as any to talk briefly about IPv6 multicast. As a general statement, it is true that the IPv4 address space is in imminent danger of exhaustion, and operators are starting to move to IPv6 with a certain amount of urgency.[*] But the IPv4 *multicast* address space is *not* congested. First things first: operators will probably keep multicast on IPv4 for quite some time. However, G.988 amendment 1 revised OMCI multicast extensively for multiple providers and previews, and it made sense to include IPv6 as well.

---

[*] Carrier-grade NAT may defer the crisis for yet another few years!

6.3 QUALITY OF SERVICE

No economically feasible network can support traffic at the maximum worst-case rate that could possibly be generated. Prioritization and rationing is therefore necessary at each of the possible congestion points in a network. Quality of service is a network-wide issue, but our focus in this book is on the ways in which G-PON contributes to the end-to-end performance. We first consider downstream QoS, then upstream. We further separate the upstream discussion into two realms: first, resource management within a single ONU and then resource management among the contending ONUs of a PON.

What are we trying to achieve?

- Traffic such as DS1/E1 pseudowires (Section 6.6) or POTS (Section 6.5) earns substantial revenue, which makes them high-priority services. Such traffic requires a *fixed rate* with low and predictable delay. A fixed-rate contract provides bandwidth even if the source generates no traffic.

- Some kinds of voice channel may alternate between silence and speech, with little or no capacity needed during the silent intervals. A similar application is a shared workplace collaboration or conference site in which only the presenter generates significant traffic, but the presenter role may change from time to time. Such traffic is suitable for a class known as *assured forwarding*, in which a contracted level of bandwidth is guaranteed to be available if it is needed but may be used for other purposes if no traffic is offered.

- A third level of QoS is called *nonassured* traffic, which takes priority over yet a fourth class, *best efforts* traffic. Service-level agreements (SLAs) specify maximum bandwidth and weight or priority for contending traffic within these latter two classes.

Conventional Internet access is a best-efforts service, with different maximum rates according to the subscriber's service tier. For example, a G-PON operator might offer a choice of Internet packages with 25-, 50-, or 100-Mb/s data rates. These would all be best-efforts services but with significantly different claims on the limited resources that might exist at network congestion points.

As to the other QoS categories—fixed, assured, nonassured—these reflect the business needs of operators as they converge all service types onto a single network. From another perspective, these other traffic classes reflect the needs of subscribers as they demand services that cannot be supported by a simple best-efforts QoS model.

Downstream QoS is largely an OLT requirement, while upstream QoS is a cooperative venture between the OLT, which allocates bandwidth among the ONUs on a PON, and the ONUs themselves, which prioritize, shape, and police their internal traffic. Different operators have different service models, which require greater or lesser sophistication in the ONU. The TR-156 model requires very little upstream traffic management capability in the single-family ONU, as long as we can dedicate one T-CONT to each traffic class—and we can.

We tend to think of QoS as an issue for upstream traffic. There is no question that upstream QoS is important, but first—why not also downstream?

6.3.1 Downstream QoS

As a general principle, traffic should be policed as early as possible in a network, ideally at the ingress network element. By the time we have incurred the resource cost of getting the traffic all the way through the network to the far end, we want to deliver it, not discard it!

That is one reason G-PON pays more attention to upstream traffic: control it at network ingress.

A second reason is that the ITU-T recommendations deal primarily with the interface between OLT and ONU, whereas downstream QoS is largely localized to functions within the OLT, or to functions within the ONU. Downstream QoS mechanisms are therefore largely invisible to the ITU-T recommendations.

Having said that, we point out that BBF, which takes a wider scope, does in fact call out downstream QoS capabilities, at least to a certain extent. TR-156 specifies that the downstream PON port in the OLT, and the downstream UNI port in the ONU, each support a minimum of four strict priority queues. Six queues are desired, and an option for weighted scheduling is desired.[*]

OMCI supports the necessary management model for the ONU's UNI, namely a set of downstream priority queues. The queue can be specified by a pointer from the GEM port network CTP (Fig. 6.4) or can be determined implicitly on the basis of P bits by the ONU vendor's MAC bridging architecture.

This is not to say that there are no issues with downstream QoS. We trust that downstream traffic was policed somewhere further upstream from ourselves. In traditional architectures, this was true to a certain extent: all data traffic flowed through the broadband remote access server (BRAS, BBF TR-59 terminology) or broadband network gateway (BNG, TR-101 terminology). Whatever its designation, this was an edge router that knew the SLA parameters of each subscriber and, in theory, knew about choke points in the access network (TR-59 calls this hierarchical scheduling), and could assure that any traffic admitted to the downstream access network would be successfully delivered.

Several factors make this assumption questionable. First, in many cases, subscribers have the ability to select their own providers. If each provider has a separate BNG, no single BNG has a view of access network congestion. This issue can arise, not only within aggregation and access concentrators (think OLTs), but even within a single ONU, for example, an MDU, whose multiple subscribers may be served from several BNGs. Layer 2 business traffic—TLS—might not flow through a BNG in the conventional way at all. TR-59 says that traffic commitments beyond the ken of the BRAS must be subtracted from the resources available to the BRAS itself, but in

---

[*]BBF specifies the same capabilities at the upstream interface from the OLT toward the aggregation network, well beyond the scope of this book.

some circumstances, this rule could require us to subtract more than 100% of the network capacity.

Further, the OLT and the ONU are distinct and physically separate equipments from the BNG, and it is not easy to instantaneously communicate their congestion status to the BNG. By the time the BNG discovered congestion at some downstream choke point, the congested packets would all have been discarded due to queue overflow. The access network control protocol (ANCP) is sometimes advocated for real-time congestion control. The IETF ANCP working group[*] is more realistic, expecting only soft real-time capabilities.

The final and most important issue with centralized congestion management (admission control) is that the access network is a tree in the downstream direction, with several points of multicast replication. Because of IGMP proxies at various points (within an MDU, e.g., and again within the OLT), even a BNG integrated into the OLT would have no way to know which multicast groups were directed to which subscribers. Multicast traffic is expected to be a very large fraction of the total, and if it cannot be controlled from the BNG, the game is lost. We sometimes hear a counterargument *not* to have proxies in the access network—but then the network does not scale.

Intractable congestion issues can be overcome by overengineering network capacity. Within some reasonable bounds of cost, this may be the best solution.

6.3.2 Upstream QoS

We have seen how we can inspect upstream Ethernet frames as they enter the ONU from a UNI, how we can classify and mark them, assign priorities, either explicitly or implicitly, and map them into GEM port network CTPs on their way to and through the OLT toward the BNG. Now we need to address the questions of capacity, congestion, and performance.

The ONU-centric OMCI information model assumes that all upstream traffic congestion points are localized at the PON interface, the ANI. The ONU vendor approximates this model as well as possible within the constraints of real hardware and software design.

We first consider QoS within a single ONU. In the second half of this section, we discuss PON capacity assignment by the OLT, including dynamic bandwidth assignment (DBA).

6.3.2.1 QoS within an ONU

The ONU is responsible for ensuring fair treatment of the various traffic classes that may be committed by the operator to each of its various UNIs. We map each flow to a unique GEM port, but, except in the simplest of ONUs, the number of flows precludes us from mapping each single GEM port to its own T-CONT. In this section, we

---

[*] See https://datatracker.ietf.org/wg/ancp/charter/.

assume the need for multiple flows (GEM ports) to map into a given T-CONT. We also assume that each T-CONT serves only one traffic class, and that the OLT deals with different traffic classes at the level of PON-wide scope.

All Right, What Is a T-CONT, Really?

We said that a GEM port is a flow identifier, whose scope is limited to the G-PON itself. A T-CONT is a *group* of flows, and it is meaningful in the upstream direction only. It is the management layer equivalent of the alloc-ID (Chapter 4), the atomic unit of capacity management by the OLT's bandwidth assignment algorithm.

In the early days of B-PON, a T-CONT was conceptualized as an aggregate of flows (GEM ports) with arbitrarily different traffic classes. According to that model, the OLT would allocate resources between ONUs, but the ONU itself would allocate resources internally. An ONU would only need to support one T-CONT. Current practice is that an ONU supports several T-CONTs, each containing GEM ports that share a single class of service. The OLT then assigns upstream capacity, not with ONU granularity but on the basis of class of service.

Figure 6.13 is the classical picture that relates GEM ports to T-CONTs: we map some number of GEM ports (flows) into a T-CONT, which represents a specific traffic class. Each ONU supports one or more T-CONTs. Something like this diagram appears in every description of G-PON. To explore QoS, suppose we have three best-efforts flows but with different service-level agreements.

Although Figure 6.13 does not show it, a real ONU would necessarily include at least one queue. Traffic cannot instantaneously flow upward from the T-CONT. It is just that if there is only one queue, and we have no desire to tune its performance, we agree to omit it from the management model diagram—yet another reminder that management diagrams are *not* signal flow diagrams. When we dig a bit deeper, we shall see that we also want queues in our OMCI model.

First, consider what happens in the ONU of Figure 6.13 if GEM port 2000 blasts a large burst of data at full UNI speed into the single implicit queue. Assume for the moment that our single, unmanaged queue does not overflow. Nevertheless, any

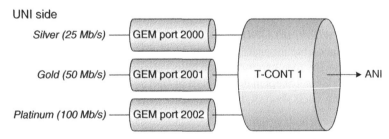

Figure 6.13 Relation of upstream GEM ports to T-CONT.

traffic that may originate during this time on GEM ports 2001 and 2002 must wait, a phenomenon called head-of-line blocking. All GEM ports, including even port 2000, will receive better service overall if their frames are metered into the queue so that everyone gets a fair share.

Further, since we hypothesized different service grades on our GEM ports, a fair assignment implies that for each byte contributed to the queue by GEM port 2000, we should be willing to enqueue 2 bytes from GEM port 2001 and 4 from GEM port 2002.

G-PON specifies the weighted round-robin (WRR) scheduling discipline for this situation.

Weighted Round Robin   Weighted round robin (WRR) is a form of scheduling in which contending traffic at a given priority level is assigned a share of resources according to the provisioned weight of each contender. Fairness is proportional to weight, so that numerically larger weights claim proportionately more resource. If R is a measure of assigned resource, and w is the assigned weight, fairness between competitors i and j is achieved when $R_i/w_i = R_j/w_j$, assuming that neither i nor j can be fully satisfied with less. Some implementations may measure R in bytes or bytes/second, while others may measure R in packets (per second).

Strict Priority   For completeness, now is the point to introduce the alternative. In strict priority (SP) queuing, a higher priority[*] client receives as much of the available capacity as it wants. Lower priority queues receive nothing until all higher priority queues have been completely emptied. Even here, however, we may want to meter the high-priority source's traffic injection rate to avoid head-of-line blocking. High priority does not imply license to exceed the traffic contract.

In a real ONU, it could well suffice to have one or more WRR T-CONTs, each T-CONT with a different strict priority or weight known only to the OLT. The next section discusses the OLT's role in upstream QoS.

How would we model our Figure 6.13 example in OMCI? Recall from Figure 6.4 that each upstream flow eventually appears on a GEM port network CTP. Figure 6.14 picks up the story there. The new MEs are the traffic descriptor, the priority queue, and the T-CONT itself.

T-CONT   The T-CONT managed entity is very simple. It contains the value of the TC-layer alloc-ID to which it responds in bandwidth maps (Chapter 4), and a discipline, either strict priority, or weighted round robin, or none, if it does not support scheduling at all. The T-CONT of Figure 6.14 need not support scheduling because it is fed by only one queue.

Traffic Descriptor   The traffic descriptor is the home of provisioned traffic parameters. It is associated with the GEM port network CTP, that is, with each

---

[*]In OMCI, 0 is the highest priority; in Appendix A of TR-156, larger numeric values have higher priorities. This is an error in TR-156.

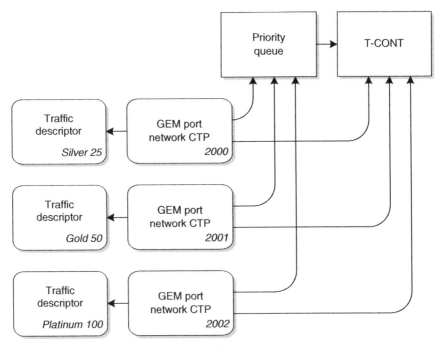

Figure 6.14 Simple scheduling.

individual flow; the same descriptor can be used by any number of GEM ports that share the same parameters. The attributes of the traffic descriptor include:

- Committed and peak information rate.
- Committed and peak burst size.
- Whether incoming upstream traffic is already color marked, and if so, how to interpret the marking. There is no explicit color field in an Ethernet frame; P bits, DEI, and/or DSCP AF (assured forwarding, RFC 2597) codes at the IP layer can be interpreted in various provisionable ways (see IEEE 802.1ad and RFC 4115).
- Whether to mark outgoing upstream traffic with the same choices. The traffic descriptor also allows for an invisible color mark to be used only within the ONU itself. If the frame survives, it goes on upstream with no alteration.
- Choice of algorithm, either RFC 4115 or RFC 2698 or the ONU vendor's default, which is probably one of these. Both describe two-rate, three-color marker (trTCM) algorithms. Although they are similar, RFC 4115 is somewhat more robust than 2698.

To understand the committed and peak attributes, suppose we have two buckets, whose sizes are, respectively, the committed and peak (= excess) burst size. These buckets are continuously but uniformly drained at the committed and peak information rates, respectively, giving rise to the alternate term *leaky-bucket algorithm*. The algorithms may also be described in terms of token buckets, with the idea that frames must be paid for with one token per byte, the tokens to be taken from buckets that are replenished at the provisioned peak and committed rates

By imposing a size on the bucket, we prevent the accumulation of arbitrarily large credits over the course of time. If the bucket overflows with tokens, well, that is just too bad—they are lost and gone forever.

When an upstream frame arrives, it is deemed to be green if it can be added to the committed bucket, or equivalently, if there are enough tokens in the committed bucket to pay for it. If it does not fit into the committed bucket, it is tried against the peak bucket. If it fits, the frame is marked yellow; if it does not fit, the frame is marked red. Red frames are discarded. A given frame goes entirely into one or the other (or neither) bucket but is not split between buckets.

An incoming upstream frame may already have been marked yellow by some previous scheduler, perhaps in the RG. Such a frame is tested only against the peak bucket: if it fits, it remains yellow, else it is marked red and discarded.

The OMCI traffic descriptor ME specifies how to mark frames with colors, and the priority queue ME includes enforcement attributes.

Priority Queue   The priority queue ME allows us to define probability thresholds that govern the discard of yellow—and green—traffic as a function of queue occupancy. We might expect that the queue should admit green traffic unconditionally, but even with green traffic, we may wish to specify what happens if the queue nears overflow, so the priority queue also provides thresholds for green traffic. Discarding a certain amount of traffic, either yellow or green, helps to signal higher layer protocols—TCP, in particular—to slow down.

We illustrate the threshold settings with the example of Figure 6.15. For traffic of a given color, as long as the queue occupancy is less than the lower threshold, frames are admitted unconditionally. If the queue occupancy exceeds the upper threshold, frames of that color are discarded unconditionally. For queue occupancy between the lower and upper thresholds, frame discard is a linear probability function, with a provisionable upper threshold probability value for each color. In Figure 6.15:

- We provision yellow traffic to be admitted unconditionally if the queue is less than 50% full.
- The probability of yellow discard increases linearly to 40% when the queue reaches 80% occupancy.
- Above 80% queue occupancy, we discard all yellow traffic.
- We admit green traffic unconditionally until the queue is 85% full.
- Above 85% occupancy, we discard green traffic with linear probability, up to 95% discard at 95% queue occupancy.
- Above 95% queue occupancy, we unconditionally discard even the green traffic. (This point is just for illustration: in real life, we would probably go all the way to 100% before discarding 100% of green traffic.)

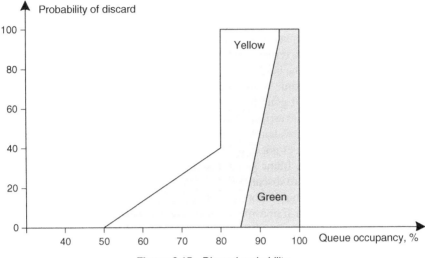

Figure 6.15 Discard probability.

Another QoS tuning mechanism is the ability to set the priority queue's depth. It would appear that we should just use all of the memory available for queuing, but this is not necessarily the right strategy. As a queue fills up, it increasingly delays traffic; at some point, service may deteriorate or even become unusable because of the additional delay. For real-time or semi-real-time traffic, if a packet is delivered too late, it might as well not be delivered at all.

Finally, we can set the priority queue to signal PAUSE frames to the incoming streams when it reaches an occupancy threshold. The user equipment's ability to recognize pause frames is of course not guaranteed, and this feature is not well-defined if several UNIs are each contributing a part of their individual flows to a particular queue.

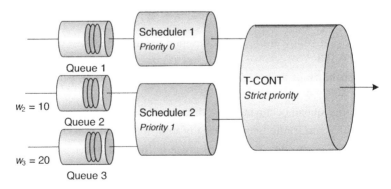

Figure 6.16 Strict priority among weighted queues.

Flexibility in QoS Resource Configuration

A few paragraphs ago we assumed that we had a WRR T-CONT. The scheduling discipline of a T-CONT, along with quite a number of other aspects of QoS, may or may not be provisionable.

From the drawing conventions, we observe that T-CONT and priority queue MEs are automatically created by the ONU itself. They are determined by the ONU's hardware design because they imply real, physical resources that are arranged in some particular way. It is correct to infer that different ONUs can be expected to have different capabilities, aligned to the operators' requirements and the market demand perceived by the ONU vendor.

According to the vendor's design, the ONU contains some number of T-CONTs, each with some number of priority queues, linked either directly or through some number of traffic schedulers. Not all of them need be used, of course, but the ONU design establishes a maximum that cannot be exceeded.

The OMCI equipment model expects T-CONTs and their related schedulers and queues to reside together on the same real or virtual circuit pack. These MEs may be fixed and permanent as part of the ONU itself; in a chassis-based MDU, they may also be created or destroyed along with their parent circuit pack MEs.

Imagining that they are wired directly into the hardware, the default legacy model has fixed associations between priority queues, traffic schedulers and T-CONTs, and their QoS disciplines. However, current silicon technology usually interrelates them through firmware, meaning that the relationships can be changed. Accordingly, the OMCI model allows for provisionable association among priority queues, traffic schedulers, and T-CONTs, subject to the constraint that they must always reside in the same cardholder (slot). The scheduling policy may also be read-only (default) or read-write.

Most OMCI models in G-PON have null defaults, but legacy OLT software may not be expecting provisionable priority queues. The ONU must therefore default to some useful preconfiguration of queues, schedulers, and T-CONTs. As well as their associations with one another, T-CONTs, priority queues, and traffic schedulers are preconfigured by the ONU vendor with a default scheduling discipline and the necessary related attributes. These associations and attributes can be used as is or reconfigured if the ONU supports flexibility and if the OLT is prepared to take advantage of it.

The OLT can determine whether the ONU supports flexible provisioning by querying the QoS configuration flexibility attribute of the ONU2-G managed entity.

We have said nothing about the traffic scheduler ME. Why not? The OMCI model does indeed allow us to construct arbitrarily elaborate trees of queues and schedulers, alternating between SP and WRR (Fig. 6.16 is a simple example). However, they appear to have no practical application, so we see no need to explore the options in detail. G.988 appendix II.3 describes the possibilities.

Why is there no need for a configuration such as that shown in Figure 6.16, in which we imagine a T-CONT that enforces strict priority multiplexing of traffic that contends with other traffic on a weighted basis? Because we can push the strict priority part into the OLT's bandwidth assignment algorithm instead. That is why we evolved to a single class of service per T-CONT.

6.3.2.2 QoS among ONUs of PON: DBA

In the previous section, we saw how upstream traffic is queued into T-CONTs (alloc-IDs). Now let us talk about assignments of capacity onto the PON.

The granularity of a capacity allocation is 4 bytes (XG-PON) or 1 byte (G-PON). Allocations can be as large or as small, as frequent or as infrequent, as the OLT chooses to make them. The ONU's responsibility is to enqueue traffic in one or more queues feeding each T-CONT, to report the queue backlog with T-CONT granularity when requested by the OLT, and, when granted permission, to transmit the traffic upstream according to the provisioned queuing disciplines.

PON Traffic Descriptor   The assignment of upstream capacity among the various T-CONTs—alloc-IDs—of the various ONUs on the PON is necessarily an OLT function. It is referred to as bandwidth assignment, and usually as DBA, because the dynamic aspects are the most challenging parts.

Following G.984.3, the XG-PON TC-layer recommendation G.987.3 defines a traffic contract in terms of a descriptor D, which is a vector:

$$D = (R_F, R_A, R_M, \chi_{AB}, P, w) \tag{6.1}$$

where

$R_F =$ Fixed bandwidth commitment. This bandwidth is assigned to the flow unconditionally, even if the source generates no traffic.

$R_A =$ Assured bandwidth commitment. The traffic contract commits to supply this capacity on demand, but if the source generates less traffic than R_A at a given time, the unused capacity may be used for other purposes. The transient response of the algorithm is important when an assured source starts and stops demanding capacity.

$R_M =$ Maximum bandwidth commitment. This defines the maximum capacity granted to the flow. According to the definition, no more than R_M is granted even if some PON capacity remains unused.

$\chi_{AB} =$ Specifies the basis for capacity assignment beyond $R_F + R_A$. It takes on values from the set {*none, nonassured (NA), best efforts (BE)*}. The NA and BE values are meaningful if and only if $R_M > R_F + R_A$.

$P =$ Priority to be used in assigning best-efforts capacity. P does not affect the nonassured contract.

$w =$ Weight to be used in assigning best-efforts capacity. Like P, the weight w is not involved in the nonassured contract.

This traffic descriptor, which is used in the context of OLT capacity assignment, is completely unrelated to the OMCI traffic descriptor managed entity of the previous section, which we used within the ONU to color upstream frames. In fact, the discussion in this section has no connection to OMCI: the action is in the OLT.

While a traffic descriptor D can model any traffic commitment anywhere, its use in G-PON is specifically in managing upstream PON capacity across the alloc-IDs in the various ONUs. Each alloc-ID i is provisioned at the OLT with a particular traffic descriptor D_i, and the OLT assigns upstream PON resources accordingly.

If the upstream PON link has capacity C, it is clear that across all traffic contracts i,

$$\sum_i \left(R_F^i + R_A^i \right) \leq C \tag{6.2}$$

Equation (6.2) may be viewed in several ways: first, it is a simple statement of physical possibility or impossibility; second, it imposes a provisioning constraint on the operator; third, it defines a business constraint on the operator's ability to sell service commitments on this particular link.

If it is straightforward to define a traffic descriptor for an aggregate flow of fixed and assured contracts, the opposite is true of nonassured and best-efforts contracts, each of which may be an aggregate, within the individual ONU, of several subscriber flows (GEM ports).

The traffic descriptor's maximum commitment must surely be as great as the largest of the individual contracts within each aggregate, but we can legitimately sell more capacity than the simple sum of the contracts. Statistical multiplexing allows for oversubscription; among other things, the level of oversubscription depends on the size of the contending group. For example, the statistics of flow i, an aggregate of 10 subscribers, each with $R_M = 10$ Mb/s, are not the same as those of flow j from a single subscriber with $R_M = 100$ Mb/s.

This problem lies beyond the scope of standardization. The operator must determine the aggregate traffic contract parameters according to its own practices, according to its own algorithm. In practice, the upstream link of a G-PON may be adequate for many years to come, such that difficult choices need not be made, at least not soon. Intractable problems can be rendered tractable through overengineering.

Now that we have some idea of the problem to be addressed, we dig further into the details of dynamic bandwidth assignment.

OLT's DBA Cycle   A capacity allocation algorithm works in consecutive blocks of 125-μs frames, frequently 8 frames (1 ms) or 16 frames (2 ms). For purposes of discussion, assume that we are on a 1-ms cycle.

Figure 6.17 illustrates how this works. During, or perhaps right at the end of, millisecond m, the OLT collects queue backlog information from all T-CONTs. In millisecond $m + 1$, the OLT plans a capacity allocation, a block of bandwidth maps that spans eight frames. During millisecond $m + 2$, the OLT broadcasts one frame of the block of allocations at the beginning of each downstream frame, and the ONUs execute the plan in transmitting their upstream traffic. Our example OLT contains a

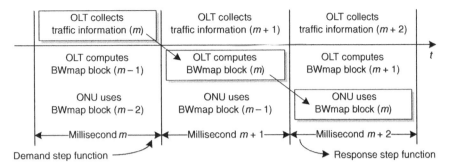

Figure 6.17 DBA cycle.

pipeline[*] engine, busy collecting additional traffic information and planning the next block of allocations.

Best case, the OLT may be able to react to a change in traffic demand in 1 ms. Suppose that the algorithm grants no overcapacity. Now suppose, best case, that a step increase in demand arrives—at an infinite arrival rate—just in time to be included in a DBA report at the end of data collection interval m. Best case, the increased bandwidth is assigned and effective at the beginning of interval $m + 2$, a bit more than 1 ms later. Response delay creates pressure to reduce the size of the DBA cycle. However, 1 ms is greatly reduced from the cycles of earlier generation products, and it may not be easy to further improve upon it.

One way to improve performance would be for the OLT to run sliding computational windows, for example, duplicating the entire structure of Figure 6.17, with an offset of 500 μs or redoubling it to duplicate the entire structure with an offset of 250 μs. With enough computing horsepower and enough heuristics in modifying previous bandwidth commitments, the response latency could be brought down considerably.

Why would we suggest sliding 1-ms windows, rather than just running DBA on a 125-μs cycle? Because not all upstream frames are equal. Let us see why not.

Assigning Fixed Bandwidth   The easiest task of the OLT is to allocate fixed capacity. It does not even need a dynamic algorithm for this. Figure 6.18 illustrates a 1-ms eight-frame allocation block, into which we insert several fixed allocations R^i_F for alloc-IDs i.

To minimize delay, we insert fixed allocations for alloc-IDs 1, 2, 3, and 4 every four frames, while by hypothesis, alloc-IDs 5 and 6 are less sensitive to delay, so we just slot them in once per block. To minimize jitter, we repeat the same allocation in every 1-ms block. We could even have fixed allocations that show up only every N milliseconds, if appropriate.

---

[*] This example illustrates the fastest way: a pipeline minimizes traffic delay and jitter and buffer memory. Overhead can be reduced by collecting status reports and computing capacity assignment levels much less often, as a tradeoff against response time.

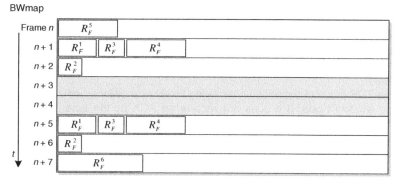

Figure 6.18 Fixed allocations.

Why every four frames, instead of every two frames, or every frame? And why are frames 3 and 4 grayed out and empty?

Recall that, from time to time, the OLT needs to open a quiet window to discover new ONUs on the PON. By reserving two frames for a 250-μs quiet window, and never putting fixed or assured allocations into these frames, we avoid jitter that would otherwise be introduced into this traffic. Fixed and assured traffic tends to have tight restrictions on delay and jitter, so it helps to be predictable. Our allocations $R_F^1 \ldots R_F^4$ occur at exactly uniform intervals; a 500-μs delay with zero variation is preferable to a 250-μs typical delay that once in a while might peak at 750 μs. The jitter buffer in the latter case would have to be deep enough to absorb the extra delay, so the best case delay of the de-jittered flow would be 750 μs.

A larger differential reach implies a wider quiet window and reduces the ability of the PON to minimize delay. The 450-μs window needed for 40 km effectively limits fixed allocations to once per millisecond, rather than twice. In the 40-km case, we might choose a 1.25-ms computational block, 10 frames. By doing so, we could maintain the practice of allocating $R_F^1 \ldots R_F^4$ twice per block, but at intervals of 625 μs rather than 500. Allocated less often, $R_F^1 \ldots R_F^4$ would be proportionately larger, and fixed-rate traffic would be delayed by an extra 125 μs.

How large should an allocation be? It is theoretically possible to allocate as little as 8 bytes of payload in G.987 XG-PON and conceivably a single byte of payload in G.984 G-PON. But the shortest legal Ethernet frame is 64 bytes long. Tiny allocations imply massive fragmentation and a great deal of potentially unnecessary burst overhead. From an efficiency point of view, it makes sense to have fewer, larger allocations.

Assignment of fixed capacity is the easiest capacity management task of the OLT, but as we see, even this is not entirely trivial.

Collecting Traffic Information   To go beyond fixed capacity assignment, the OLT needs information about the actual traffic load on the T-CONTs of the PON. It can obtain this in either or both of two ways.

Efficiency Is Not an Unmixed Blessing

Large grants at greater intervals are indisputably more efficient in terms of upstream PON efficiency. Efficiency is only one of our goals, however. Delay is important in real-time services; even some non-real-time services are affected by delay.

The TCP protocol depends on receiving timely acknowledgments (ACKs) on a sliding window basis. If it does not receive an ACK by the time it reaches its maximum retransmission window, the TCP server pauses and begins to time out. If it times out before it receives the ACK, it retransmits the last window. A constraint on upstream grant spacing is a limit on downstream speed and may lead to unnecessary retransmission.

This remains true even if the maximum upstream rate is provisioned high enough to convey all the ACKs, if the DBA algorithm grants bandwidth only in large chunks at comparatively wide intervals, say once per millisecond. The TCP server, which has been downloading at hundreds of megabits per second, hits its retransmission window, stops, and waits for ACKs, which arrive late and in bursts. TCP stacks differ in terms of timeouts and window sizes, so this effect is more pronounced on some stacks than on others.

As a general principle of layered architecture, it is desirable for nested feedback loops to have time constants that differ considerably. An order of magnitude is a good rule of thumb. For example, if the transport layer can do a protection switch within 50 ms, higher-order service restoration might hold off for a few hundred milliseconds before it attempts to reconfigure the network. Unfortunately, TCP and PON resource allocation have feedback loops with comparable timing, and neither has a lot of room for maneuverability.

- Status reporting DBA (SR-DBA) is based on queue occupancy reports, which are sent by the ONU in response to queries from the OLT.
- Traffic monitoring DBA (TM-DBA) is based on the ability of the OLT to observe the presence or absence of idle GEM frames in the payload of each alloc-ID (T-CONT).

There are advantages and disadvantages in each method. In TM-DBA, there is no way for the OLT to know whether the continuing absence of idle GEM frames indicates a trivially small demand in excess of the current supply or whether it indicates a dramatic surge in demand. The OLT may choose to ramp up the assignment gradually, hoping to satisfy it by degrees, or it may take the safe and simple—but potentially wasteful—approach of assigning capacity on the assumption that the T-CONT is offering traffic at rate R_A, or even R_M. If a status report is available, the OLT can respond in a more nuanced way.

A queue occupancy report provides better information, but it still does not indicate precisely how much capacity is needed. In G.984 G-PON, it is not even specified whether the status report includes or excludes the capacity contained in the current grant, the grant that requests the status report itself. The ambiguity is recognized, the

ONU is required to choose, and the OLT is required to deal with either option. G.987 XG-PON specifies that the current BWmap's allocation is to be subtracted from the reported queue occupancy.

But that is not the whole story. In G.984 G-PON, the status report is a nonlinear approximation, basically a logarithmic estimate of the queue backlog (Section 4.1.5), with quantization errors that round upward. In G.987 XG-PON, the status report is a linear measure, but it does not include GEM frame overhead. If the OLT grants capacity equal to the reported backlog, it will surely turn out not to be precisely correct. The OLT can apply some rules of thumb to grant a bit more capacity than is requested, but the only way for the OLT to know that it has drained the queue is to check for idle GEM frames in the upstream transmission.

Of course, all of this assumes a conveniently discrete traffic model: a block of traffic appears in an ONU queue. All of it gets reported to the OLT. Its T-CONT receives a transmission grant and transmits the complete queue, whereupon the cycle begins anew. The continuing real-world incremental arrival of additional upstream traffic, with asynchronous DBA report timing, means that everything we have said is an approximation: still, a good way for the OLT to know that it is keeping up with demand on a continuing basis is to check that there are a few idle GEM frames—but not too many—in the upstream transmission.

Assigning Variable Capacity As we discussed above, the OLT assigns fixed capacity on the basis of provisioned SLAs before it considers dynamic assignment at all. Then, given an estimate of traffic demand, the DBA algorithm assigns capacity to assured, nonassured, and best-efforts contracts in that order. But how much do we assign to a given alloc-ID? Above, we assumed that we would attempt to satisfy the demand completely.

The DBA algorithm does indeed satisfy any and all demands of assured contracts—that is what *assured* means. Because of the inherent delay in response, it may be desirable for DBA to overshoot when it observes an increase in the demand of an assured contract, helping to assure that any backlog is cleared expeditiously. Although the standards assume work conservation—that is, no intentional wastage of resources—there is clearly an opportunity for heuristics that may improve performance, especially if the upstream capacity is not oversubscribed in the current cycle.

After all of the assured commitments have been met, the OLT considers non-assured demand. Fairness in the nonassured class is defined to mean that whatever capacity remains in the PON is assigned to unsaturated nonassured contracts in proportion to their fixed plus assured commitments. That is, eligible alloc-IDs i and j receive nonassured capacity R_{NA} such that

$$\frac{R_{NA}^i}{R_F^i + R_A^i} = \frac{R_{NA}^j}{R_F^j + R_A^j} \qquad (6.3)$$

If an alloc-ID saturates, that is, if its fair assignment either satisfies its demand or reaches R_M, that alloc-ID receives its saturation assignment, and the remaining capacity is distributed among the other nonassured alloc-IDs.

The traffic descriptor parameter $\chi_{AB} \in \{$none, NA, BE$\}$ means that a given alloc-ID participates in at most one of the nonassured or best-efforts classes, never both. After all nonassured commitments have been fully satisfied, the algorithm considers best-efforts demand.

Whatever capacity remains in the PON is assigned to unsaturated best-efforts contracts in proportion to the headroom between their maximum rates and their fixed plus assured rates. That is, for any two eligible unsaturated alloc-IDs i and j, the best-efforts assignment R_{BE} ideally satisfies

$$\frac{R_{BE}^i}{R_M^i-(R_F^i+R_A^i)} = \frac{R_{BE}^j}{R_M^j-(R_F^j+R_A^j)} \tag{6.4}$$

For pure best-efforts service, that is, service in which $R_F = R_A = 0$, alloc-IDs receive capacity in proportion to their provisioned maximum rates R_M, as would be predicted.

As before, if an alloc-ID saturates, it receives its saturation assignment and the remaining capacity is distributed among the remaining unsaturated best-efforts alloc-IDs.

Recall that the traffic descriptor D includes optional[*] priority and weight components. These pertain only to best-efforts traffic, where they override the default best-efforts description, as follows.

Best-efforts contracts at the highest priority level are served first. Best-efforts contracts at each successive lower priority level are served only if all higher priority demands are completely saturated.

Within a given priority class, unsaturated best-efforts contracts are granted resource R_{BE} such that, ideally,

$$\frac{R_{BE}^i}{w_i} = \frac{R_{BE}^j}{w_j} \tag{6.5}$$

thereby divorcing the best-efforts assignment from its dependence on R_F and R_A. This is not to say that the total assignment is independent of R_F, R_A, or R_M: for alloc-ID i the total assignment would be $(R_F^i+R_A^i+R_{BE}^i)$, which may not exceed R_M^i.

Optimizations and Complications As mentioned above, it may be desirable to adjust the size of the BWmap computational block, depending on the size of the quiet window, which defaults to 20 km equivalent differential reach but may be altered by provisioning.

---

[*] Optional in G.984.3 and G.987.3. Mandated by TR-156.

The OLT DBA algorithm needs to know whether a quiet window is to appear in the upcoming BWmap block. If not, it can use the quiet window frames for dynamic allocations. Of course, quiet windows need not occupy an integer number of frame times; the 450-µs window for 40 km is a case in point.

In Figure 6.18, we dropped capacity allocations into the BWmap framework uniformly but otherwise more or less ad hoc, and we imply that the allocation of variable capacity would follow the same model. The natural result of this would be a large number of small bursts from each ONU, incurring a great deal of unnecessary burst overhead. So our DBA algorithm should rearrange allocations such that alloc-IDs belonging to a given ONU fit together into contiguous bursts, subject to the other restrictions, including the minimization of jitter for fixed capacity.

By now it should be clear that the answer to the G-PON efficiency question is the same as that given by the proverbial good accountant: "What would you like the answer to be?" And this is even before we consider FEC and the overhead for PLOAM and OMCI.

We mentioned other restrictions. What other restrictions? Well, for one, G.984 G-PON does not permit an upstream burst to cross a physical layer frame boundary. G.987 XG-PON removes this restriction.

Another factor for the capacity assignment algorithm to consider is that every alloc-ID needs some level of background assignment, especially if TM-DBA is in use, and also if the OLT and ONU are participating in a low-power mode (Section 4.5). As to the default alloc-ID (Chapter 5), the OMCC should receive allocations often enough to at least report OMCI alarms in a timely fashion.

Recall from Chapter 4 that the actual time occupied by the burst on the PON does not coincide precisely with the values specified in the BWmap. The DBA algorithm must therefore adjust the BWmap parameters to provide for the burst overhead, including a choice of profile. Time must be allocated for an upstream PLOAM message and (DBR), if requested, and for the possible expansion of the burst by FEC parity bytes. Reshuffling allocations from separate bursts into contiguous bursts saves overhead but implies additional bookkeeping in computing the proper overhead.

This is a significant amount of computation to be repeated for dozens or hundreds of alloc-IDs every millisecond. Given the intrinsic uncertainties associated with random traffic arrival rates, we would surely forgive a DBA algorithm that incorporated a few shortcuts.

Finally, there are rules for the construction and interpretation of the BWmap. The following list comes from G.987 XG-PON. There is no explicit list for G.984 G-PON.

- The BWmap must be structured such that start time values increase monotonically. This gives the ONUs the maximum possible time to structure their upstream bursts, knowing that no reordering of start times will be needed.
- The maximum start time is 9719 words, the last word of the 125-µs upstream frame).

- The maximum allocation size is 9720 words. Combined with the latest possible start time, we see that a burst based in frame N could in fact consume all of upstream frame $N + 1$, if the OLT so desired.
- Potentially concatenating several allocations, the maximum burst size is 9720 words.

The remaining G.987 rules exist to bound the complexity that may need to be supported by an ONU. The values are arbitrary, but are thought to be large enough that they will not cause problems in real-world applications.

- No BWmap may contain more than 512 allocation structures.
- In a given BWmap, no ONU may receive more than 64 allocation structures.
- No BWmap may assign more than four bursts to a given ONU.
- No burst may contain more than 16 contiguous allocation structures.
- If the OLT allocates consecutive bursts to the same ONU, the bursts must be spaced by at least as much as would be required for bursts allocated to different ONUs, plus an additional processing margin of 512 bytes. This is a not very subtle incentive to combine allocations into single bursts whenever possible.

As to the ONU, it interprets BWmaps with a few rules of its own, which we omit. A properly debugged OLT should never send illegal BWmaps, so the ONU should never need to enforce the rules. The interested reader is referred to G.987.3.

One note on open issues related to QoS: we have assumed that QoS parameters are statically provisioned, both into the OLT and the ONU. But when a fourth-generation wireless device opportunistically connects to a G-PON ONU and demands several tens or hundreds of Mb/s of capacity, we will need dynamic adjustment of that ONU's QoS, not only satisfying the wireless user's contract but balancing the new demand against the resource commitments that already exist on the PON.

6.3.3 Congestion Points within an ONU

At the time of writing, discussions were beginning on the validity of the assumption that the only congestion points in the ONU were at the ANI (upstream) or the UNI (downstream). ONUs such as MDUs may well have PON MAC chips connected through comparatively narrow [gigabit media-independent interface (GMII)] channels to Ethernet switch chips on the UNI side.

The proposal under discussion is to define some form of bridge interworking model (Fig. 6.19), which would represent the ANI from the viewpoint of the Ethernet switch on the UNI side. At the same time, it would represent a UNI to the bridge contained in the PON MAC device. In such a model, everything we have said about QoS would remain true, but could be applied twice on each flow as it traverses the

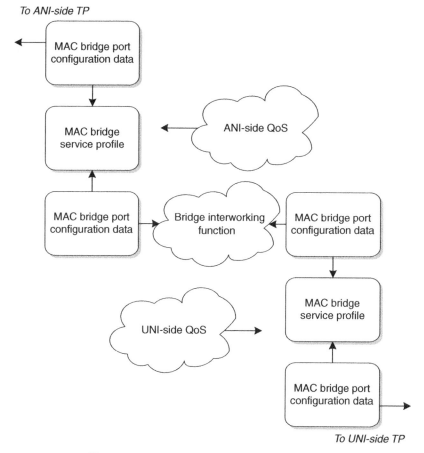

Figure 6.19 Separating ANI- and UNI-side bridges.

ONU, and thereby give us some control over the congestion point at the internal interface.

This is similar to the configuration addressed by BBF TR-167, on a G-PON-fed DSLAM, but differs because here OMCI manages both the ANI and UNI sides.

6.4 IP SERVICES

We started this chapter with Ethernet because, to a first approximation, an ONU is a layer-2 device. That is, it cares mostly about Ethernet frames, MAC addresses, and VLANs and less about higher layers of the stack, in particular the layer-3 Internet protocol IP. Certainly, an ONU may contain a residential gateway function, which is usually an IP device, but such functions are not managed via OMCI. Either they are managed by the service provider by means such as Broadband Forum's TR-69 suite or they are free for the subscriber to manage directly.

Residential Gateways

A residential gateway is a somewhat open-ended device that evolved from its original function as a home router connected to a DSL modem. In G-PON, a simple RG may be nothing more than a MAC bridge from a single PON-side Ethernet service to several Ethernet UNIs. More common RGs also include routing, with NAT between UNI and ANI, with DHCP and DNS servers, and possibly with additional kinds of physical interfaces such as 802.11 wireless. Such RGs are often purchased and managed by the subscriber. If the RG also contains ATAs, some parts of it are also managed by the service provider.

The RG is evolving into something of a centralized information appliance that may also contain a print server or a storage server. Other functions that could be centralized in RGs—or not—include coordination of home security monitoring, smart meter reading and other machine-to-machine (M2M) applications.

Telecommunications operators may offer such devices as remotely managed products, but there is no enthusiasm to use OMCI to manage them. Broadband Forum publishes RG requirements in TR-124 and publishes a series of information models that describe how to manage various RG features with the TR-69 protocol.

This functional partition is not the result of a grand architectural plan but simply represents the way in which the market has evolved. Indeed, OMCI does include a model intended for managing an embedded router, although it must be said that the model would require further study if it were actually to be used. Instead, the lack of community interest caused the router model to be left behind when the previous OMCI recommendation G.984.4 was migrated into the current recommendation, G.988.

What do we mean when we say that an ONU is mostly not a layer-3 device? We have already touched on a few applications in which the ONU must be aware of IP, most recently in the multicast section—IGMP and MLD—and the mention of DSCP priorities above. From Section 5.1.3.2, we know how to represent IPv6 addresses in OMCI, either as hybrid IPv4/v6 addresses or in their own right through new attributes. But the real application for IP functionality lies in the ONU's option to provide integrated VoIP telephony. If we are to have voice over IP, we cannot rely on the ONU doing no more than just recognizing IP packets for snooping, discarding, or expediting; the ONU needs its own IP stack.

The core of an ONU's layer 3 functionality is the IP host config data managed entity, which represents an IP protocol stack.

IP Host Config Data The original OMCI model for an IP stack was designed for IP version 4. It is represented by the IP host config data ME, which is sufficiently specific to IPv4 that a separate ME definition is warranted for IPv6. This is reasonable; if an ONU supports coexisting IPv4 and IPv6 stacks, it will want separate MEs for them anyway. The IPv6 stack is new software, so a new model is not surprising.

How is the IP stack used? Several other OMCI MEs can point to an IP host config data ME.

- TCP/UDP config data—this is the most meaningful associated ME. It models one of the two transport protocols TCP or UDP, along with the port in use by a particular application. The application in question—VoIP, for example—links to the TCP/UDP config data, which in turn links to the host.
- MAC bridge port config data—modeling the mapping from IP onto Ethernet.
- 802.1p mapper.
- (Extended) VLAN tagging operation config data, allowing us to classify and tag the Ethernet server layer underlying the IP client.
- IP host PM.

With the exception of the IP host PM ME, which is implicitly linked, each of these MEs contains a pointer to an IP host. Prior to the advent of IPv6, this meant a pointer to an IPv4 stack. If the ONU also contains an IPv6 stack, modeled with a separate ME class, the pointer potentially becomes ambiguous. It would have been possible to create a new code point in each of the associated MEs to indicate whether it was associated with an IPv4 or an IPv6 stack, but a client ME such as a TCP/UDP (transmission control protocol/user datagram protocol) config data rarely or never cares what kind of an IP stack it is using (evidence of a nicely layered architecture!).

Rather than requiring explicit new provisioning of a TP-type attribute, G.988 specifies that the ONU create IP and IPv6 host config MEs within a single ME ID namespace. Because the addresses never coincide, an IP host config pointer can never be ambiguous.

Both versions of IP host config data allow the IP address either to be manually provisioned or to be assigned by a DHCP server. IPv6 always assigns itself a link-local address. IPv6 is also capable of listening to router advertisements (RAs) and is capable of assigning its own globally unique address, based on DHCP or RA information. Following the self-assignment of an address, IPv6 performs duplicate address detection (DAD), and iterates on its choice of address if there is a conflict. An IPv6 interface can theoretically have several global addresses, but, if the address is to be manually provisioned, OMCI assumes that one suffices.

In both versions of the IP host config data, a set of read-only attributes exposes the current values of information such as IP address(es), gateway or default router, DNS servers, and IPv6 on-link prefixes. In a manual provisioning regime, this read-only information matches the manually provisioned values.

Testing The OMCI *test* and *test result* commands provide ways to test an IP route, either by requesting a series of pings (ICMP echoes) or with a traceroute. The far-end IP address may be explicit in the test request or may be provided by way of a string to be interpreted by a DNS server. This latter feature verifies DNS connectivity and functionality as well as the ability of the network to transfer IP packets.

6.5 POTS

We said that the IP stack was primarily needed for the support of VoIP, delivering POTS.* Despite competition from cell phones, the need, and consequently the ability, of G-PON to offer POTS goes without saying.

6.5.1 Overview

Speech Signals   A telephone set is a transducer that converts sound waves to baseband analog current variations in the transmit direction. In the receive direction, it converts current variations into sound waves. For many years, these analog waveforms have been digitized by encoder–decoder (codec) circuits and conveyed through the telecommunications network in the form of digital streams. To conserve bandwidth, back in the days when bandwidth was expensive, the coding laws (specifications) are nonlinear (see ITU-T G.711). In North America and Japan, the μ law predominates, whereas the A law is used in most other places. In both domains, the standard format is a stream of 8-bit samples derived at an 8-kHz sampling rate (thus the universal 125-μs frame), for a flow rate of 64 kb/s.

A single pair of wires conveys analog baseband energy in both directions simultaneously to and from the telephone set. If the two-wire circuit's impedance does not match that of the telephone set, energy is reflected. Speech energy at zero delay is intentionally fed from transmitter to receiver as so-called sidetone, which gives the phone a live sound, but delayed (far-end) reflections are audible as echo. For a given amplitude, echo is less tolerable as delay increases. Because of the packetization delay common to VoIP, most current telephony circuits include echo cancellation features. Absolute delay is still a problem, as we all know from the problems of talking on top of one another.

All G-PON-based telephony is VoIP. Speech signals are often carried over a real-time protocol layer (RTP, RFC 3550), which assists in maintaining timing in networks that cannot guarantee synchronicity.

Signaling   As well as power for speech, telephony requires power for signaling. A conventional telephone set appears as an open circuit when it is on-hook. When the subscriber lifts the handset, the switchhook closes and current flows. This provides the bias current that is modulated by the speech waveform; at the same time, the central office end of the loop recognizes the onset of current flow as an off-hook event and provides dial tone.

The subscriber signals the desired destination in one of two ways. Legacy telephones interrupt the loop current with a series of dial pulses. The number of

---

* The term *POTS* is sometimes used in the sense of an RJ-11 jack with remote power feed, with *VoIP* used to suggest an ATA that is visible to the subscriber, possibly terminating in a device that is not a traditional telephone (e.g., headset plugged into a PC). This is a false dichotomy—all G-PON telephony is VoIP. The real distinction is between best-efforts VoIP and VoIP with a QoS contract.

interruptions conveys digit information; digits are distinguished from dial pulses by timers. Dial pulse telephones are still supported, but most telephones today generate tone signaling, in which a digit is determined by the combination of two tones, one each from two groups of four frequencies. The dual-tone multifrequency (DTMF) spectrum was carefully designed to avoid harmonic relations between any of its tones, which is why DTMF signals do not sound musical.

At the called end, the central office equipment (in G-PON: the ONU) generates a ringing or alerting signal. A common alerting pattern in North America is a 2-s ringing burst, followed by 4 s of silence; national and regional standards vary widely. The ringing signal is a high-voltage low-frequency wave, capable of delivering a substantial amount of power, which was at one time necessary to operate a mechanically vibrating bell. At the same time, the network may send calling party name or number information to be displayed on the called telephone set.

While the called party is being alerted, the central office equipment generates an audible ring[*] signal to the calling party. When the called party answers, the telephone network cuts through an end-to-end speech path and may create an accounting record for billing.

From this very brief summary, it should be apparent that, as well as a speech path, we need signaling to and from each end of the connection. In the pre-VoIP telephone network, subscriber signaling is carried in the bit stream to a so-called class 5 switch, which would today be regarded as a POTS application server. That is, the class 5 switch contains the intelligence to interpret and generate subscriber signaling, route the call to the next hop, apply features such as calling party identification or three-way calling, and record call information for billing or traffic analysis.

As legacy equipment, class 5 switches are not directly capable of terminating VoIP. ITU-T H.248 is a signaling protocol well adapted to class 5 switches serving VoIP through so-called media gateways.[†] For some time, H.248 will be a common signaling protocol, but as legacy class 5 switches are retired—they are a sunk cost to the operator and will be kept in service until they no longer generate enough revenue to pay for their own upkeep—SIP signaling is expected to predominate.

SIP is the session initiation protocol (RFC 3261). As well as for voice telephony, SIP can be used for other media types; in fact, it underlies IMS, the IP multimedia system. SIP supports the idea that the applications intelligence resides in the endpoints (the ONUs or external ATAs), although it does allow for SIP proxies at intermediate points between the end stations. As might be expected, the OMCI model for H.248 is much simpler than the model for SIP.

Be it H.248 or SIP, VoIP carries signaling in its own logical channel.

VF Specials A wide variety of voice frequency (VF) special services has been defined over the years, coin phones being the best-known example. VF specials that

[*] This tone is often called ringback. To the grandparents of the telephone pioneers, ringback was a quite different signal with a quite different purpose. Times change.
[†] H.248 is sometimes known as media gateway control protocol (MGCP), and the MGC acronym appears in OMCI and elsewhere.

cannot be served by simple adaptations of POTS are unlikely ever to become directly available from G-PON systems. This is not a limitation of G-PON's capabilities, merely a reflection of the cost of developing and supporting a specialized interface whose total market might be 50 units. The alternative is that such services may be delivered over G-PON in the form of fractional DS1 or ATM pseudowires, ultimately to be supported on private automatic branch exchange (PABX) or other legacy hardware. Pseudowires are the topic of Section 6.6.

If the ATA is external to the ONU, the only concern of the G-PON equipment and the OMCI management model is to properly classify and prioritize the VoIP stream for the necessary QoS. In best-efforts VoIP, the network may know nothing about the underlying service at all.

An ONU may itself contain one or more ATAs through which POTS telephones are connected directly. If the ATA is integrated into the ONU, questions of power, heat dissipation, surge protection, drop testing, and such must be resolved by the ONU designer, as well as software for signaling and management. Without wishing to minimize the importance of hardware design, we focus our discussion here on the configuration and management of POTS service. An integrated ATA may be managed by BBF TR-69 and TR-104, just as if it were separate, or it may be managed by OMCI. We first consider the OMCI case.

6.5.2 POTS UNI

Different operators in different countries have their own unique requirements, both at the hardware and software levels, and an ONU intended for world markets must cope with these variations. Fortunately, components have evolved to the point that this is today primarily a matter for software and is therefore a demand for configurability.

As with all other user–network interfaces in G-PON, OMCI represents the POTS UNI with an ME called a physical path termination point, in this case, the PPTP POTS UNI. Significant attributes of the POTS UNI include:

Impedance, provisionable such that the ONU UNI matches the expected sub-scriber set as well as possible. A good impedance match reduces reflections and echo, but because subscribers can, and do, connect arbitrary numbers and types of equipment to telephone drops, this attribute may be mostly useful for troubleshooting of exceptionally bad cases.

On-hook transmission, full-time or part-time. A speech path may exist even when the telephone set is on-hook to support services such as intrusion monitoring. Otherwise, an on-hook speech path may only be cut through during the alerting phase of call setup, when it is needed for features such as calling number delivery, and then be torn down when the call disconnects.

Receive and transmit gain settings. Switches introduce their own loss, but it may be necessary to add or reduce the amount of loss in particular cases. Again, this attribute is more likely to be useful in responding to subscriber complaints about the telephone never being loud enough, or always being too loud, to the far-end party. Direction is defined from the viewpoint of the telephone set, so the receive direction is from the network toward the subscriber, while the transmit direction is upstream from the subscriber. This convention is consistent with Telcordia GR-909, a common reference in remotely served POTS applications.

Holdover time. Intrusion alarms commonly rely on continuous loop voltage on a copper pair to the central office, the equivalent of a POTS circuit. If an intruder cuts the copper pair, the intrusion detector times out and activates a local alarm. An ONU may emulate this behavior by holding the loop voltage up for a provisionable amount of time after losing its connection from the PON, to avoid false alarms, then releasing it so that a local alarm circuit can operate.

6.5.3   H.248 Signaling

Figure 6.20 illustrates the major MEs and their relationships in an H.248 VoIP service.

As we see, several profiles govern VoIP, including one called *VoIP config data.* It is automatically created by the ONU itself, and is intrinsically associated with all VoIP services on the ONU. Its significant capabilities include:

Signaling protocols supported by the ONU, one or more of
- SIP
- H.248
- MGCP.

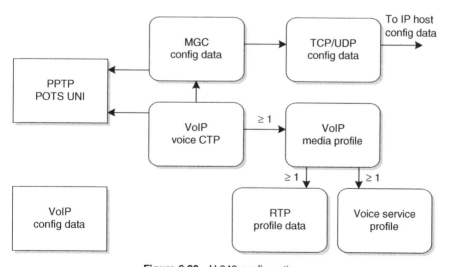

Figure 6.20   H.248 configuration.

Signaling protocol currently in use.

VoIP configuration methods supported by the ONU, one or more of
- OMCI
- Configuration file retrieval, further discussed in Section 6.5.6
- Broadband Forum TR-69
- IETF sipping config framework (RFC 6080)

VoIP configuration method currently in use.

VoIP configuration file server uniform resource ID (URI)

VoIP configuration file state:
- Inactive: configuration retrieval has not been attempted.
- Active: configuration was retrieved.
- Initializing: configuration is now being retrieved.
- Fault: configuration retrieval process failed.

There is only one VoIP config data in an ONU, but there can be any number of the other profiles, and different POTS ports can associate with different profiles.

Significant attributes of the *voice service profile* include:

Announcement type—what happens when a subscriber goes off hook but does not attempt a call within the dial-tone timeout interval.
- Nonspecific: ONU vendor's default
- Silence
- Reorder tone
- Fast busy
- Voice announcement

Jitter target—target value of the jitter buffer in milliseconds.

Jitter buffer maximum depth

Echo cancellation—on, off

PSTN protocol variant—the ITU-T E.164 public switched telephony network (PSTN) country code. More on this later.

DTMF digit levels and duration—the power level and duration of dual-tone multifrequency digits that may be generated by the ONU toward the subscriber set. These digits are delivered to a telephone that is capable of displaying information about the calling party.

Hook flash minimum and maximum time settings—the minimum (maximum) duration to be recognized by the ONU as a switchhook flash. A shorter interruption in loop current may be recognized as a dial pulse or ignored, depending on call state; a longer interruption is recognized as a disconnect.

The remaining attributes are recent additions to the OMCI model and are based on BBF TR-104.

Tone Pattern Table   A table, each of whose entries specifies a complex tone (or silence), a duration, and an optional reference to another table entry. By linking tones and silence together, possibly cyclically, it is possible to define continuous, varying, or interrupted tone sequences, repetitive or not. Not only can the tones themselves be composites of up to four frequencies, but they can be modulated to create a warble effect.

Tone Event Table   A table, each entry of which links a telephony event with a tone pattern. As an option, the event may be linked with a file, for example, to play a voice announcement. Events include:

- Busy tone
- Confirmation tone
- Dial tone
- Message waiting
- Off-hook warning (idle, receiver off hook, timed out without having originated a call)
- Ringback (really: audible ring)
- Reorder
- Stutter dial
- Call waiting
- Alerting signal
- Special dial
- Special info
- Release
- Congestion
- Intrusion
- Dead tone

Ringing Pattern Table   A table, each of whose entries specifies a ringing pattern, a duration, and an optional reference to another table entry. By linking ringing and silence together, possibly cyclically, it is possible to define continuous or interrupted ringing sequences, repetitive or not.

Ringing Event Table   A table, each of whose entries specifies a telephony event for which a ringing sequence is defined. As with the tone events, an entry may point to a ringing sequence or to a file that plays, in this case, a ring tone. In addition, a text string may be specified, to be displayed on the subscriber's telephone set. The ringing events are default ring and splash ring.

The *VoIP media profile* ME includes capabilities to specify fax mode, one of upward of a dozen voice coding standards, with the ability to provision an order of preference in case the first choice is not supported by the far end, packetization time in milliseconds (recommended value 10), whether to suppress transmission when the codec detects silence, and whether DTMF digits are carried in-band or in a parallel signaling channel.

The last of this set of profiles is the *RTP profile data* ME. It specifies the range of UDP port numbers to be used, how to set the DSCP bits on outgoing RTP packets [expedited forwarding (EF), recommended], and it enables or disables piggyback events, tone events, DTMF events, and channel-associated signaling (CAS). RFCs 4733 and 4734 are the place to learn more. It also provides a pointer that permits voice traffic to be carried in a separate channel from signaling.

We need yet a few more MEs to set up our H.248 service.

The *VoIP voice CTP* mostly just points to other MEs, but this is also where we specify how the subscriber originates a call:

- Loop start—the usual method, in which the switchhook simply closes the circuit between tip and ring leads
- Ground start
- Loop reverse battery
- Pay phone, coin first
- Pay phone, dial tone first
- Multiparty

And, yes, although still lumped in with POTS, these capabilities go beyond conventional POTS into the realm of VF specials. And yes, most of these options may never be implemented in real ONUs.

The *MGC config data* ME specifies addresses for redundant media gateway control (MGC) servers and their connection parameters.

Finally, the *TCP/UDP config data* contains the IP parameters needed to set up the IP connection. We look further into that aspect later on.

6.5.4 SIP Signaling

From the introductory remarks of Section 6.5.1, it is not surprising that SIP configuration is more flexible and more complex than H.248 provisioning. H.248 features are largely supported by the remote class 5 switch, while SIP is based on a distributed model, many of whose functions are expected to be supported by the ONU itself.

The ONU model includes a SIP agent that represents the VoIP endpoint. In theory, it can communicate directly with a far-end SIP agent, but the practical needs of large telecommunications networks usually imply a proxy server and a SIP registrar. The proxy acts on behalf of the endpoints and knows how to route calls and where to find servers. The registrar is responsible for authentication and billing.

Figure 6.21 illustrates the MEs involved in a SIP voice service. The lower part is the same as in Figure 6.20. But where the VoIP voice CTP pointed to an MGC config data ME in the H.248 case, it now points to a SIP user data ME.

Along with the VoIP voice CTP and the authentication security method, the SIP user data ME specifies the unique details of each individual SIP termination. Most of the other MEs are profiles, so-called or not, which may be shared by any number of clients.

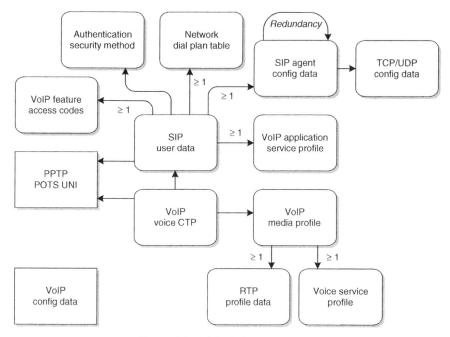

Figure 6.21 SIP VoIP configuration.

In the *SIP user data* ME, we have the user address of record (AOR)—possibly the directory number—that identifies this user to the SIP server, a few timers and a number of pointers to tie the management information together. This ME also contains a pointer to the URL of a possible voicemail server.

The *authentication security method* contains login information for the SIP agent and could be unique to a SIP client instance.

The *SIP agent config data* contains URIs that specify SIP proxy servers, SIP DNS servers, registrars, registration timers, and related information. We can also specify a redundant SIP agent to be used if the first one is unavailable. The primary agent is the one pointed to by the SIP user data; the secondary agent is the one pointed to by the primary agent; and the secondary agent may point to yet another backup agent.

The *VoIP application service profile* includes attributes to enable various options such as caller ID, call waiting, call progress, or transfer features, and call presentation features. There are also URIs for a direct connect remote endpoint, a bridged line agent, and a conference factory.

The *VoIP feature access codes* ME is a list of the digit sequences to be recognized as invocations of a variety of features such as call hold, caller ID activate/deactivate, intercom service, and the always essential emergency number.

The *network dial plan table* contains partial and critical dial timeout values, along with a dial plan. A dial plan is a set of digit patterns; when the ONU sees a pattern match, possibly with a timeout, it recognizes that the subscriber has completed dialing some particular kind of number and forwards the number to its server.

Dial Plans

A dial plan is the template by which a dialed digit receiver knows when a given string of digits is complete, so it can translate them into the address of a far-end termination somewhere in the world. The alternatives are:

- To time out on the dialed digit string before forwarding any of the digits, which would delay response and annoy subscribers, or
- To forward every digit as it is received, which adds unnecessary signaling traffic to the network and presupposes that all signaling goes to the same signaling server, or
- To have a start-dial signal that permits the subscriber to explicitly indicate when the digit sequence is complete[*]

The preferred solution is none of the above but, instead, pattern definition and recognition by way of the dial plan.

To use North America as an example, three digits ending with -11 (411, 511, 911) are usually interpreted as special numbers, and are forwarded without waiting for additional digits. Seven digits (555–1212) refer to a specific called number within the same area code, unless 10 (or 11-) digit dialing is in effect. In the workplace, we frequently dial 4 or 5 digits for internal company numbers, the first digit of which is predefined by the PABX, with an escape code such as 8 or 9 to get into the public numbering plan.

All of these possible strings are either unambiguously definable as patterns or can be recognized as patterns after a timeout that implies the subscriber does not intend to dial further digits.

As an example of the format, "0T" is a dial plan table entry that would usually appear in North American dial plans. The string matches the single dialed digit 0 after a timeout that allows for possible additional dialed digits.

Dial plans have existed deep in the heart of the telephone network since the obsolescence of progressive signaling. SIP wants the client to be intelligent, so the dial plan comes out to the periphery.

6.5.5 IP and Ethernet

The previous section showed how to define POTS from the telephony point of view. Using SIP as an example, Figure 6.22 provisions the stream into IP and Ethernet, and onto the PON. We pick up the thread at the TCP/UDP config data ME, which specifies a protocol [UDP in this case, usually augmented with real-time protocol (RTP)], a port, a DSCP field, and a pointer to the IP host config data. Observe that any number of POTS ports can use the same SIP agent, thereby the same UDP port.

---

[*] The # key evolved from MF signaling to DTMF as a start-dial signal from the days when operators manually set up long-distance connections. In many cases, it still works today.

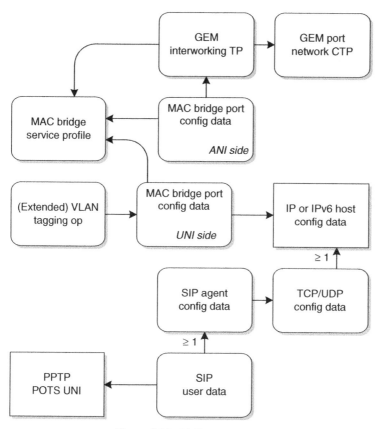

Figure 6.22   VoIP connectivity.

From the IP host config data, the connection toward the T-CONT is exactly what we would expect, including a possible extended VLAN tagging operation ME that permits us to add VLAN tags; we always add at least one tag. The most common voice service model is 1:*N* tagging, in which VoIP traffic from several ONUs, each with a separate C tag, is aggregated into a single VLAN at the OLT, which adds an S tag that directs the VoIP flow into the core network.

The model of Figure 6.22 assumes that voice and signaling traffic coexist on the same IP address. Voice traffic—particularly peer-to-peer SIP-based VoIP— necessarily exists on a quite public network because it can flow anywhere in the world, while for security reasons the operator may wish to segregate signaling traffic into a private network. OMCI includes an optional attribute in the RTP profile data ME that permits voice traffic to be directed to a different IP host config ME, thence to a different destination. Parallel construction would suggest that the pointer be to a TCP/UDP config data, but voice traffic is always carried over UDP, and the UDP port number is already specified in the RTP profile.

6.5.6   Configuration Beyond OMCI

The OMCI model includes a large set of standardized POTS variations, but many operators around the world have defined unique features of their own. An attribute of the voice service profile allows the country code to be provisioned, which was intended to permit variations to be coded into the software.

But some large and diverse countries have more than one standard. In addition, vendors like to take their products into new markets without first having to go through a complete software release. And finally, a chip set may require bits and bytes that just do not lend themselves to standardization.

So OMCI has an escape mechanism that allows custom files to be downloaded and applied to VoIP terminations, using the same MEs and procedures that apply to software upgrade (Section 5.2.9). This might be described as a push model, with persistent storage. In contrast, the config file described in the VoIP config data ME is understood to be available and downloaded afresh whenever the ONU initializes (pull model, cached storage).

OMCI also defines two config portal MEs, one for H.248 signaling and one for SIP. Their purpose is to provide read-only windows of VoIP configuration information. The content and format of the readback are unspecified. It could be, but is not required to be, a readback of the config file itself, for example, from a TR-69 ACS or of some vendor-specific file format.

If VoIP is configured through a separate file, most of the MEs we have previously discussed become unnecessary. Surviving MEs include the PPTP POTS UNI, of course, and the VoIP config data—the tool that sets up config file mode in the first place. It is, of course, permitted, and probably preferred, that the OMCI model survive intact, with the configuration file used merely to populate the attributes of the model. That would permit a standard OMCI-based manager to at least query the configuration of the VoIP service.

We have not talked about VoIP performance monitoring. Suffice it to say that OMCI defines five MEs to collect PM on a number of aspects of VoIP, as well as a VoIP line status ME. These MEs are still valid if POTS is provisioned by way of a config file, so that the OMCI manager can see what is going on for purposes of monitoring and troubleshooting. Three of the PM MEs are not tied to the PPTP POTS UNI; they can only report on the ONU as a whole. There is no possibility of granular views, for example of separate SIP agents. To date, this has not been an issue since no one has needed to proliferate SIP agents.

6.5.7   ATA External to ONU

If the analog telephone adaptor is logically or physically separate from the ONU, OMCI cannot be used to provision the service. We mentioned Broadband Forum's TR-69 as one of the common non-OMCI provisioning methods, together with the corresponding information model in TR-104.

From the viewpoint of the G-PON ONU and OMCI, the interface to an external VoIP service is a flow contained in a real or virtual Ethernet UNI. The G-PON ONU side of the real or virtual Ethernet interface is expected to classify, tag, and prioritize the IP flow associated with VoIP, just as it would for any other Ethernet-based service.

6.5.8 Loop Test

POTS testing is one of the more complicated functions of an ONU, but also one of the more important functions. There are two perspectives and tests for each: looking out and looking in.

The outward-looking set of tests evaluates the drop wire and the telephone set. It measures hazardous and foreign potential, both AC and DC. It measures resistance between tip and ring or the resistance from either to ground and assesses the presence and magnitude of a ringing load. It can return measured values or declare pass–fail against thresholds included in the test command.

The inward-looking set of tests is intended to evaluate the ONU's own ability to generate telephony signaling and confirm that the signaling is in fact interpreted correctly at the far end. The dial-tone make-break test generates an off-hook signal, checks for the presence of dial tone, and measures dial-tone power and delay. Then it generates a tone or pulse digit to break dial tone, measures dial-tone break delay, and measures quiet-channel power. Clearly, these tests are less significant in a SIP signaling environment, where most of these activities are entirely local to the ONU itself. A variation on this test specifies that the ONU dial a complete digit string, presumably to a test line.

6.6 PSEUDOWIRES

POTS is certainly a TDM service, but it is universally treated as an application in its own right, as we do in this book. Business customers may need what are usually designated TDM services, typically full or fractional DS1 or E1. These are carried over the packet network, and over G-PON in particular, in the form of pseudowires. The standards—and OMCI—also define Ethernet and ATM pseudowires. Any of these pseudowire types may be needed for another important application, namely mobile backhaul.

This section opens the door to a great deal of information that is necessarily beyond the scope of this book. We try to fill in a certain amount of background and indicate references, but the newcomer will have to do considerable external homework to fully grasp this material.[*]

---

[*]Especially for readers with a North American focus, Telcordia's GR-499 is an excellent introduction to many of the legacy TDM telecommunications interfaces.

Pseudowire Standards

A brief summary of a few of the standards follows. Additional standards are mentioned in the text and listed in the references section at the end of the book.

Standard	Topic
RFC 3916	Pseudowire requirements
RFC 3985	Pseudowire architecture

Mappings	
MEF 8	TDM over Ethernet, Metro Ethernet Forum implementation agreement
RFC 4553	Structure agnostic TDM over UDP, MPLS, layer 2 tunneling protocol L2TPv3*
RFC 5086	Structured TDM over UDP, MPLS, L2TPv3. In-band or out-of-band signaling
RFC 5087	TDM streams encapsulated in ATM adaptation layer AAL1 or AAL2 over UDP, MPLS, L2TPv3, Ethernet
RFC 4717	Any payload encapsulated in any AAL, over MPLS
RFC 4448	Ethernet pseudowires over MPLS
Y.1413	TDM over MPLS; in-band and out-of-band signaling
Y.1453	TDM over UDP; interworks UDP with ATM or MPLS pseudowires
Y.1415	Ethernet, IP over MPLS

Let us first consider what we are trying to achieve; then we show how the OMCI management model makes it possible.

6.6.1 Time Division Multiplex Services

Digital transmission really began with the DS1, sometimes called a T1. DS1 is a TDM protocol that multiplexes twenty-four 8-bit channels, 8000 times per second, for a payload rate of 1.536 Mb/s. The framing and embedded operations, administration, and maintenance (OAM) overhead is limited to 1 bit per frame, bringing the line rate to 1.544 Mb/s. The 125-μs frame therefore contains 24 bytes plus 1 bit, 193 bits. The industry has had to work around that prime number 193 since the beginning.

OMCI models a TDM UNI with an ME called the *physical path termination point CES UNI*. We discuss circuit emulation service (CES) below, but, in fact, the PPTP CES UNI focuses on the legacy TDM interface. Its attributes include:

- *Basic Line Type* DS1, E1, J1 (Japan version of DS1), and choices for DS3 or E3, higher rate PDH protocols that we predict will never be needed in the G-PON world.

---

*The G-PON community has not needed the layer 2 tunneling protocol L2TP server option; it is not supported by OMCI.

- *Framing Choices* For DS1 (unframed, superframe, extended superframe, JT-G.704), and for E1 (ITU-T G.704 options with combinations of time slot 16 multiframe and CRC-4).

- *Line Coding* Choices include bipolar with suppression of strings of three or more or eight or more consecutive zeros (B3ZS, B8ZS), alternate mark inversion (AMI), and high-density bipolar, order 3 (HDB3).

- *DS1 Line Length or Build-out* To accommodate various wiring distances from the terminal equipment to the DSX-1 (digital cross-connect level 1), various amounts of attenuation need to be introduced in the line. In days of yore, physical attenuators were installed; today, it is a matter of software.

- *DS1 Mode* With or without smartjack features, facility data link (FDL), or power feed.

Rather than full DS1s, business customers often purchase fractional DS1 (fractional T1) service, also known as $N \times 64$. For example, a 6×64 service at 384 kb/s might be leased as a program channel by a broadcaster or used to feed a limited and, therefore, less costly, number of trunk lines to a small PABX. Fractional DS1 is also used for nonvoice services such as frame relay.

This is true in North America, also Japan. The rest of the world uses the 2.048 Mb/s E1 instead. In an E1, the 125-μs frame contains 32 time slots, 2 of which are typically reserved for framing and signaling. Fractional E1 services are also available.

The OMCI model for fractional DS1 or E1 is the *logical Nx64 CTP*. It is nothing more than a bit map of the time slots to be included in this particular service definition.

6.6.2 Circuit Emulation Service: CES

As packets began to take over the network, TDM began to shrink as a percentage of total traffic. Though it remained lucrative, the day was clearly approaching when the sustaining cost of a separate TDM network would exceed the available value. It was becoming necessary to carry TDM services over a packet network, until such time as they could be eliminated completely.

Circuit emulation was developed as a way to carry full or fractional DS1s, E1s, or other full-time TDM services, originally over ATM. The crucial requirements for CES to emulate TDM circuits were:

- Transparency, within stated bounds of cost and performance. This means more than merely fidelity. For example, voice services are very sensitive to delay, and packetization delay was, and is, a major issue.

- Transparency to the emulated service could imply active intervention on the part of the emulator, for example, to detect the failure of the physical layer at a client interconnect point, and convey it in some way [alarm incoming signal (AIS)] to the corresponding client at the far end.

- Signaling is an intrinsic aspect of voice telephony. Whether in-band or out-of-band, there had to be a way to carry signaling across an emulated circuit. Other TDM services require overhead for OAM or PM, if not for signaling.
- Frequency synchronization is necessary to avoid frame slips and the consequent degradation of service quality. Although the legacy telecommunications network is synchronous, the packet network may not be frequency synchronized. Further, large business or institutional customers may operate their own local timing domains, so that network timing cannot be used, even if it is available. Somehow, then, accurate timing had to be conveyed across a packet network.

Although CES was originally defined on ATM—and the term *CES* sometimes implies ATM—it became apparent around the turn of the millennium that ATM was no longer the technology of the future. CES therefore needed to be updated for Ethernet or IP or MPLS server layers.

The term *pseudowire* emerged as a generic industry term for packetization and transport of a variety of legacy services. A pseudowire can encapsulate a TDM* bit or byte stream over a generalized packet network, be it one or more layers of Ethernet, IP, MPLS, or others. The client's native QoS requirements are approximated with a match to the closest QoS available in the packet network.

Although ATM and Ethernet are themselves packet protocols, it ultimately turned out that, for reasons of scalability and network uniformity, ATM and Ethernet were also two of the possible pseudowire client types. However, ATM and Ethernet pseudowires are noticeably different from TDM pseudowires. We discuss them separately below.

The OMCI model for pseudowire is built around a managed entity called the pseudowire termination point. Figure 6.23 illustrates how a TDM pseudowire is configured in OMCI. We will not be surprised to discover a MAC bridge when we look inside the cloud, but there are other options. We will come to them below.

The interesting attributes of the *pseudowire termination point* include:

Service Type   Structured, structure agnostic, or an octet-aligned structure-agnostic DS1 mapping that carries the extra 193rd bit in each frame along with 7 bits of padding. The minimum level of circuit emulation is provided by simply considering the client to be a bit stream at a defined rate, the structure-agnostic service type. However, if we provision the TP to recognize the client's underlying structure, we can gain optimizations and additional features. For example, the emulator can carry a separate signaling stream, or it can deal with client failure in one of several possible ways.

Signaling   A choice of: no signaling, CAS carried in-band, or signaling carried in a parallel channel.

---

*There are also frame relay pseudowires, Sonet/sdh pseudowires, fiber channel pseudowires, and PPP/HDLC (high-level datalink control) point-to-point protocol/pseudowires. To date, these have not been of interest in G-PON markets; they are not included in the OMCI model.

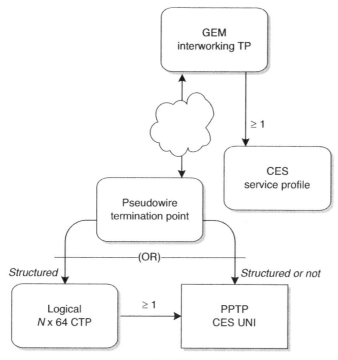

Figure 6.23   CES model.

Packet Size and Delay   The network is more efficient if more client data can be included into each packet, but the delay can become excessive. Given the capacity of modern networks, the comparatively low rate of TDM pseudowires, and the substantial revenue generated by legacy TDM services, performance may be regarded as more important than network efficiency.

Timing Mode   Network timed, adaptive, differential, or loop-timed.

Emulated Circuit ID (ECID) Sent and Received   This allows checking for the correct connectivity. We discuss ECID in the following section on payload mappings.

The *GEM interworking TP* is formally responsible for defining the interworking function, in this case between circuit emulation and GEM. It effectively does this by pointing to the CES service profile. Recall that the GEM interworking TP will also lead us into the ANI-side structures: the GEM port network CTP, priority queues, and T-CONT, where there are no surprises.

The *CES service profile* specifies the jitter buffer depth and the signaling protocol from the set

$$\{E1\,CAS,\ SF\,CAS,\ DS1\,ESF\,CAS,\ J2\,CAS\}$$

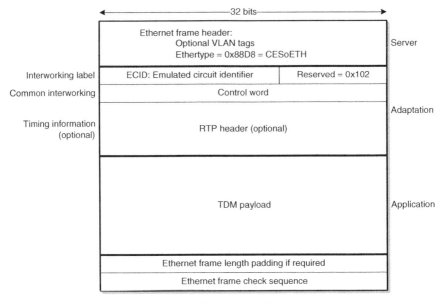

Figure 6.24 Mapping into Ethernet.

Mapping Payload into Server Layers Before going further into the OMCI model, we need to understand a certain amount of detail about pseudowire mappings from the TDM client into packet payloads. Figure 6.24 shows how a TDM application payload is mapped into an Ethernet frame. The Ethernet frame (server layer header) is completely standard, with optional VLAN tags, padding if necessary to a minimum length of 64 bytes, and a trailing frame check sequence. The application layer is just the TDM payload itself.

What is new, what we have not seen before, is the adaptation layer.

Figure 6.25 shows the same encapsulation in MPLS, and Figure 6.26 shows how client payload is mapped into IP.

In Figure 6.26, we recognize that we have seen the adaptation layer information blocks before, but notice that if the RTP header is present, it comes *before* the control word, rather than *after* it.

The similarities among these mappings are, of course, not accidental.

Clearly, the server layer header needs to be specific to the particular server, so we would expect nothing more than abstract similarities in the server header. The client payload blocks are the same in all cases. Let us examine the adaptation layer.

Adaptation Layer The adaptation layer consists of an interworking label, a control word (sometimes optional), and an optional RTP header.

In the Ethernet case of Figure 6.24, the interworking label comprises the ECID and a reserved field. Figure 6.27 expands the interworking label into its details. The reserved field with value 0x102 in the Ethernet mapping happens to coincide with

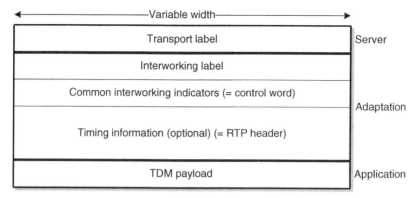

Figure 6.25 Mapping into MPLS.

meaningful fields in other mappings. As a matter of fact, Figure 6.27 illustrates a standard MPLS label (RFC 3032).

The 20-bit ECID can be arbitrary but needs to be unique and known to both endpoints, such that multiplexed client flows can be properly identified. The ECID permits several pseudowires to be multiplexed within a single Ethernet VLAN or under a single MPLS transport label.

The semantics of the traffic class field TC (formerly EXP—see RFC 5462) are unspecified. The field is set to zero.

The S bit is always set to 1 to indicate that this label is at the bottom of the (MPLS) label stack.

If the pseudowire is point to point, the time to live (TTL) field is set to 2. A multisegment pseudowire has a TTL either specifically chosen for the network it will traverse or selected by policy to be larger than necessary, but not dramatically larger. TTL expiration is the tool to remove possible looping packets during MPLS network reconfiguration.

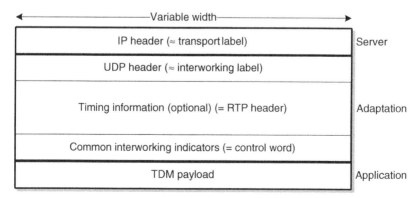

Figure 6.26 Mapping into IP.

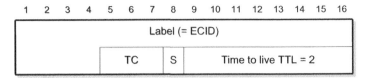

Figure 6.27 Interworking label.

Figure 6.28 depicts the control word. Mandated by some pseudowire definitions, especially those related to MPLS server layers, the control word is optional in others.

Reserved These bits matter and must be set to zero to avoid equal cost multipath (ECMP) routing in MPLS networks, which can cause frames to be interpreted as IP packets and be delivered out of order. RFC 4928 explains the issue for those who wish to dig further into the topic. RFC 4385 also uses these bits to define a pseudowire associated channel header (ACH). The ACH is not presently needed by G-PON nor is it supported by OMCI.

L *(Local)* As shown in Figure 6.29, the CES adaptor in ONU 1 sets the L bit to indicate a failure on the upstream TDM circuit. In structure agnostic mappings, this can only mean loss of signal; in structured mappings, the CES adaptor recognizes the framing structure of the client signal, so that loss of frame and incoming upstream AIS are also recognizable failures in the TDM circuit.

While sending L = 1, a CES adaptor (ONU 1) may or may not continue to forward TDM payload packets across the network, as provisioned in OMCI. Because a TDM transport circuit is typically provisioned with a fixed bandwidth assignment, it saves no network capacity to suppress invalid packets.

If the far-end adaptor (ONU 2) sees the L bit set, it may transmit service-specific AIS or trunk conditioning onto the TDM circuit. The OMCI provisioning model permits the operator to specify the desired behavior.

R *(Remote)* If the pseudowire fails in the direction from ONU 1 to ONU 2, as shown in Figure 6.30, ONU 2 sets the R bit in the reverse direction, attempting to inform ONU 1 that it is receiving none of its traffic. ONU 1 may in turn generate an RDI indication toward its TDM attachment equipment (also provisionable in G-PON systems through OMCI). To be precise, ONU 2 sets the R bit upstream whenever its TDM interworking circuit is in downstream LOF state.

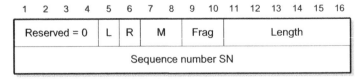

Figure 6.28 Control word.

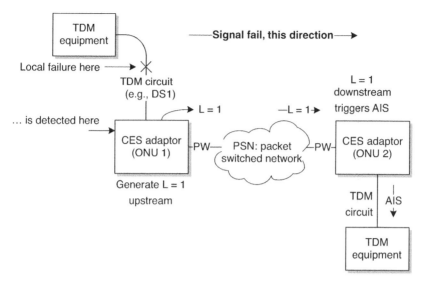

Figure 6.29 L bit behavior.

M   The two modifier bits can be used to extend the semantics of the L bit. Only M = 00 is mandatory, and not all pseudowire definitions include values for the M bits.

When L = 0:

00   No problem.

01   Reserved.

10   RDI is appearing at the TDM equipment from somewhere yet further downstream.

11   MEF 8: indicates that the current frame contains non-TDM data, for example, signaling.

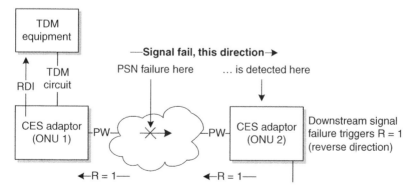

Figure 6.30 R bit behavior.

When $L = 1$:

00   TDM defect that should trigger far-end AIS.

01   Y.1413: idle TDM data; declare no alarms but generate an idle data
stream toward the far-end TDM circuit if defined. In MEF8, this
code point is not used.

10   Y.1413: corrupt but possibly recoverable TDM data. The semantics
of this value are unspecified. In MEF 8, this code point is not used.

Frag   The two fragmentation bits indicate that the packet either contains an
entire multiframe (or that no multiframe is defined), or that the packet contains
the first, last, or an intermediate segment of a multiframe. DS1 and E1 and the
$N \times 64$ packets of interest to G-PON should never need to be fragmented
because their multiframes (superframes) are small enough to fit into a normal
packet.

Length   In CES over Ethernet, the length field is used to indicate the presence of
padding if the unpadded Ethernet frame would have been shorter than 64 bytes.

Sequence Number   This is a circular counter that increments with each new
frame. It is used to detect lost and misordered frames. Some pseudowire
definitions—especially TDM pseudowires—allow for the possibility of reordering frames at the receiving end, but at the expense of increased delay.

Figure 6.31 illustrates the RTP (real-time protocol, RFC 3550) header. The RTP layer
is optional, but RTP is the standard way to support absolute or differential timing,
so an RTP header is to be expected for emulated TDM services, in which
synchronization is vital. Recall that POTS is also normally accompanied by RTP
for the same reason. Adaptive timing is also possible in which the output clock
frequency is based on the average arrival rate of downstream data, but adaptive
timing often proves to be unsatisfactory.

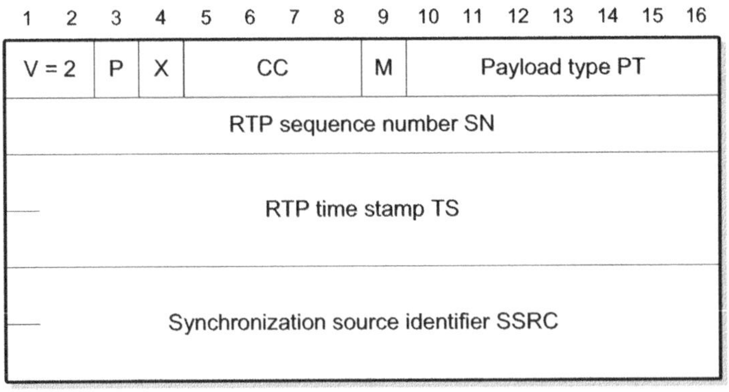

1	2	3	4	5	6	7	8	9	10	11	12	13	14	15	16
V = 2	P	X		CC				M			Payload type PT				
RTP sequence number SN															
RTP time stamp TS															
Synchronization source identifier SSRC															

Figure 6.31   RTP header.

V   The version number is set to 2 (RFC 3550).

P, X, CC, M   These bits are not used and are set to 0.

Payload-Type PT   Some value must be allocated from the range of dynamically assignable RTP types reserved by IANA, determined by ONU vendor default or by provisioning. The value is used only for possible checking of misconnection at the far end, and does not in fact distinguish different types of payload at all.

RTP Sequence Number   This field takes on the same value as that of the sequence number in the control word.

RTP Time Stamp   Time stamps are determined and processed in accordance with RFC 3550. MEF 8 requires the support of at least absolute timestamp mode, using an 8-kHz reference. Other absolute reference rates are optional, as is differential timestamp mode.

Synchronization Source Identifier SSRC   RFC 3550 defines the rules for generating and processing SSRC. It is also optionally used by the receiving end to detect misconnection, and despite the implication of its name, it is not related in any way to any particular source of synchronization.

Returning to OMCI Model for TDM Pseudowires. . .   When we looked at the MAC bridge model, we discovered ancillary MEs that were necessary to flesh out the basic model. The same is true in TDM pseudowires, as illustrated in Figure 6.32.

The *RTP pseudowire parameters* ME specifies the way in which RTP may be used in pseudowire transport. As we just discussed, the purpose of RTP is to convey timing across a packet network that may or may not have a common reference frequency. Management attributes include:

Clock reference rate

Time stamp mode—absolute, differential, or unknown/not applicable

Payload type (PT) and SSRC, sent and received.

The *pseudowire maintenance profile* plays a necessary supporting role since the standards specify that many of the policies be configurable. Its attributes include:

Jitter buffer parameters. If it is known that the network will introduce large packet delay variations, the jitter buffer depth can be increased to soak them up, but at the expense of increased delay.

Fill policy—what to do if no packet is available at the time the interworking circuit needs a packet to generate the next TDM output. It may be best to repeat the previous packet, for example, or to play out AIS or an idle pattern.

Alarm thresholds for misconnected packets, lost packets, or buffer underrun or overrun. Underrun and overrun occur when the TDM service requires a payload packet that has either not yet been delivered (underrun) or when the

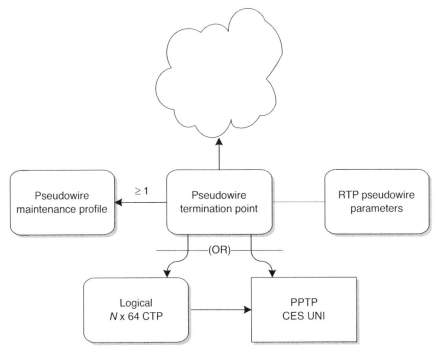

Figure 6.32 TDM pseudowire ancillary MEs.

network delivers so many packets in such a short time that the playout buffer cannot retain them all (overrun).

Severely errored seconds (SES) declaration criterion.

L-bit receive policy. Recall that the L bit in a downstream packet indicates a fault in the far-end TDM connection (Fig. 6.29). When the ONU sees an L bit, it can be provisioned to send AIS to its local client, to repeat the last packet, or to idle out the service.

R-bit transmit and receive policy. The R bit indicates loss of packets from the far end (Fig. 6.30), more specifically, loss of frame in the TDM reconstruction process. When the ONU crosses a configurable threshold of lost downstream packets, it sets the R bit in its upstream direction. When it receives an R bit, our ONU—at the far end, the transmitting end—can be provisioned through this ME to do nothing, to transmit an RDI or other indication, or to perform some configurable form of trunk conditioning toward its own TDM connecting circuit.

TDM services have a long and rich history of performance monitoring, and OMCI supports this feature with several PM MEs, not shown in Figure 6.32. Rest assured that we can count and collect code violations, errored seconds, severely errored seconds, unavailable seconds, controlled slip seconds, failures, and many other classical measures of TDM facility performance.

6.6.3 CES and Pseudowires in OMCI

The MEs of Figure 6.32 take care of the TDM and the packetization part of the pseudowire service. Now we look inside the cloud, the same cloud we saw in Figure 6.23, but working upward one layer at a time. We will get to the GEM interworking TP in just a bit.

In OMCI, a pseudowire may be carried over IP (layer 3), directly over Ethernet (layer 2), or via MPLS. Figure 6.33 illustrates the managed entities involved. In comparison with Figure 6.32, we see that the pseudowire termination point is the interface between the packet server layer and the TDM interworking function.

We already know about the TCP/UDP config data. The *Ethernet flow TP* is an Ethernet interworking point. Its attributes include:

MAC Addresses   Source and destination addresses of the pseudowire endpoints. The local (source) address is read-only, in the expectation that it is built into the ONU hardware. It is not implied that this MAC address be unique; indeed, an ONU may have a single MAC address for everything.

VLAN Tag Info   Whether the pseudowire frames are tagged, and if so, the value of the tag.

If the pseudowire is carried directly over Ethernet or via IP over Ethernet, the cloud of Figure 6.32 contains a MAC bridge model, as shown in Figure 6.34. All of the

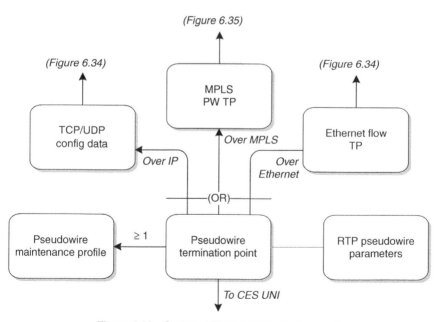

Figure 6.33   Choice of TDM* pseudowire transport.

*Emphasis: this is for TDM pseudowires. ATM and Ethernet pseudowires have different management models, discussed below.

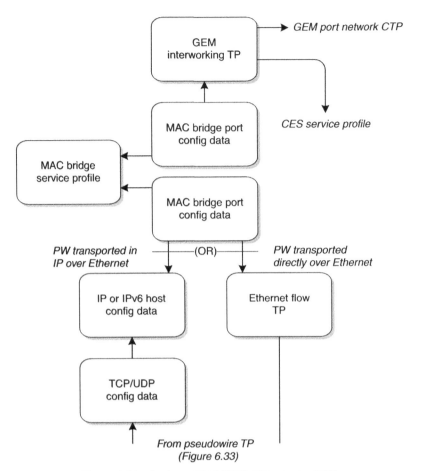

Figure 6.34   Getting a TDM (CES) PW onto the PON.

MAC bridging options are available, including tagging, filtering, and forwarding choices. The GEM mapping and the connection to the T-CONT is exactly as would be expected, with all of the QoS capabilities we discussed in Section 6.3.

Figure 6.35 shows the set of MPLS variations for TDM pseudowires. We carry the pseudowire over MPLS, and MPLS can in turn enlist the support of Ethernet as a server, via the Ethernet flow TP.[*] MPLS can also use UDP over IP over Ethernet, and if the OLT is capable of decoding MPLS directly from GEM, that option is available, too.

In Chapter 4, we described the mapping of MPLS directly over GEM. To clarify this in Figure 6.35: the MPLS Pseudowire (PW) TP invokes this mapping by omitting

---

[*]RFC 3985 says the server layer can be MPLS or IP, but does not mention Ethernet. MEF8 is an implementation agreement for circuit emulation over metro Ethernet.

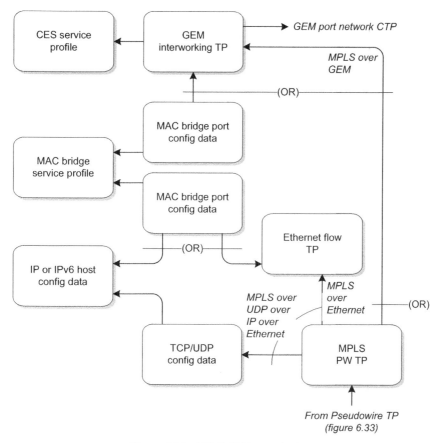

Figure 6.35 MPLS TDM pseudowire.

the MAC bridge completely, and pointing directly to a GEM interworking TP instead.

The *MPLS pseudowire termination point* controls the MPLS aspects of the pseudowire tunnel. Label provisioning on the ONU is static; there is no perceived need for the ONU to support an LDP stack.

Pseudowire Type Only Ethernet and various kinds of ATM pseudowires are defined in this ME. TDM pseudowires are defined in the associated pseudowire termination point ME.[†]

Directionality Upstream only, downstream only, or bidirectional. Strictly speaking, a bidirectional circuit comprises two pseudowires, but they are required to follow the same path through the network, and OMCI provisions both directions together.

[†] There is no way to provision the MPLS PW TP to be transparent so that the pseudowire TP can shine through. This is an error in the model.

Single or Double Label Option Also label values, including traffic class (TC).

Control Word Preference Whether each packet includes a control word. This attribute is only meaningful in mappings in which the control word is an option.

ATM Pseudowires Fixed bandwidth services such as $N \times 64$ may be mapped into ATM via AAL1 adaptation. AAL2 can be used when there is a high-priority commitment for variable bandwidth, for example, for compressed speech; AAL5 is appropriate for data services. These are of interest because legacy equipment may offer a DS1 or E1 interface whose next layer is ATM. Although structure-agnostic mapping would get the DS1 or E1 across the network, it would arrive at the far end looking like a DS1 or E1, while the legacy far-end terminating equipment might prefer to see a stream of ATM cells.

Quite a number of mappings are defined in the standards by which the various ATM adaptation layers can be mapped into pseudowires, transparently or not, one-to-one or N-to-one, either at the virtual circuit or virtual path connection (VCC, VPC) level, concatenated or not, and with OAM cells if needed. We omit the extensive details on grounds that ATM, though widely deployed, really is not the technology of the future. RFC 4717 is a good reference for those who wish to dig deeper; see also RFC 4816.

Figure 6.36 shows the OMCI management model of an ATM pseudowire. Rather than the pseudowire TP of the earlier models, the ATM pseudowire starts with a specialized ME called the PW ATM configuration data. As well as representing a number of ATM options, based on the MIB of RFC 5605, it includes the ability to swap ATM VPI and VCI values both upstream and down.

Observe that the ATM pseudowire model links back to the B-PON OMCI model of G.983.2. ATM definitions were not carried forward into G-PON, under the assumption that legacy interfaces were, well, legacy. When support for ATM interfaces in mobile backhaul was proposed, the community agreed not to reimport all of the ATM definitions from G.983.2 into G.988, but simply to refer back to the B-PON standard, whose definitions remain valid. The most interesting ATM UNIs for mobile backhaul are those at 1.544 and 2.048 Mb/s, our DS1 and E1 friends.

An ATM pseudowire may be carried over UDP, over MPLS, or over Ethernet, and if the choice is MPLS, the MPLS frames may themselves be mapped either to UDP, Ethernet, or directly to GEM.

Figure 6.36 illustrates another error in the present OMCI model. The GEM interworking TP wants to point to a service profile, but none is defined for an MPLS ATM pseudowire (or an MPLS Ethernet pseudowire either, for that matter—see the next section). In fact, the necessary management associations are present, if not in the standard way, so it can be made to work.

Ethernet Pseudowires Finally, Figure 6.37 illustrates the OMCI management model of an Ethernet MPLS pseudowire (Y.1415), which can also be transported over Ethernet (!), over IP, or directly over GEM.

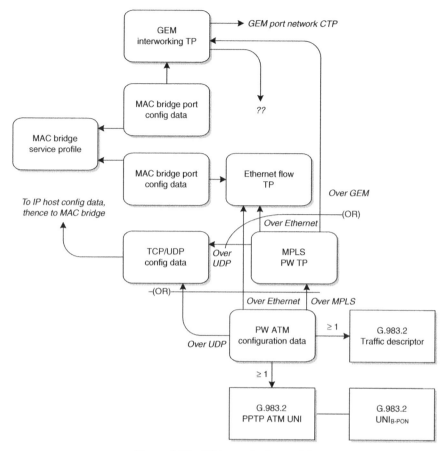

Figure 6.36 ATM pseudowire model.

There is no provision in this model for an RTP layer because none is needed. Ethernet encapsulation is represented by the PW Ethernet configuration data ME, whose only function is to link together the other MEs.

6.7 DIGITAL SUBSCRIBER LINE UNIs

While a pseudowire represents a complete service from UNI through ANI, DSL provisioning is primarily a matter of customizing UNIs.

Two forms of DSL are common today, asymmetric ADSL2+ and very high speed VDSL2. Their management views share many features; to avoid acronym conflict, they are known in OMCI as xDSL, where *x* can be either *A* or *V*, depending on context.

At its lowest level, a digital subscriber line is just a twisted pair that carries digital signals rather than the baseband analog signals of a classical telephone service, and

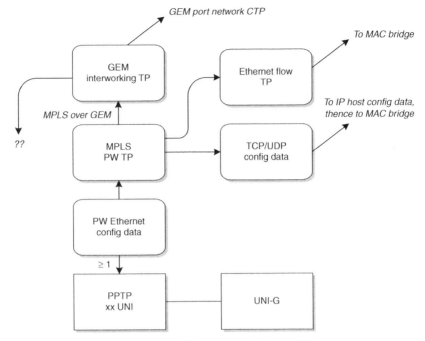

Figure 6.37 Ethernet pseudowire model.*

*See also the update to the MPLS model in the recently-consented G.988 amendment 2.

often in parallel with classical analog telephony on the same copper pair. DSL has been around for a number of years and has evolved, especially as silicon technology has brought advanced digital signal processing (DSP) into the realm of large-scale inexpensive feasibility.

Some forms of DSL are of little interest in PON applications, for example, high-speed HDSL and symmetric high-speed SHDSL (ITU-T G.991 series). There exists a G.993.1 VDSL standard, sometimes called VDSL1 to disambiguate it from VDSL2. Although OMCI includes managed entities to support it, it has largely been superseded, and it is not documented in G.988. This turns out to be convenient because G.993.1 VDSL does not align well with the common xDSL management model defined in G.997.1.

The ADSL family is widely deployed in one version or another (the ITU-T G.992 series), with ADSL2+ dominating applications with longer loop lengths. When the copper drop is shorter, VDSL2 (G.993.2) is the technology of choice, capable of speeds up to 100 Mb/s and more. The secret to high bandwidth is to shorten the copper loop, which is the business justification to push fiber deeper into the network.

The G.998 and G.993.5 series of recommendations describe ways in which DSL speed can be increased through techniques called vectoring and bonding. Vectoring uses complex signal processing algorithms to cancel far-end crosstalk between pairs that share a common cable or binder group. Bonding increases speed by simply using more pairs. Vectoring and bonding can be combined for dramatic performance over short to moderate loop lengths. We hesitate to mention numbers because the field evolves so quickly, but several hundred megabits per second over distances of several hundred feet are possible with two bonded pairs.

xDSL service is often delivered over the same twisted copper pair as POTS. Electrical filters separate out voice frequencies and reserve higher frequencies for data. Common xDSL frequency bands go up to 8 and 12 MHz, with 17 and 30 MHz also available in applications that require more bandwidth. Within these bands, data is modulated onto a comb of subcarriers, spaced at intervals of 4.125 or 8.25 kHz, a modulation technique called discrete multitone (DMT).

The xDSL modem at the remote (subscriber) end is called an ATU-R (ADSL) or a VTU-R (VDSL2), or in OMCI, just an xTU-R. The equivalent modem at the central office end is called an xTU-C. It is still an xTU-C, even if it resides in an ONU only a few hundred feet from the subscriber's premises.

There is a pervasive myth that OMCI cannot provision DSL. The grain of truth behind the myth is that OMCI has no need to provision xTU-Rs. That is not its purpose. But the largest single section of G.988—about 70 pages—is dedicated to the provisioning of the xTU-C, and to single- and double-ended loop test (SELT and DELT).

Management of all pertinent flavors of xDSL is standardized in ITU-T G.997.1, and the OMCI model follows G.997.1 as closely as possible.[*] However, at the time of writing, OMCI did not yet include management models for vectoring or bonding. The continuing rapid development of xDSL features, with their associated complexity, may be well suited to the vendor-specific file download feature described in Section 5.2.9.

Figure 6.38 illustrates the MEs associated with an xDSL UNI. For the sake of simplicity, we omit the half dozen PM MEs that are also part of the complete picture. At the top, we see the usual connection into a MAC bridge. This model assumes that the flow across the xDSL UNI is direct Ethernet, that the modems are in so-called packet transfer mode (PTM). We discuss Ethernet over ATM over xDSL below.

Notice that there may be as many as four MAC bridge ports associated with the UNI. An xDSL UNI supports several bearer channels, and OMCI allows up to four upstream, four downstream. One of the four is selected through the two most significant bits of the xDSL UNI managed entity ID. And yes, this violates all kinds of uniform structuring rules, but it exists. OMCI is nothing if not pragmatic.

The PPTP xDSL UNI part 1 had more than 16 attributes, the maximum available in an OMCI ME, so a PPTP xDSL UNI part 2 was created, simply as an extension.

Each bearer channel has its own profile, for a maximum of eight. The purpose of several of the profile MEs is to limit signal amplitude at different frequencies, thereby minimizing radiative interference with other users of the spectrum, for example, amateur radio. Other profiles establish speed, power, and SNR; objectives and bounds; if the modems are capable of it, and provisioned to do so, they may

---

[*] OMCI always follows other standards as much as possible; it only invents new models where nothing exists already.

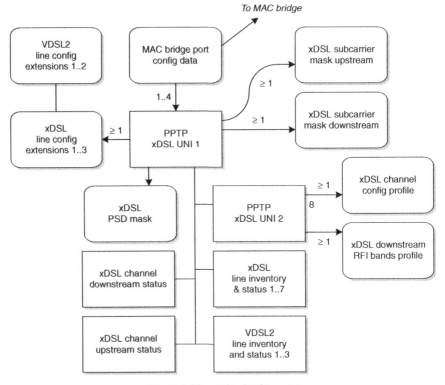

Figure 6.38 xDSL OMCI model.

renegotiate dynamically to optimize performance within these bounds. The profiles also specify the negotiation time constants.

The idea of a profile is that it defines a set of parameters that can be used by any number of clients. Unfortunately, historic creep has rendered the line configuration profiles unsuitable for use by more than a single UNI because the profiles also control line diagnostics execution, SELT (single-ended loop test), and DELT (double-ended loop test). The results, as well as basic inventory information, are available through the line inventory and status MEs.

Other scars of history are apparent in the names of the VDSL2 line configuration extension MEs. When they were originally created, their attributes pertained only to VDSL2; subsequently, the ADSL2+ community adopted some of them into G.997.1 as well. OMCI has preserved the original name in the interest of continuity, although it is no longer an accurate description.

ATM Over xDSL As mentioned, the basic xDSL provisioning model assumes direct Ethernet carriage at the xDSL UNI. Ethernet over ATM is slowly fading away in DSL technology, but it is still widespread. Figure 6.39 illustrates the OMCI model, where we omit the details of Figure 6.38 to focus on the difference in ATM provisioning.

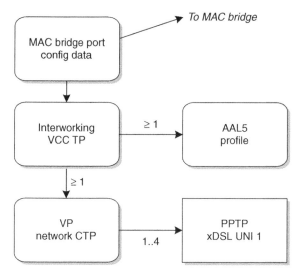

Figure 6.39   ATM interworking.

ATM offers considerable generic flexibility. Several ATM profiles are defined in the standards, and ATM allows several virtual circuits (VCs) to be contained within a virtual path (VP). But existing ATM-DSL modems only need VPs, not VCs, and they only need AAL5 encapsulation, so that is all that the OMCI model supports. Although they are documented in G.988, the MEs themselves are inherited from the ANI side of B-PON (G.983.2), when the PON itself was based on ATM.

6.8 RF VIDEO

In earlier days of PON, it was not uncommon to find RF video on the same fiber with a B-PON or G-PON, at its own separate wavelength, 1550 nm. This is a downstream analog or quadrature amplitude modulated (QAM) signal, independent of the digital PON, except for control and monitoring, and often with an upstream signaling adapter on the PON for uses such as channel changing.

This made perfect sense at the time because IPTV standards were not complete, device technology had not reached the point of large-scale inexpensive manufacture, digital set-top boxes were not widespread, and the downstream bandwidth of B-PON was comparatively limited. All of these constraints have been removed or relaxed over the past few years. RF video overlays will continue to exist for a number of years to come, but they are gradually fading away as IPTV becomes the mainstream technology.

7

OTHER TECHNOLOGIES

In this chapter:

- *Ethernet PON: EPON*
- *Wireless*
- *Copper*
- *Point-to-point fiber*
- *WDM PON*

In this book, we have discussed G-PON in detail. In this, the final chapter, we turn to a high-level comparison of some of the other access technologies, existing or potential. Some of them compete with G-PON; others are complementary.

- *EPON* Ethernet PON is the technology most often compared with G-PON. It addresses the same point-to-multipoint topology as G-PON, but in a distinctly different way. EPON will be of interest to students of G-PON, so we treat it first and in considerable detail.
- *Wireless* The single most important feature of wireless networks is mobility. It is therefore tempting to consider wireline and wireless networks to occupy completely different spaces, and indeed the introduction of this book takes that position. But the fixed network is heavily involved with wireless networks, and

Gigabit-capable Passive Optical Networks, First Edition. Dave Hood and Elmar Trojer.
© 2012 John Wiley & Sons, Inc. Published 2012 by John Wiley & Sons, Inc.

dynamic wireless handoff, and provisioned wireless backhaul will become increasingly important in the future.

- *Copper* Because there is a large amount of copper already in place—a sunk cost—incumbent operators are strongly motivated to reuse it if they can. In subscriber drops, copper nicely complements G-PON.

- *Ethernet, Point-to-Point* Especially for business customers, large radio base station installations, and remote DSLAMs, dedicated fiber carrying GbE or 10GE is without doubt the simplest solution. If plenty of fiber is available, point-to-point access is tempting. As with any technology, however, it is not without its issues.

- *WDM-PON* If point-to-point runs are conceptually the most straightforward distribution architecture, it could make sense to combine a number of them on a single fiber through WDM. While this is often considered the ultimate access architecture, there remain challenges, some of which will persist even if the optical technology becomes cost equivalent.

Let us examine these options.

7.1 ETHERNET PON, EPON

As well as G-PON in its 2.5G and 10G variations, there is a widely deployed alternative known as EPON, Ethernet PON, running at 1 Gb/s in both directions. Just as G-PON has evolved to 10 Gb/s downstream, so has EPON, as so-called 10GE-PON. 10GE-PON standardizes both 1G and 10G upstream rates. The China Communications Standards Association (CCSA) also recognizes a version of 1G-EPON with a downstream rate of 2^* Gb/s.

This section discusses the differences and similarities of the G-PON and EPON technologies. We are comparing families, so unless otherwise specified, the term *G-PON* is understood to refer generically to both G.984 and G.987 systems, while the term *EPON* refers to both 1G-EPON (IEEE 802.3ah) and 10G-EPON (802.3av) systems.

Comparing these technologies is a bold venture into the no-man's-land of marketing claims. Although we attempt to present the case for EPON fairly, it will be recognized that the authors of this book are familiar with the intricacies of G-PON more than with those of EPON.

A further issue arises in the evaluation of EPON. Many of its aspects are not documented in the released IEEE standard 802.3, for example, the usage of queue backlog reports. Topics such as these are within the scope of the P.1904.1 service interoperability for EPON (SIEPON) project, but at the time of writing, this existed only in draft form. Further, the various interests that were active in the SIEPON work

---

*The downstream symbol rate is 2.5 Gb/s; the line coding is 8B10B. The resulting capacity is twice that of a 1G-EPON, whose symbol rate is 1.25 Gb/s.

had differing views about how some of these functions should be performed, with the result that the draft formalizes a number of options.

Finally, it is not the purpose of this book to explain EPON in detail; we limit ourselves to EPON features that can be compared and contrasted in some meaningful way to G-PON. It will be seen that the two technologies address similar issues, often in somewhat different ways, and that neither technology has a fatal flaw. The choice of technology is likely to be a function of an operator's specific needs, rather than accumulated points on an abstract score sheet.

7.1.1 Perspectives

Although they end up in the same space, G-PON and EPON arrive there from somewhat different directions. It may be instructive to step back for a moment.

The G-PON family is standardized by the ITU, whose telecommunications arm ITU-T has well and faithfully represented the telecommunications industry for many years. The ITU-T perspective understands the rigors of every outdoor environment imaginable, the need for continuing incremental upgrade and maintenance, and the realities of subscriber churn. Equipment and facilities represent a large investment with a modest rate of return; they are left in place as long as they continue to provide revenue service to subscribers. Subscribers expect uninterrupted service, even during disasters—especially during disasters!

Upgrade is a matter, not of technological advance per se, but of market and competitive demands beyond the capability of existing infrastructure, and of increasing operations and maintenance cost that eventually tips the scale in favor of replacement. As well as the cost of new equipment or facilities, the operational difficulties of an upgrade are daunting. Large numbers of subscribers must be rolled from their existing infrastructure onto new infrastructure with little or no service disruption. As part of the job of relocating subscribers and services onto new infrastructure, all of the massive and interrelated databases must be updated, and the continuing flow of service orders and repair orders must not get lost in the transition. Migration en masse is the exception, not the rule.

The 802 LAN/MAN standards committee of the IEEE standardizes the EPON family. This body has given the data communications industry a continuing series of solid and reliable standards that are universally used, the best known of which is commonly referred to as Ethernet. The historic IEEE 802 perspective is that of a data center or information technology (IT) manager, not that of a telecommunications operator. IT installations are largely indoors, at worst buried between the buildings of a campus, or networked between a finite number of corporate sites over facilities provided and operated by third parties. The corporate IT manager is responsible for the premises network, which is usually not true of a telecommunications operator.

In the corporate environment, the latest technology is barely adequate to cope with the insatiable demand for performance. Equipment is an expense, not an investment, to be depreciated as rapidly as possible to make way for the next generation: bigger, faster, better. And with comparatively small networks, with comparatively few services, with comparatively simple databases, often with slack periods overnight and weekends, the operational cost and difficulty of upgrade are modest.

To the telecommunications operator, the long lifetime of the network, coupled with incremental upgrade and evolution, makes standards compliance and interoperability vital. On the other hand, the corporate IT manager expects to replace the equipment within a few years anyway, may have an equipment contract with only one or two vendors, and is more interested in the early availability of the next step in technological evolution than in how religiously it conforms to standards and interoperates with other vendors' equipment.

At this point, we freely admit that we have intentionally drawn large-scale caricatures for the purpose of contrast. The days of best-efforts IT are long gone, as are the days of POTS-only telecommunications service. Even historically, neither camp was entirely one way or the other, and as their interests converge, both camps recognize the importance of all of these factors. We nevertheless argue that these perspectives help to understand the different approaches and emphases of the technologies.

As a trivial example of the difference in perspectives, the IEEE camp would never think to impose a repetitive 125- μs frame on a data flow. What on earth for? To those whose background includes everything from DS1 to SDH, the question would also not arise—it would simply be taken for granted. How else would you do it?

An interesting difference is that the telecommunications side of the industry tends to evolve bandwidth technology in multiples of 4, looking for robust cost effectiveness, while the data communications side evolves in multiples of 10, in search of leading-edge performance.

There is a historical difference in perspectives of efficiency. Again to exaggerate for effect: the perception of the IT manager, considering a premises or campus network, is that the medium has essentially unlimited bandwidth and very low cost. In the tradeoff between performance and efficiency, performance wins without question. It is okay if the medium sits idle for most of the time, with statistical bursts of traffic, because we have only short runs of comparatively inexpensive media. If the medium becomes congested, replacement or augmentation is reasonably painless. The telecommunications perspective is that of a large long-term investment in transmission facilities that span substantial distances, facilities that, for economic reasons, must carry as much traffic as can possibly be devised, both in the technology of the endpoints and in the statistical traffic engineering of the load.

Reliability and recovery time were less crucial in early datacom networks—in the early days, companies did not live or die on the strength of their IT—and troubleshooting was easier and less costly. Datacom networks might scale to a few hundred or a few thousand users spread over at most a few hundred sites, while telecom networks were interconnected to serve millions of subscribers over thousands of towns and cities.

It was okay for a file transfer to take an indeterminate amount of time as long as the data arrived error free. Data communications protocols discard frames if they contain errors, and TCP simply retries if data get lost or damaged in transit. In contrast, a telephone subscriber can deal with bit errors (voiceband noise), but excess delay can

TABLE 7.1   PON Family Standards

Family	Standard	Standardization Date
ITU-T G-PON	G.984	2003–2004
ITU-T XG-PON	G.987	2010
IEEE EPON	802.3ah[a]	2004
	P.1904.1	2012
IEEE 10G-EPON	802.3av	2009
	P.1904.1	2012

[a]Both 802.3ah and 802.3av are amendments to IEEE 802.3. 802.3ah has been merged into the parent document, and 802.3av will be merged when 802.3 is next republished.

make a voice circuit completely unusable. In the brave new converged world, both factors are vital: delay must be minimized, and subscribers do not tolerate pixelation on their video screens. We, the industry, have raised the bar: everything matters!

In the early days, a data center could be shut down for maintenance and upgrade possibly every night or at least every weekend. But telecommunications networks have long been committed to 24×7 availability with an absolute minimum of scheduled maintenance downtime.

It must also be said that telecommunications operators are restricted far more than corporate IT managers by union work rules and government regulation, restrictions that affect the boundary between what is rapid, economical, and possible and what is not.

7.1.2   Standards

Table 7.1 summarizes the G-PON and EPON families and the standards that define them.

The most significant competitive advantage of the EPON family is that of early market availability. The standardization dates do not appear to be significantly different, but first-generation EPON products were available before the corresponding G-PON products, and the same is true of the 10G second generation.

Early standardization reflects an earlier start, but more importantly, it reflects the fact that 802.3ah and 802.3av represent only the lower layers of 10G-EPON, equivalent to limited parts of the XG-PON recommendations G.987.2 and G.987.3. In XG-PON, G.987 and G.988 specify essentially everything up to the service layer. Work on higher layer aspects of EPON continues in the IEEE P.1904.1 working group, *Standard for service interoperability in Ethernet passive optical networks*, commonly known as SIEPON,[*] a document whose formal release is expected in 2012.

---

[*] Many people pronounce it sigh'-pon, but it has a German look and we like to give it a German pronunciation zee'-pon: *haben Sie SIEPON gern?*

The differing scope of the standards helps understand one of the other claims and counterclaims:

- G-PON standards are claimed never to be complete. In fact, any living technology evolves into new applications and needs to solve new problems—PON-fed mobile backhaul is a case in point. If a standard is static, it is safe to conclude that it describes only a legacy technology. From the standpoint of 2012, think of the ATM standards, for example, or SDH. The G-PON standards do indeed evolve, but they are hardly unstable—the G-PON community is careful to retain backward compatibility as the standards move forward. G-PON standards are extended but rarely superseded or deprecated.
- To the extent that the EPON standard is indeed frozen, nonstandard implementations are encouraged. EPON at 2 Gb/s and optics that support 28–30-dB-loss budgets are examples of market opportunities that have been addressed by the EPON community without benefit of IEEE standardization. In the G-PON community, both of these would have been the subject of amendments to the standards and, it must be admitted, an extra year before products were available. The IT manager is less concerned about multivendor interoperability than is the telecommunications operator.

7.1.3 Optical Budget

We need not go into details of launch power and sensitivity here. Information about the G-PON values appears in Chapter 3, and the reader interested in that level of detail about EPON is encouraged to consult IEEE 802.3 clause 75. More useful in an overview is Table 7.2, which compares the target ODNs. Both technologies define several loss budget classes, with a view to market-based tradeoffs between the severity of an operator's technical requirements and the cost of components.

Both 10G-EPON and XG-PON were developed with the perspective of considerable deployment experience. As we see, both technologies recognize the need for higher budgets in defining the second generation, but they differ significantly. EPON is especially popular in high-density and MDU-centric markets, markets that require only modest loss budgets, so the 10G-EPON community stopped at 29 dB loss. In contrast, the XG-PON market, which tends to focus on the long reach needed for universal geographic coverage and central office consolidation, and the high split ratio needed for fiber to the home (FTTH), *started* at 29 dB. Coupled with the higher overhead efficiency expectations of XG-PON, about which we shall have more to say below, this helps to explain cost and market availability differences.

7.1.4 Rates

Table 7.3 shows the nominal speed of the PON contenders. We also show wavelength assignments, to be discussed in the next section.

Together with cost, the rates comparison may be the most important criterion against which operators make their decisions. As of today, our perspective is that

TABLE 7.2 PON Optical Budget Classes

1G-EPON,^{a} 1Gb/s Symmetric

Designation	PX10 down	PX10 up	PX20 down	PX20 up
Optical loss, max, dB	19.5	20	23.5	24

10G-EPON,^{b} 10 Gb/s Downstream, 1 or 10 Gb/s Upstream

Designation	PR(X)10		PR(X)20		PR(X)30
Optical loss, max, dB	20		24		29

G-PON,^{c} 2.5 Gb/s Downstream, 1.2 Gb/s Upstream

Designation	A	B	B+	C	C+
Optical loss, max, dB	20	25	28	30	32

XG-PON,^{d} 10 Gb/s Downstream, 2.5 Gb/s Upstream

Designation	N1	N2	E1	E2
Optical loss, max, dB	29	31	33	35

^{a}*1G-EPON*—Deployment experience indicates that larger (nonstandard) optical budgets are desirable. A variety of optical devices exists in the market to satisfy loss budgets from 25 to 30 dB or more.
^{b}*10G-EPON*—PRX designations refer to 10G downstream, 1G up. PR designations are for symmetric 10G systems.
^{c}*G-PON*—Class B+ evolved in 2005 based on experience with planning and deployment and is today the norm; there is little or no deployment of class A, B, or C G-PON. Class C+ was defined in 2008 as a way to add 4 dB to ODNs equipped with class B+ ONUs. C+ OLT transceivers are being produced and qualified at the time of writing. Their market penetration is yet to be seen, but it would not be a surprise if C+ were to become the de facto standard for G.984 G-PON.
^{d}*XG-PON*—Two nominal classes and two extended classes are defined. N1 is compatible with G-PON class B+ ODNs, allowing an extra 1 dB for wavelength-splitting filters. Cost-effective devices for the extended classes, especially 10G devices with an E2 budget, are understood to be a challenge.

G-PON and XG-PON offer a better match between the cost of technology and the needs of most operators. 1G downstream is becoming inadequate to meet subscriber demand for increased capacity. As mentioned, there also exists a 2G downstream de facto version of EPON: does it have the market strength to continue and extend the

TABLE 7.3 PON Rates and Wavelengths

	G-PON Family		EPON Family	
	G-PON	XG-PON	1G-EPON	10G-EPON
Downstream				
Rate, Gb/s	2.5	10	1	10
Wavelength, nm	1490	1577	1490	1577
Tolerance, nm	±10	−2, +3	±10	−2, +3
Upstream				
Rate, Gb/s	1.2	2.5	1	1, 10
Wavelength, nm	1310	1270	1310	1310, 1270
Tolerance, nm	±20	±10	±50	±50, ±10

first generation? As to G.984 G-PON, its 2.5G downstream rate should suffice for yet a few more years; it is enough to dedicate about 40 Mb/s to each of 64 subscribers, enough for several high-definition video channels.

There also exists a market pull for 10 Gb/s downstream, from operators who wish to put as many as 500 subscribers on a PON, sharing the optics cost and often reusing copper drops via MDUs.

As to upstream, 1G EPON is vulnerable to increasing market demand for user-created content, while 10G is more difficult and expensive than is warranted by currently known applications. The XG-PON community argues that 2.5G upstream is a better fit for the technology and economics of the next few years. There was no enthusiasm for a 2010 proposal to develop a 5-Gb/s upstream variant of XG-PON.

7.1.5 Coexistence and Upgrade

Table 7.3 also shows the wavelengths assigned by the PON standards. There is considerable commonality between comparable generations of G-PON and EPON. Convergence toward common wavelengths, especially in the 10G standards, is expected to increase component volumes and reduce cost for both camps. However, the 50-nm tolerance of EPON's 1-Gb/s upstream rate (1260–1360 nm) presents a problem for upgrade, because it overlaps the 1270-nm spectrum assigned to 10 Gb/s. This prevents the independent coexistence of 1G and 10G traffic on different wavelengths of the same ODN.

The spectral inefficiency issue was recognized early in the G-PON days and addressed with G.984.5 before any significant amount of G-PON deployment had occurred. While one of the G.984.5 options allows the full 100-nm spectrum— ITU-T tries never to make existing implementations nonstandard—the actual practice is that all G-PON deployments adhere to the narrower ±20-nm spectral tolerance. G-PON and XG-PON can independently coexist at different wavelengths on the same ODN.

Because simultaneous coexistence on separate wavelengths is not possible, 10G-EPON standards and products support upstream coexistence via TDMA. That is, all ONUs on the PON, regardless of generation, receive grants from a single pool.

Assume for the moment that all of our upstream traffic, from whatever generation, is at 1G, so that we do not have to worry about mixed rates. Even so, our bandwidth scheduler needs to know about all ONUs of both generations. Migration to 10G-EPON buys us no increased upstream capacity. Upstream resource sharing also means that the OLT blade must be replaced; it can hardly be expected that the DBA engine can be shared across independent circuit packs.

Now, operators just hate to scrap a perfectly good investment and are rightly concerned about the software transparency of a drop-in replacement blade that must continue to support existing subscribers seamlessly. Vendors, on the other hand, would be perfectly happy to sell combo 1G/10G-EPON OLT blades, but there does not necessarily exist a win–win price level for both operator and vendor. In terms of demand, a PON may serve only one or two subscribers who are prepared to pay for service that implies 10G-EPON, so the cost of the new OLT blade is frontloaded. All of this complicates the business case for migration from 1G to 10G-EPON.

But let us take one step further. Suppose the problem with the existing 1G-EPON was that its upstream capacity was inadequate. Suppose our 10G-EPON upgrade includes the 10-Gb/s upstream rate option. Suppose that the 1-Gb/s upstream capacity of our legacy EPON is 90% full. We upgrade to 10G-EPON, with 10 Gb/s upstream. But because 90% of our shared upstream capacity is used for 1G legacy service, we have only 10% of the capacity available for 10G. We have effectively doubled our upstream capacity, not increased it by an order of magnitude. We can give our brave new 10G subscriber 1 Gb/s upstream—undoubtedly enough, right?—but not 10 Gb/s. If a second 10G subscriber comes online, each is good for 500 Mb/s uncontested bandwidth. Our premium customers are contending for shares of a resource that is significantly smaller than would first appear to be the case.

To look at this another way, dual-rate TDMA complicates, not only optical receiver design, but also traffic engineering. The resource cost of some bits is 10 times that of others. Perversely, the resource cost of low-value legacy 1G bits is 10 times as high. In the best of all worlds, this incentive would cause us to migrate to 10G-EPON as fast as humanly possible; in the real world, most of our incentives are to retain the legacy 1G-EPON network as long as we can.

Finally, if the OLT chassis was designed for 1G-EPON, it may not have the backplane capacity to be the correct platform for combo blades carrying a migration into universal 10G service.

Now, how does this play in the G-PON space? Because of wavelength separation, G-PON and XG-PON coexist, completely independently, on the same ODN. An XG-PON upgrade opens an entire 2.5-Gb/s upstream resource for use, not merely an increment on the previous resource. An XG-PON upgrade does not render the existing G-PON OLT obsolete; the XG-PON blade can occupy the same OLT or a new one, and the new one need not be from the same vendor (operators also dislike being locked into a single vendor). Traffic engineering and network planning can be done on the basis of two independent access networks that just happen to share the same fiber, rather than one dual-rate network.

7.1.6 Yes, But Is It Ethernet?

In the marketing competition between G-PON and EPON, it is sometimes claimed that EPON is just plain Ethernet, while G-PON is, well, something else. IEEE 802.3 is the body that standardizes Ethernet, so this is of course trivially true as a matter of definition. It is worth pointing out, however, that the preamble of an EPON Ethernet frame differs from that of the same frame on an Ethernet LAN, by containing the logical link ID (LLID), effectively a PON-level address. The 10G-EPON line code and FEC arrangement are unique, as compared to other IEEE Ethernet formats. And while EPON retains the standard Ethernet interframe gap at the client layer, it is not apparent that this is a benefit.

In comparison, G-PON encapsulates Ethernet frames into GEM frames, stripping their preambles, which are easy to re-create if they should ever be needed. Including Ethernet interframe gaps, which do not exist in G-PON, GEM frame headers are more efficient than Ethernet frame overhead. And unlike 802.3 frames, GEM frames can be fragmented, which makes G-PON less vulnerable to lost capacity at burst boundaries.

Yes, G-PON is not a pure 802.3 PHY; it is better in some ways and equivalent in others. Nor is EPON the same Ethernet that would be found on another 802.3 PHY. It does not matter: both convey standard Ethernet frames transparently.

7.1.7 Fragmentation

G-PON is built on the long-standing telecommunications tradition of 125-μs frames. In the downstream direction, a distinct and recognizable framing bit pattern repeats at precisely this interval. As well as providing a convenient location for the downstream bandwidth map, a PLOAM channel per Chapter 4, and FEC and scrambler resynchronization, the repeated frame-start instant is the reference time for upstream burst generation and optionally for time of day.

Because the G-PON framing pattern recurs at precise times, it necessarily preempts any GEM frames that might be in the process of transmission. There are several ways in which this issue could have been addressed, but the chosen solution is GEM frame fragmentation. Unless the remaining scrap of capacity is unusably small, a client frame is simply broken into two GEM frames, which are reassembled at the other end. The same phenomenon occurs in the upstream direction, where the boundary occurs at the end of the specific grant allocation.

With its datacom heritage, EPON has no concept of a repetitive frame. Downstream fragmentation is therefore not an issue. However, the upstream channel faces the same issue as G-PON: what if the next frame is too long to fit into the remaining grant time?

EPON does not fragment frames. It addresses this issue through the concept of queue sets, in which the upstream queue occupancy report proposes one or more values for the next grant size. If the OLT grants exactly the capacity requested by one of these queue sets, and depending on the queue discipline, there is no fragmentation and no wasted capacity. Observe that, for this to be true, the EPON ONU's queue report must—and does—include the proper overheads for interframe gaps and frame preambles—ten 100-byte frames require a larger allocation than one 1000-byte frame.

As to G-PON, there is no interframe gap. GEM frame headers are *not* included in the queue reports, and in G.984, the report is only an approximation anyway. Even under the best of circumstances, a G-PON allocation is therefore imprecise. We discuss the G-PON case in Section 6.3.

If the upstream resource is congested, the OLT may not be able to grant as much capacity as is requested. EPON allows the reporting of multiple queue sets, with different grant suggestions in each, and the OLT is invited to pick one. This policy limits the granularity—or the efficiency—of EPON grants and consequently the fairness of resource allocation. In contrast, the fine-grained grants of G-PON are able to achieve better fairness.

Both G-PON and EPON DBA algorithms may include heuristics that grant some level of capacity beyond the reported queue backlog, in anticipation of newly arrived traffic or unanticipated messages or overheads. If there is no backlog, such additional capacity is wasted in both cases, but if it turns out that there is additional traffic to send, G-PON fragmentation uses the extra allocation more efficiently than does nonfragmenting EPON.

Another factor is that, in one of the proposed SIEPON options, newly arrived high-priority EPON traffic preempts lower priority frames in the ONU's queue. From an abstract point of view, this is the right thing to do, but it invalidates all commitments about efficient grants.

7.1.8 Efficiency

We revert a moment to the historical perspective, intentionally exaggerated. The telecommunications operator wants equipment, and especially transport facilities, to be loaded with as much traffic as possible all the time. Idle investment earns no return. Efficiency is a key criterion. In contrast, the corporate IT manager over-engineers inexpensive campus and building transmission facilities, so they do not become bottlenecks. Idle facilities are a good thing; the IT manager is evaluated on network performance.

G-PON optical components have historically been somewhat more expensive than EPON components, as much as anything because G-PON has gone for efficient use of the medium, particularly of upstream bandwidth. G-PON's interburst gaps are recommended to be short, and the burst header, necessary for level setting and clock alignment, is also suggested to be fairly aggressive. (All PON variants include ways to adjust gaps and burst headers to accommodate early technology that may be less capable than subsequent, more mature implementations.)

IEEE 802.3 requires the full interburst gap and preamble on every frame transmitted, while in G-PON, GEM frames are continuous. The difference is less dramatic than it sounds in 10G-EPON, because 10G-EPON uses the extra time for FEC and line coding. In comparison, XG-PON shrinks the payload rate to accommodate FEC.

The question always arises: which is more efficient, G-PON or EPON[*]? There are so many variables that it is impossible to specify a simple numeric efficiency for any one of the four alternatives, and the variables differ in ways that make it difficult to set up fair comparisons. We have mentioned some of the factors above and in Section 6.3.2.

Just to take the single largest efficiency factor, consider FEC overhead. In 1G-EPON, FEC overhead not only comes out of the payload reservoir but also increases the interframe gap and preamble to improve byte delineation statistics in the presence of high BER. In 10G-EPON, the overhead of FEC is built into the raw bit rate, along

---

[*] An equivalent question: what is the subscriber payload rate of the various families and generations?

TABLE 7.4   FEC Efficiency Cost

	1G-EPON	10G-EPON	G.984 G-PON	G.987 XG-PON	
Direction	Both	Both	Both	Down	Up
Algorithm	255, 239	248, 216[a]	255, 239	248, 216	248, 232
Efficiency	93.7%	87.1%	93.7%	87.1%	93.5%
Optional?	Yes	No	Yes	No	Yes

[a]10G-EPON combines 64/66b line coding with FEC. Computed on 66-bit vectors, the FEC algorithm is (255, 224), including some zero padding, with a bitwise efficiency of 87.7%. The table entry shows the equivalent code in bytes.

with the line coding. In the G-PON family, FEC overhead is subtracted from the raw bit rate (the resulting downstream rate is almost the same between XG-PON and 10G-EPON). In some cases, FEC is optional; in other cases, mandatory. The protocol efficiency varies, but we must also remember that FEC is not deadweight overhead. It is there to improve the optical budget. Which is the best bargain? You decide! Table 7.4 summarizes just the cost due to FEC overhead.

Arguably the largest remaining factor contributing to overall efficiency is upstream burst overhead, which of course depends on the number of bursts and their sizes. This varies as a function of the number of ONUs (or LLIDs), the OLT's DBA capacity allocation strategy, and the level of optical components technology, which shows up in the size of the necessary interburst gaps and the length of the burst headers.

As a parting word on this topic, recall from our discussion in Section 6.3 that efficiency is not always a good thing.

7.1.9 Addressing Granularity

There are three specialized addressing constructs in G-PON, the ONU-ID, the T-CONT/alloc-ID (upstream only), and the GEM port. EPON has only one PON addressing construct, the LLID, but also offers visibility of priority queues and therefore of traffic classes.

The G-PON TC-layer ONU-ID is assigned during the activation process. It is used during the ONU's session lifetime as its identifier for management messages, both at the PLOAM and OMCI levels. The EPON ONU's primary LLID—additional LLIDs may exist—is likewise assigned during the activation process, and serves to identify the physical ONU during its session lifetime.

The G-PON T-CONT and the EPON LLID are the quanta in which the OLT manages contention for upstream bandwidth. Each is capable of reporting its backlog to the DBA process in the OLT, and each is capable of receiving and acting upon an upstream grant.

In this sense, a G-PON ONU with a single T-CONT is approximately equivalent to an EPON ONU with a single LLID. The EPON LLID is typically fed by eight priority queues, and its DBA queue occupancy report includes values for each of the eight, so that the OLT can evaluate traffic contracts across all ONUs on the PON at the level of individual queues. If we had a single T-CONT in a G-PON ONU, we would not have that level of visibility into its component traffic classes.

And indeed, it is partly for this reason that the usual G-PON model is for an ONU to have multiple T-CONTs, one for each traffic class. The OLT's DBA process then arbitrates capacity with T-CONT granularity on the basis of these traffic contracts. The approximate equivalent in EPON would be an ONU with up to eight LLIDs, each fed by a single queue.

Why the number eight? Because EPON prioritizes traffic on the three 802.1Q priority bits. As it turns out, there is not that much need for all eight—Broadband Forum TR-101 only requires four classes of service, with an objective of six. TR-156 and TR-200, dealing with G-PON and EPON, respectively, repeat these numbers as minimum requirements and objectives. In the development of TR-200, it was observed that Ethernet chipsets usually have eight queues, one for each P-bit code point, but the requirements for four and six were retained because the requirements are intended to state market need, not technology capability.

A G-PON ONU could also have eight T-CONTs, if desired. It could have more than eight, if desired. More than eight? In an application with multiple VLANs and multiple providers, even the same P-bit values might need separate traffic contracts, or the P bits in some VLANs might be ignored altogether. An ONU designed for business subscribers is indeed likely to offer more than eight T-CONTs, some of which may be used to segregate Ethernet business services from other traffic. Could we do the same thing in EPON? Of course, with additional LLIDs in the ONU.

As to flow identification, we observed in Chapter 6 that the primary benefit of the GEM port concept is uniformity in rapidly classifying incoming traffic.

In data communications system architectures, it is common for frames to acquire an extra flow identifier as they traverse the internals of the system, so that frames need be classified only once, and successive engines can act without needing to repeat the classification. In many ways, the GEM port simply extends this internal system concept across the ODN.

The corresponding primary benefit in the EPON specification is the preexistence of silicon that can classify flows on the basis of MAC addresses, VLAN or priority tags, and more. If wire speed packet inspection is readily available, there is no need for G-PON's extra layer of streamlining. However, the EPON classification rules may need to be provisioned in both ONU and OLT, so that frames can be reclassified after they cross the PON.

7.1.10 Management

G-PON ONUs are managed via OMCI, which is discussed in detail in Chapters 5 and 6. IEEE 802.3 does not deal in any depth with the question of management, and several models have evolved in different markets. At the time of writing, this topic is under consideration in IEEE P.1904.1, SIEPON. Options include OMCI, DOCSIS (data over cable service interface specification, proposed by the cable TV industry, and "extended Ethernet OAM," which is based on IEEE 802.3 clause 57. Clause 57 describes only a limited number of maintenance functions, but it is indeed extensible. For purposes of complete ONU management, the extensions are the important part.

The EPON community may agree in SIEPON not to specify the full range of functionality encompassed by G-PON's G.988, in which case the rest of the problem must be addressed in other ways, by adapting complementary existing models such as OMCI or the BBF TR-69 family or by creating a new model.

The likely outcome is that different management models will be developed for different markets. Full service-level intervendor interoperability is a continuing challenge in both technologies, but G-PON is arguably further along this path than is EPON.

7.2 WIRELESS BROADBAND

We now turn our attention to an access technology that is about as different from EPON as it is possible to be. The wireline and wireless networks complement one another in several ways, sometimes as competitors, sometimes with one as client, the other as server.

Wireless as Client At one extreme is the IEEE 802.11 wireless support built into a home gateway G-PON ONU. Such a device certainly adds value to an ONU product line, but it poses no technical or management issues of concern to G-PON. To G-PON, it is just a subscriber UNI, the functional equivalent of a copper drop.

At perhaps the opposite extreme is the G-PON ONU colocated at a mobile base station hotel and used to backhaul mobile traffic. To the ONU, the client is a DS1 or an E1 or increasingly often, an Ethernet feed. We have noted several times the importance of a time-of-day reference from the ONU to aid in synchronizing new-generation base station equipment. Other than that, the ONU perceives standard clients with fixed-rate QoS expectations or comparable high-priority QoS in the case of Ethernet.

Mobile backhaul is expected to become a major application of G-PON, especially as mobile networks evolve through technologies such as high-speed packet access (HSPA) and long-term evolution (LTE and LTE-A) into multimegabit IP-based mobile services. An ONU that specializes in this function is called a cellular backhaul unit (CBU).

When a willing ONU is available, a roaming user equipment (UE) may connect dynamically, thereby offloading traffic from the mobile to the wireline network. The G-PON ONU may find itself acting in the role of 802.1X authenticator, assisting to identify and authorize the UE. Once the UE is accepted, the question arises of what to do with its traffic.

It may be possible to preprovision VLAN or possibly a pseudowire specifically for the purpose of dynamic mobile connection to a specific ONU, although scalability concerns would arise if this needed to be implemented on every ONU.

We also ask whether, on the basis of UE authentication, the access network will dynamically reassign QoS, including bandwidth and priorities, to align with the UE's privileges and usage. The sudden appearance of a 100-Mb/s upload device on an ONU can hardly be hidden in the granularity uncertainties of a PON's QoS provisioning. At the time of writing, the topic was in the early days of discussion.

When it is available, landline backhaul is generally the preferred alternative; it costs more to transport a bit over radio than over fixed plant, and it is desirable to conserve radio resource for its unique added value, mobility.

Wireless as Competition   As an alternative to placing long runs of new cable, fixed wireless broadband may be the most economical way to serve small numbers of subscribers in rural areas. IEEE 802.16, the standard for worldwide interoperability for microwave access technology (WiMax), can deliver several tens of megabits per second and can span distances of up to 50 miles, although it cannot do both at the same time. Next-generation WiMax may be able to deliver rates up to 1 Gb/s.

Wireless as Partner   Fixed wireless technology may also be a good choice for PON protection. In practice, there is little redundancy in fiber cable routes, and misadventure may well affect all wirelines to a given area. Even with reduced bandwidth, wireless protection of a G-PON could be a way to maintain high-priority services such as emergency telephony.

Wireless as a Ship Passing in the Night   The common public radio interface (CPRI) is a specialized protocol that allows the separation of radio antenna electronics from radio base stations. Signals are optically carried between the two nodes on what is essentially a backplane extension protocol. Phase skew and delay are critical, the payload bears no resemblance to Ethernet, and bit rates lie in the range 600–2500 Mb/s. CPRI is clearly not a good candidate to be mapped onto a G-PON.

It could make sense, however, to put CPRI signals onto other wavelengths of the same optical distribution network with a G-PON, simply as a way of getting them from point *a* to point *b*. The G-PON equipment might be called upon to assist with maintenance, monitoring, and alarming.

7.3 COPPER

Returning to landline technology, we cannot ignore copper, the oldest technology in the access network. Twisted pairs delivering POTS and/or DSL from central offices or remote terminals still exist in vast numbers. The good thing about copper is that it exists and has been paid for. The downsides of copper include:

- Broadband speeds are difficult to support over long loops. Typical DSL speeds lie in the range 1–10 Mb/s, depending on reach.

- To support DSL, loops must have had their bridged taps removed at considerable operational expense. This is less of a concern looking into the future than it was before, because much of that work has already been done.
- Distributing frame congestion is a problem, which we discuss in more detail below.
- Equipment connected to copper plant is vulnerable to lightning and power line induction, and copper plant can radiate energy that interferes with other services.
- Copper plant, whether aerial or buried, eventually deteriorates in the rigors of an outdoor environment.
- Copper is expensive, even to the point that cable theft is a problem.

For these and other reasons, the edge of the network has been moving outward toward the subscriber for many years, originally in the form of digital loop carrier (DLC), usually sized for 48–384 POTS subscribers and today usually combined with DSL. Fiber to the home is the ultimate realization of this trend, but the economics do not always justify FTTH. When it is a question of cost–benefit, new fiber versus preexisting copper plant versus subscriber demand, the optimum tradeoff is often to install fiber to a neighborhood, a fiber-fed DSLAM, or to the curbside or to an MDU. The final service drop remains copper.

The economically achievable rate and reach of DSL have increased dramatically over the years, thanks to mathematical innovations made feasible through Moore's law.[*] New DSL technologies such as vectoring are capable of boosting DSL speeds on single pairs, while bonding increases speed by using two or more pairs in conjunction. At the time of writing, it appears to be feasible to deliver 100 Mb/s on a drop perhaps as long as 500 m; whether gigabit-per-second service can be delivered over the same loop is an open question, but every year sees advances in technology that would have seemed improbable the year before.

High-performance DSL over short copper drops is a natural client of the optical access network. Copper from the central office will continue to decline in importance as time goes by, but copper in the last 500 m will remain a major player for many years to come.

7.4   ETHERNET, POINT TO POINT

If an operator has an abundance of fiber, point-to-point optical Ethernet is at least conceptually the simplest form of access. Each subscriber gets a dedicated fiber; there are no questions of traffic isolation or of QoS in the distribution network. Some of the security issues disappear. Unbundling among multiple service providers happens at the physical layer. Service flexibility is absolute. If 1-Gb/s Ethernet is not enough, we can change the service to 10 Gb/s by replacing the subscriber

---

[*] A continuing issue is the cost of power for these advanced algorithms. Especially in remotely powered equipment, every milliwatt counts. The cost of power may dominate the choice of technology.

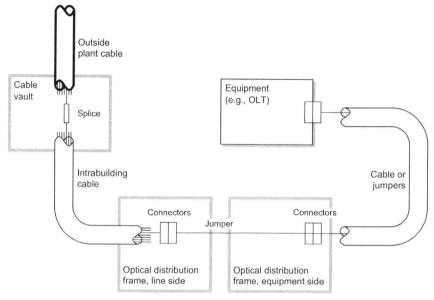

Figure 7.1   Fiber routing.

terminal and patching the central office end to a new port, possibly on a different equipment unit. If the remote endpoint requires CPRI rather than Ethernet service—fine! Install the correct terminal and patch to the appropriate equipment at the radio base station hotel.

There is no worry about power levels, neither the maximum launch power of our service nor crosstalk from foreign signals. We can easily deal with one subscriber who is willing to pay for redundant fibers, other subscribers who require rings, and yet other subscribers in the same area who are not prepared to pay for redundancy at all.

Everything has its price, however. In the traditional world of copper outside plant, large cables left the central office and were fanned out into smaller cables at cross-connect points in the field. This was a form of statistical oversubscription: while any given subscriber could get two or even three pairs if necessary, there might be only 20% headroom by the time a cable had been cross connected back to the CO.

On the other hand, the cross connections in the field had to be continually updated as subscriber services churned over the years. This particular form of churn is less likely in a fiber network, because there is no need to dedicate more than one working fiber to a subscriber.[*]

Churn at the CO is harder to avoid. Of course, we could assume that there is only one service provider, that 1 Gb/s will be the right-size service for all subscribers forever, and that we have essentially 100% take rate. Even under these assumptions, however, we are faced with a distributing frame problem. Every fiber that enters the

---

[*] Unless unbundling is done at the physical fiber level and a customer subscribes to more than one service provider.

CO appears on an optical distributing frame (ODF). At the rate of one fiber per subscriber, our jumper management problem is, for example, 20 times as bad as if we were able to serve 20 subscribers with each fiber.

Figure 7.1 shows a typical arrangement in which an outside plant (OSP) cable is spliced in or near the cable vault to an intrabuilding cable, which is easier to route to the ODF. By the way, these are a few of the components that do not show up on high-level PON network diagrams but that must be included in optical budget planning.

From the ODF, a jumper may go directly to the equipment, but as illustrated in Figure 7.1, it is more likely that all optical ports of all equipment in the CO will be likewise cabled or jumpered to separate terminations on the ODF, and the two ODF appearances interconnected with jumpers. The ODF is itself complex, comprising a number of line-side racks and a number of equipment-side racks. The jumper between line and equipment sides goes from one ODF rack to another, which may be some distance away, potentially on another floor of the central office. Another essential reason for an ODF, not shown in Figure 7.1, is test access to the outside plant fibers.

As well as protecting the fiber, the cable and jumper arrangement must ensure that the fibers' rated bend radius is not violated and that there is space to store slack cable or fiber. A significant industry exists devoted to ODFs and their associated components and practices.

All of this would be easy if the jumpers were static. But over time—and we are talking about many decades here—subscribers move around, they change their minds about their service needs, and on the CO equipment side, we will incrementally extend, upgrade, and replace network elements. This churn magnifies the problem. In comparison with a single fiber that serves 20 (for example) subscribers, not only does point-to-point fiber require 20 times as much expensive floor space, not only do we have the continuing operational expense of rearranging 20 times as many jumpers—a recordkeeping expense as well as a direct labor cost—but if the history of the copper plant is a guide, it will prove difficult and eventually impossible to remove buried jumpers or cables from the mass of newer fibers. Over time, the compost heap of dead fiber will overwhelm the distributing frame or the raceways throughout the office. Fiber in a raceway is recommended not to exceed 50 mm in depth, such that old jumpers can be removed; maintaining this policy over the decades will require discipline in a large CO with a lot of fiber.

Even if the raceway is only 50 mm deep in fiber, the various destinations of fiber cables and jumpers will inevitably result in interwoven tangles. Conceivably, an old fiber that was tangled in the mass of newer ones could be extracted by cutting off and pulling out manageable lengths one at a time. But this is expensive and risky: expensive in terms of ODF labor, risky because disturbing the mass of existing fiber could disrupt service, either through fiber or connector damage or simply because of bend radius violations. And, of course, there's always the risk of cutting loose the wrong fiber—they all look alike.

Ribbon cables, microducts, and air-blown fiber are all ways in which to mitigate the problem, although they introduce their own issues—the limited number of mating cycles of an optical ribbon connector, for example. Assuming that these approaches can be made to work, their efficacy in avoiding long-term ODF congestion remains to be seen.

Vendors would like to solve the churn problem with automated optical cross-connect systems, which could also provide test access. This solution requires substantial improvements in insertion loss and reliability and especially in cost. Until then, manually placed fiber jumpers are used, and test access is attained by manually patching in an OTDR at the ODF when needed.

Dedicated fiber has undeniable advantages, but it comes with its own problems.

One way to serve 20—or 100—subscribers with a single fiber is to use it to feed a remote multiplex terminal, be it a remote DSLAM for DSL access over vastly shortened drops, be it a remote G-PON OLT, or be it an Ethernet switch or router on the premises of a business customer. Dedicated fiber is also the current favorite for mobile backhaul and, if not the favorite, is at least a strong contender for DSLAM feed. Such applications also minimize the churn problem.

7.5 WDM PON

At the time of writing, there was an emerging consensus in the industry that after the various versions of 10G PON, and maybe in competition with 10G PON, the next generation would be WDM PON. There are at least as many definitions of WDM PON as there are contributors to the standards forums. This section explores a few of the issues and options but does not attempt to predict the winners.

A certain amount of WDM PON is already deployed in venues such as Korea. Generally speaking, however, current technology is too costly to compete with G-PON or EPON. Vendors of WDM PON systems argue their merits in specialized applications, such as symmetric high-capacity (high-revenue) business services, or carrying specialized protocols such as CPRI, or wavelength unbundling.

The idea of WDM PON is that the single trunk fiber can carry a number of wavelengths—all the flexibility of point to point, with fewer fibers. Thirty-two wavelengths in each direction is perhaps the low end of the scale, and some proposals go as high as 96 or even more. Each wavelength carries its own signal independently of the others, meaning that every subscriber can have, not only a dedicated service instance, but a dedicated protocol: fast Ethernet here, GbE there, 10GE there, sonet/SDH or OTN on this wavelength, RF on that wavelength, a G-PON OLT or reach extender on some other wavelength, and CPRI on yet another.

This is true in theory.

Just as with point-to-point fiber, flexibility requires that the OLT end be broken out into separate wavelengths on an ODF. Each wavelength, now on a separate fiber, is then cross-connected to an individual equipment termination here and there around the central office. In Figure 7.2, we show a single fiber as far as the line-side ODF—so we do indeed reduce the fiber problem on the outside plant side. Then we separate out the individual wavelengths with an arrayed waveguide (AWG), which may or may not be mounted on one of the ODF racks—we show it on a separate passive components rack. Each wavelength is then jumpered through

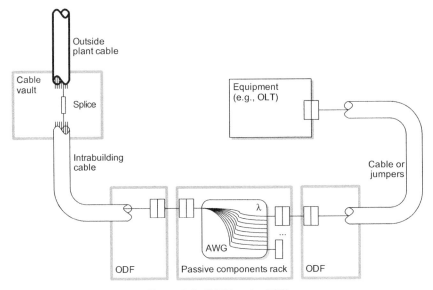

Figure 7.2 WDM at the ODF.

the equipment-side ODF to the proper equipment port. Within the CO, it is the same as a point-to-point network and must address the same issues of churn and congestion.

Figure 7.3 illustrates an alternative to the fully flexible option. Fiber congestion is minimized when the AWG is at the OLT, possibly breaking out into an optical ribbon

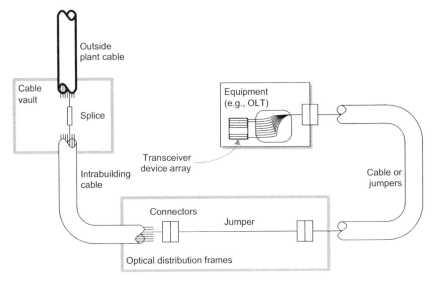

Figure 7.3 WDM on the circuit pack.

cable that terminates via a single optical connector on an array of transceivers, all of whose capabilities are the same. The AWG may be on the same circuit pack as the transceiver devices or on an optical termination subassembly installed in the equipment chassis. Capital cost is substantially reduced in this scenario: an array of transceiver devices is far less expensive than a separate (plug-in) optical transceiver for each wavelength.

Now, we are prepared to believe that a parallel MAC array could terminate your choice of either fast Ethernet or gigabit Ethernet. Also 10GE? Maybe. But at some point, we have just moved the traffic congestion point from the fiber to the OLT's backplane interface; if a significant number of subscribers migrate, over time, to higher speeds, we will probably need to relocate their services onto an OLT with higher throughput. It is not clear how that will work incrementally, and without service disruption. And, certainly, we cannot terminate other transport protocols on this OLT blade. It is Ethernet for everyone! Too bad about that CPRI link.

We could take the position that GbE per subscriber is enough forever. Twenty or 30 years from now, we might look back at this view with amusement or frustration, or we might agree that we got it exactly right.

Another oft-cited advantage of WDM PON is that it permits unbundling at the wavelength level. But if all parties are constrained to run the same protocol and terminate on the same OLT, it smells more like layer 2 unbundling. Some operators who base their businesses on unbundling are reluctant to go with layer 2.

These problems will either be solved or they will be deferred. Maybe special, different protocols and upgrades just go onto another fiber. All things considered, this is not at all a bad solution. Hypothesize a default fiber with default GbE services and a second, parallel fiber with a fully flexible DWDM spectrum for the high-value, high-revenue specials. On this second fiber, we can put 20—or 100—subscribers onto each wavelength through a remote multiplexing terminal, be the terminal a remote DSLAM for DSL access over vastly shortened drops, be it a remote G-PON OLT, or be it a 10G or 40G or 100G Ethernet switch or router on the premises of a business customer. Such a parallel fiber could even absorb overflow if the default fiber fills up, addressing a scalability issue that characterizes all PONs.

How would we distinguish this from an existing DWDM system or, with a few additional features, a reconfigurable optical add–drop multiplex (ROADM)? As we mentioned above, everyone has a different definition of what is and what is not a WDM PON. We could take the position that WDM PON is here today! Or if this is not it, we could challenge anyone to tell us how we will know when the real thing arrives. (*Hint*: try colorless ONU optics.)

A Question of Color Colorless optics are expected to be a universal requirement for WDM PON. The inventory and logistics cost of a separate device type for each wavelength would essentially preclude deployment. The question is then how to determine the wavelength of a service.

There are two approaches. With a fixed filter such as an AWG at each end of the shared-wavelength fiber, physical connection to one of the filter ports determines the service wavelength. Optical receivers are broad-spectrum devices and care little about their precise wavelength assignment, but the transmitter wavelength must match the passband of the filter. The optics in such a transceiver necessarily tune their transmitters by exploiting the wavelength of the received signal in one way or another.

In a technology called self-seeding, a broadband light source at the OLT is filtered by an AWG to deliver a narrowband pilot wavelength to each ONU. The ONU uses the pilot wavelength to lock the wavelength of its upstream laser. A similar arrangement tunes the OLT, often using the same broadband light source. The line width is fairly broad, and chromatic dispersion limits the achievable rate to something like 1 Gb/s per wavelength, over distances of something like 20 km, perhaps as much as 40 km.

Another proposal is called wavelength reuse, in which the downstream wavelength is stripped of its modulation, then essentially reflected back upstream at a power level that differs between the ONU's 0 and 1 transmit levels.

A third proposal uses tunable lasers, although the mechanism for automatically tuning the laser remains open to innovation. Because their line widths can be very narrow, tunable laser WDM PON promises the best rate-reach characteristics. However, tunable lasers are the most expensive of the current alternatives.

The other approach to colorless optics is to combine a tunable filter with the tunable transmitter. This could be appropriate if a WDM PON were overlaid on a (colorless) G-PON ODN, so that the entire WDM PON spectrum was visible to each ONU.

Our present view of WDM PON technology is that there will be several sweet spots that combine technical requirements, economics, and subscriber markets. WDM PON will find applications where its higher cost is justified by the improved optical budget that facilitates longer reach, by its protocol agnosticism, and possibly by its ability to unbundle at the physical layer. Residential service (wavelength to the home) will prove to be the most difficult.

7.6 ACCESS MIGRATION

The perceptive reader will have noticed that this book never strays far from concern about real-world feasibility, also known as the business case. It matters not how wonderful a technology is if it adds no value to the vendor, operator, and subscriber. Especially from this chapter, it should be apparent that there are many ways to add value, with no single right answer for all environments.

We conclude this book with a few thoughts about open issues. Their theme is that we should try to work on the important problems, not just the interesting technology.

Congestion Above, we described the pending issue of ODF congestion. This problem needs to be solved before it occurs; if we let it get out of control, a tangled ODF will be very difficult to recover. Fortunately, a subindustry is diligently working on this already.

Power PON technology intrinsically requires a powered device at the far end.

In the case of the SFU, the subscriber provides the power and is responsible for the backup battery. It remains to be seen whether subscribers eagerly grasp their new responsibility, but we, ourselves, would not be particularly keen.

Chapter 2 discusses some of the issues of remotely powering an MDU. In many environments, fiber to the curb with VDSL2 drops is an ideal solution, except for the power. Add powering to the equation, and it becomes so difficult and expensive as to preclude deployment.

Both of these use cases illustrate the fact that power, and power reduction, is vitally important. It has nothing to do with political correctness: it is a matter of real costs, here and now, that can make or break the case for deployment.

This perspective suggests that intermittent power conservation intervals are not particularly useful. What we require is low power, all the time. Just to mention two topics that come to mind:

- Traditional POTS telephones are power hogs: could an operator offer free replacement electronic telephone sets as part of a G-PON roll-out? No more 5-REN loads, no more high-voltage ringing, no more 24-mA loop current. Our cell phones require none of these—do we really have to support rotary-dial 500 sets forever? In terms of power consumption, could an ONU be designed to be more like a cell phone?

- DSL achieves dramatic speeds with vectoring and bonding, but at the expense of more and more complex DSP algorithms and considerable power on the wire. Are there better ways? (As we went to press, G.fast was gaining a lot of attention in the community. Back-powered, with speeds up to 1 Gb/s over a single pair, it will be extremely valuable if it can be done.)

Could we get MDU power from the subscriber, back down the drop? If subscribers are to power an MDU, we need ways to assure that each active user contributes an equal share and that the MDU does not fail if some users disconnect or lose power. These seem like solvable problems.

Power is not exciting, but it is important. Power issues can be show-stoppers.

Upgrade We have talked a lot about coexistence of current and next-generation technologies on preexisting optical distribution networks. At risk of heresy, we propose a different perspective.

Suppose that G-PON turns out to be a good match for most subscribers' demands for many years to come—the plain black telephone of the twenty-first century. Suppose that there are a few pioneers here and there, with the need for advanced service—and the willingness to pay. What if the take rate for next-gen never exceeds, say, 10%? Never is a long time; make it, say, 10 years? How do we make a business case for access network upgrade?

The primary upgrade challenge may not be coexistence on installed ODNs, but finding ways to cost-effectively serve sparse populations of advanced users. Ninety-six wavelengths at the end of a feeder fiber is only useful if we have, say, 70 customers nearby.

In terms of PON distribution networks, this might mean locating many splitters at a single point, with connectorized drops so that individual subscribers could easily be relocated onto a new service. There is no issue of upgrading all ODNs; we will upgrade one ODN and move the pioneers onto it. Shifting the high-demand users onto a different service has the side effect of allowing better service to the remaining subscribers, those who do not migrate. This would fit nicely into the more usual scenario for the bulk of subscribers: gradual evolution, gradual uptake of new services, and gradual increase in revenue.

And maybe we do not even need to multiplex the new service onto an existing trunk fiber—just use one of the spare fibers in the feeder plant. The operations issues are surely easier if we use a new fiber. Just to mention one, our power user is more likely to be willing to pay for protected access.

APPENDIX I

FEC AND HEC IN G-PON

I.1 REDUNDANCY AND ERROR CORRECTION

There are two basic ways that redundancy can help to keep errors under control in a communications system. If redundancy is used merely to detect errors, the receiver can request that erroneous data be sent again by the transmitter until error-free transmission is achieved. Obviously the transmitter needs to get feedback from the receiver to know which data have to be retransmitted. This requires a full duplex data connection and a protocol capable of coordinating forward data transmission and backward error notification, generally known as automatic repeat request (ARQ) protocols.

The advantage of an ARQ scheme is that fully reliable transmission can be achieved as long as retransmission is repeated until the receiver detects no errors—and assuming that it is indeed possible to detect the complete absence of error. Error detection requires that some form of redundant content be transmitted along with the payload data.

The downsides of an ARQ protocol are additional protocol overhead, potential additional delay and jitter due to retransmission, on the order of multiple round-trip times, and large buffers at both ends of the link. The transmitter needs to retain data in case it is needed for retransmission, and the receiver may need to buffer data until a full frame can be assembled.

In some applications, delay is secondary to reliability. The ubiquitous TCP is an example in which the integrity of a file transfer matters far more than delay, and buffering is not a problem because we assume file storage devices at each end. On the

Gigabit-capable Passive Optical Networks, First Edition. Dave Hood and Elmar Trojer.
© 2012 John Wiley & Sons, Inc. Published 2012 by John Wiley & Sons, Inc.

other hand, some applications rely on timely delivery—voice and video, for example—and retransmission is of no value whatever. If errors are to be corrected at all, the function must be part of the mainstream flow.

The way to exploit redundancy in real time is through error-detecting and error-correcting codes, often just called error-correcting codes (ECCs). An error-correcting code with sufficient redundancy enables nearly error-free data transfer without feedback. The choice of code is a tradeoff between the performance of the worst-case channel and receiver, the demands of the application, and the overhead cost of redundancy. A prime example is FEC, which is used in G-PON. Another example from G-PON is HEC, originally the acronym for header error control, but generalized in XG-PON to hybrid error control or correction.

I.2 FORWARD ERROR CORRECTION

G-PON and XG-PON provide FEC capabilities to achieve transmission quality better than the performance of the given optical link budget. G-PON requires residual BER not worse than 10^{-10}, while XG-PON requires 10^{-12}. The FEC code is chosen to deliver this performance in the face of a raw error rate that in XG-PON may be as bad as 10^{-3} downstream, 10^{-4} upstream. For G-PON systems with class C+ optics and FEC, the pre-FEC upstream BER needs to be not worse than 10^{-4}.

A FEC encoder at the G-PON transmitter introduces redundancy in the form of additional parity data added to GTC frames. A FEC decoder in the electrical portion of the receiver utilizes this redundancy to find and correct errors. Its performance is limited by the FEC code's error detection and correction capabilities. Clearly, the added redundancy reduces the capacity of the system for client data, as illustrated in Figure I.1.

With the help of Figure I.2, let us examine how FEC improves link quality. The figure depicts a hypothetical relation between received power or optical SNR and the resulting BER of an optical receiver, with and without FEC. Assume that the channel delivers performance along the curve WP1–WP2 without FEC. Received power level P_H delivers the specified service quality BER_L. If the received power decreases, the receiver walks up the curve toward WP2, where it delivers higher BER_H, which does not satisfy the quality requirement. If we now switch FEC on, the FEC decoder exploits the added redundancy to correct transmission errors, which moves us onto the curve for a FEC-enabled receiver, from operating point WP2 to WP3. We now achieve the required BER_L at the lower received power P_L. As long as the actual

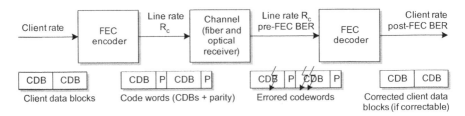

Figure I.1 Communication system with channel encoding and decoding.

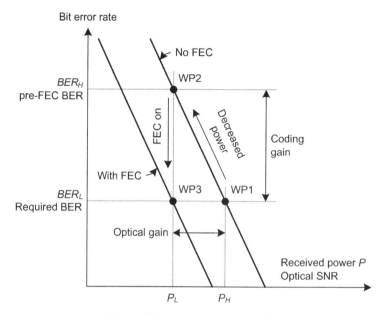

Figure I.2   Improvement with FEC.

received power remains above P_L, the actual BER does not exceed BER_L. Thus, FEC relieves the requirements on received power or optical signal-to-noise ratio (OSNR) for a specified BER.

We designate $P_H - P_L$ as the optical gain, the difference in received power required to achieve BER_L with FEC on and with FEC off. The coding gain is the vertical distance between BER_H and BER_L and is determined by the mathematics of the FEC algorithm and the statistical error distribution. Observe that optical gain is *not* the same as coding gain. Optical gain is what we really care about, because it affects the ODN budget. The nearer the BER curve is to being vertical—and many real-world optical receivers have very steep response curves—the less optical gain is possible, regardless of FEC.

For a receiver without FEC, it suffices to specify the maximum acceptable error rate BER_L. For a receiver with FEC, the pre-FEC requirement BER_H determines the necessary optical parameters. We then specify the FEC algorithm such that it delivers post-FEC-quality BER_L.

Reed–Solomon codes are a good choice of FEC algorithm.

I.2.1   Reed–Solomon Codes

G-PON and XG-PON use Reed–Solomon (RS) block codes, which are a subset of more general BCH* codes with the following properties:

---

* From the initials of their developers: Bose, Chaudhuri, and Hocquenghem.

Nonbinary Block Code RS codes operate on symbols rather than on bits, where a symbol is defined to be a block of bits, in our case 8 bits. Regardless of the number of incorrect bits within the symbol, the symbol can either be corrected or not, depending on the correction capability of the code. This makes the RS code especially suitable for burst error applications.

Systematic Code Systematic RS codes do not alter the input data they work on. Computed FEC parity bytes are appended to input data blocks to form code words, keeping the original data intact. The original intention in G.984 G-PON was to allow for inexpensive ONUs that did not support FEC. Such ONUs would nevertheless need to understand FEC-encoded data, if only to ignore the parity bytes. Today all ONUs implement FEC and this objective is not relevant. Another consideration is that in G.984 G-PON downstream framing, the frame delimiter is part of the first FEC code word. The ONU needs to be able to synchronize on the downstream frame in order to delineate FEC code words prior to decoding. The frame delimiter must therefore remain unchanged in position and content whether FEC is used or not. In G.987 XG-PON, the physical layer overhead is not part of the first FEC code word, and all ONUs and OLTs are capable of generating and decoding FEC. Systematic coding would therefore not be necessary in XG-PON. The systematic coding property is nevertheless inherited by reuse of the G-PON RS code in XG-PON.

Code Performance An RS code performs well compared to other linear codes with an equal amount of redundancy.

Hardware Efficiency Efficient hard-decision decoding algorithms exist for RS codes. Even long codes can be cost-effectively implemented in hardware, as needed for the high data rates found in optical communications networks.

We start with a semiformal description of the properties of an RS code. Refer to Figure I.3.

An RS(n, k) code word comprises n symbols derived from k input message symbols and $p = n - k$ parity check symbols. Each of the $q = 2^m$ possible symbols comprises m bits of input data, where m is a positive integer greater than 1.

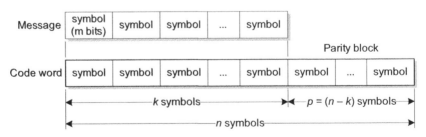

Figure I.3 RS code terminology.

Symbols are elements of a finite field.[*] When $m = 8$, a symbol equals a byte, which can represent 256 unique values.

RS(n, k) codes on m-bit symbols exist for all n and k for which

$$0 < k < n < 2^m + 2 \tag{I.1}$$

This yields a wide range of RS codes with fully flexible parity size p, directly related to the number of correctable errors t, which is given by

$$t = \begin{cases} \dfrac{n-k}{2} & \text{for } p = n - k \text{ even} \tag{I.2} \\[2mm] \dfrac{n-k-1}{2} & \text{for } p = n - k \text{ odd} \tag{I.3} \end{cases}$$

Most RS(n,k) codes used in communications systems are $(n,k) = (2^m - 1, 2^m - 1 - 2t)$ with $m = 8$. For $t = 8$ and $t = 16$, the codes are therefore RS(255, 239) and RS(255, 223). These two codes form the basis for FEC in G-PON and XG-PON.

The code rate R, redundancy overhead O, and minimum Hamming distance $D_{\min}$ are given by

$$R = \frac{k}{n} = \frac{k}{k+p} = \frac{1}{O+1}$$

$$O = \frac{p}{k} = \frac{n-k}{k} = \frac{1}{R} - 1$$

$$D_{\min} = n - k + 1 \tag{I.4}$$

The code rate R represents the efficiency of the code. It describes the proportion of user data symbols in a code word. For a given line rate R_c measured in bits per second, the user rate is reduced from R_c to $R_u = R \cdot R_c$ by enabling FEC.

The redundancy overhead O indicates the proportion of parity bytes.

The Hamming distance between two bit blocks of equal size is defined to be the number of locations in which bits differ. A similar definition is used when working with symbols. That is, the Hamming distance between two symbol blocks is a count of the number of symbols that differ. The minimum Hamming distance $D_{\min}$ reflects the minimum distance between any two of the q^k valid code words in an RS code.

Figure I.4 interprets this. All possible q^n words containing n symbols are located at the vertices of an n-dimensional space. Out of q^n potential words, q^k code words are chosen in such a way that equal spheres[†] centered on them contain as

---

[*] This is where we abruptly become informal. Those interested in further exploring the underlying mathematics are encouraged to start with ITU-T G.987.3 Annex B and G.709.

[†] One of our reviewers points out that there are no smooth curves in an n-dimensional discrete space. Rather than *sphere*, perhaps the word should be *squere*. As shown in Figure I.4, nonoverlapping really means that no discrete code word lies within two spheres.

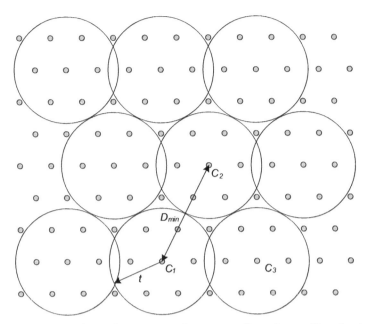

Figure I.4 Code words as centers of error correction spheres with radius t.

many other words as possible without overlapping other spheres. The spheres have radius t.

Any received word that falls within the sphere of a given code word is magnetically converted by the decoder to that code word. A received word that lies outside any of the spheres is detected as an uncorrectable error. If an excessive number of errors changes a received word from within one code word's sphere to some other code word's sphere, a decoding error occurs and the decoder outputs a wrong value.

The number of correctable errors is given by the radius t. If incorrect outputs are more undesirable than uncorrected errors, the radius t may be reduced. Fewer errors can then be corrected, but the probability that a badly errored code word may fall inside another sphere is reduced.

The value t is usually chosen to be as large as possible. An RS code is guaranteed to correct up to t symbols per code word. RS codes achieve the largest possible minimum Hamming distance for any linear code with the same (n, k):

$$t = \left\lceil \frac{D_{\min} - 1}{2} \right\rceil = \left\lceil \frac{n - k}{2} \right\rceil \tag{I.5}$$

The number of correctable errors t is half the number of parity symbols needed to achieve that performance.

If an RS decoder is capable of correcting a code word, it can easily determine how many incorrect bits were present as well as their positions.

I.2.1.1 *Symbol Error Rate*

In RS-protected systems, the symbol error rate SER, that is, the symbol error probability SEP for statistically independent errors, is relevant.

Figure I.5 shows performance curves for RS(255, k) codes for different values of $t = 1/2 \, (255 - k)$, with input symbol error probability on the horizontal axis and the probability of uncorrectable symbol errors on the vertical axis:

Slope   For large values of t, the slope of the curve is very steep: a slight variation in input SER results in a large variation in output SER. If the G-PON receiver cannot deliver its specified performance, either because of inadequate optical input power or because of component or design deficiencies, the output performance deteriorates dramatically.

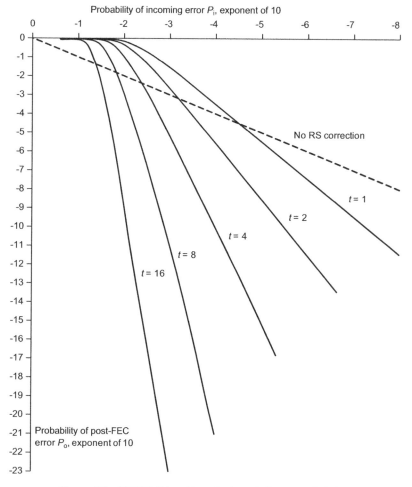

Figure I.5   RS(255,k) input to output symbol error probability.

High SER Performance   For very high input SER, RS codes perform poorly. The dotted line in Figure I.5 shows a threshold. For input SER above this threshold, the RS code introduces more errors than were present in the original input sequence. Higher t codes behave better in this respect.

An RS(255, 239) code with $t = 8$ has the ability to convert a symbol error rate $P_i = 10^{-3}$ at the input into an output symbol error rate of 8.7×10^{-12}, nine orders of magnitude better. A symbol error rate of 10^{-3} translates to a bit error rate of about 10^{-4}, as specified for XG-PON upstream transmission. An RS(255, 223) code with $t = 16$ improves $P_i = 10^{-3}$ to $Po \approx 10^{-22}$, 19 orders of magnitude better. The relation between symbol error rate and bit error rate is explained below.

I.2.1.2 Bit Error Rate

Assume a binary channel with bit error probability p_b corresponding to the long-term BER. With probability p_b for a single bit error, $(1 - p_b)^m$ denotes the probability of zero errors in a block of m bits. Therefore the probability of one or more errors in the symbol is

$$P_i = 1 - (1 - p_b)^m \qquad (I.6)$$

G.987.2 XG-PON expects a maximum BER of 10^{-3} at the ONU input (downstream). A p_b of 10^{-3} results in P_i of 8×10^{-3} for 8-bit symbols ($m = 8$). This results in an output symbol error rate of 4.7×10^{-11} for $t = 16$. Thus, on average, one symbol in a sequence of 2.1×10^{10} received symbols cannot be corrected. With $135,432$ bytes (symbols) per XGTC frame, an uncorrectable symbol occurs roughly every 20 s in the worst-case channel.

The corresponding number for XG-PON upstream, assuming 100% full bursts, is one uncorrectable symbol error approximately every 47 min. The same value is true for class C+ G.984 G-PON downstream. As to G.984 class C+ upstream, again assuming 100% burst occupancy, an uncorrectable error occurs about every 90 min. These are the values at the limit of the specified budgets; because the curves are so steep, links even slightly better than the worst-case specification will perform much better.

I.2.1.3 Burst Error Correction Capability

Regardless of the number of bit errors, an RS code can correct t symbols in a symbol of size m. Thus, an error burst of $(t - 1)m$ consecutive bit errors in a received word is always correctable, and maybe a burst of as many as tm bit errors, depending on how the burst aligns with symbol boundaries.

Figure I.6 shows how it works. The RS(255,239) code, with $t = 8$, can correct eight symbols, which means it can always correct an error burst of 57 bits, and possibly as many as 64 bits, if the 64 bits happen to align with symbol boundaries, always assuming that the code block contains no additional errors. The RS (255,223) code can always correct a burst of 121 bits and sometimes as many as 128 bits.

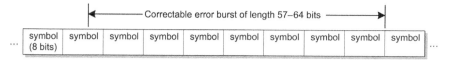

Figure I.6 Burst error correction capability, RS(255,239), $t = 8$.

I.2.1.4 Truncated RS Codes

RS(n,k) codes can be constructed for various positive integer combinations of m, k, n as long as $0 < k < n < 2^m + 2$, with $m > 2$. This gives great flexibility in designing codes. However, the risk of development and test of new hardware can be reduced by using specific standardized and well-tested RS codes. This is true even if the FEC function is required to work on shorter message blocks. At the cost of some loss of efficiency, the client data block can be truncated while retaining the same number of parity bytes. This approach can be used to construct any smaller RS(n', k', $p = k' - n'$) code from a given RS(n, k, $p = n - k$) code, where $n' < n$ (and $k' < k$), but with $p = n - k = n' - k'$ unchanged, and therefore with the same error correction performance as the longer code word.

Figure I.7 outlines the procedure at the transmit end. An appropriate number $k - k'$ of leading zero symbols is inserted prior to an ordinary (n, k) FEC operation. The zero pad is discarded prior to transmission. At the receiving end, the same process occurs: a zero pad of size $k - k'$ is prepended to the code word, a completely ordinary (n, k) FEC decoder operates on the padded word, and then the padding is

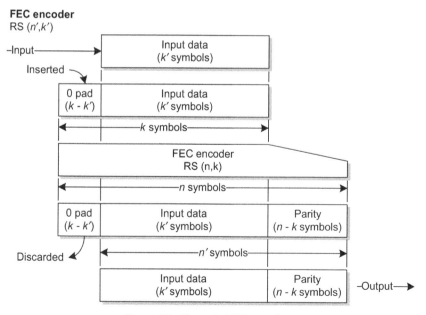

Figure I.7 Truncated RS encoding.

discarded. This technique relies on the fact that RS codes are systematic, that is, they do not alter the original data but merely append parity symbols.

Because the zero pads inserted at the encoder are never transmitted, they can never be errored. Thus, symbol errors can never occur at those positions at the decoder. In theory, the performance could be improved by taking advantage of these known values, but the purpose of truncating a standard RS code was to reuse a known and proven design. The resulting RS(n', k') code therefore has the same error correction performance as its RS(n, k) equivalent.

It is clear that the number of zero padding bytes can be whatever we choose, subject to agreement on both ends. When a transmission interval is too short for a complete FEC code word, we create what is called a shortened last code word with an appropriate pad. G.984 G-PON uses the shortened last code word principle in the downstream direction, with a pad of 135 zero bytes. The XG-PON downstream frame is designed to contain an integer number of full-size code words and therefore does not need this feature. Both G-PON and XG-PON rely on this principle in the upstream direction when the available space at the end of a grant allocation does not permit a full-size code word.

I.2.2 FEC in G-PON and XG-PON

This section peels one further layer from the details of FEC encoding and decoding of G-PON and XG-PON in both upstream and downstream directions. We begin with an overview of each.

I.2.2.1 G.984 G-PON FEC

G-PON FEC is based on the common RS(255, 239) code used in many telecom applications. The same code is used in the upstream and downstream directions; it can correct 8 bytes per 255-byte code word, regardless of the number of errored bits per byte.

G.984 mandates that OLTs support FEC in both directions, but it is the operator's option whether to switch it on or not. The ONU is not required to be able to utilize FEC to correct errors, but it must be capable of at least recognizing and ignoring the FEC parity bytes. Upstream FEC can be enabled/disabled for each ONU individually by the OLT.

Although the standard specifies FEC as an optional feature for ONUs, it is widely supported.

FEC is often not used in either direction in actual G-PON deployments. The G-PON standard B+ power budget of 28 dB can be achieved by available optics. Even the extended 32 dB C+ budget, whose standard requires FEC, can almost be achieved by available optics without the need for FEC. In any event, FEC can be used to gain one or two additional decibels for budgets similar to B+. This allows either extending the reach by a few additional kilometers or doubling the split ratio, possibly enabling 64 or even 128 users on a single PON port. For C+ and highly optimized optics, FEC gain is quite limited, since the BER curves are very steep, as shown in Figure I.2.

I.2.2.2 G.987 XG-PON FEC

While FEC in G.984 G-PON is an option, the situation is different in XG-PON, whose downstream rate is four times as great and whose upstream speed is twice as great. At 10 Gb/s, dispersion can be a problem on real-world ODNs. FEC is needed to achieve even the nominal budget classes of 29 and 31 dB. In XG-PON, a truncated version of the RS(255, 223) code is used downstream. The code is called RS(248, 216); it introduces blocks of 32 bytes of parity to correct up to 16 bytes per 248-byte code word. The reason for truncation was to fit an integer number of code words into the 125-μs XG-PON downstream frame; the PHY header is excluded.

In the upstream direction, XG-PON chose a truncated version of the RS(255, 239) code, called RS(248, 232). The use of upstream FEC is dynamically controlled by the OLT, as determined by the individual channel quality of each ONU. If an ONU's optical budget is such that good error performance is possible without FEC, we can turn the FEC off.

I.2.2.3 Summary

Table I.1 summarizes the algorithms and the choices for FEC in the G-PON and XG-PON world.

The FEC coding parameters for G-PON and X-GPON together with overhead burden and the achievable coding gain are summarized in Table I.2.

To better explain how FEC overhead affects the available data rate, Table I.3 compares line rate R_c with user data rate R_u and lists the difference. The numbers account only for FEC overhead.

TABLE I.1 FEC in G-PON and XG-PON

Protocol	Direction	Code	Support	On/Off
G-PON	Downstream	RS(255, 239)	OLT mandatory,	Configurable per PON
	Upstream		ONU optional	Configurable per ONU
XG-PON	Downstream	RS(248, 216)	Mandatory	Always on
	Upstream	RS(248, 232)		Dynamically controlled by OLT

TABLE I.2 G-PON and X-GPON Reed–Solomon Code Characteristics

Protocol	Direction	Code	Number of Parity Bytes	Overhead Percent	FEC Coding Gain Expected
G-PON	Downstream	RS(255, 239)	16	6.27%	3 dB
	Upstream				2 dB
XG-PON	Downstream	RS(248, 216)	32	12.9%	4 dB
	Upstream	RS(248, 232)	16	6.45%	2 dB

TABLE I.3   FEC Effect on Available Rate

Protocol	Direction	Line Rate, Mb/s	User Rate, Mb/s	Difference, Mb/s
G-PON	Downstream	2488.32	2332.19	156.13
	Upstream	1244.16	1166.09	78.07
XG-PON	Downstream	9953.28	8668.99	1284.29
	Upstream	2488.32	2327.78	160.54

I.3   HYBRID ERROR CORRECTION

The normal payload flow in a G-PON is expected to be nearly free of errors, either because it is protected by FEC or because the optical link is comfortably within the bounds of good performance. However, some errors have comparatively serious consequences. As an example, an error in decoding the payload length of a GEM frame causes the loss of that GEM frame and all subsequent GEM frames until resynchronization occurs.

Especially error-sensitive fields in G-PON and XG-PON are protected by hybrid error correction (HEC), a two-part error detection and correction scheme. The first part uses another BCH error-correcting code. In contrast to the symbol codes used in G-PON and XG-PON FEC, this code operates on bits. It is designated BCH (63, 12, 2), meaning that the code word is 63 bits long, including a 12-bit check field, and that the code can correct two errors. It is generated by the polynomial $x^{12} + x^{10} + x^8 + x^5 + x^4 + x^2 + 1$.

The protected field, including the polynomial remainder itself, is 63 bits long. All of the pertinent fields in G-PON are an integer number of bytes long, for example, 8, so there is an extra bit beyond the protected field. The BCH algorithm corrects up to 2 bits, and a simple parity bit in the final bit position reliably flags an uncorrectable third error.

In summary, with an overhead of 13 bits, we can correct two bit errors and detect three.

We use the same algorithm everywhere, even for fields shorter than 64 bits. The XGTC header described in Section 4.1.3, for example, is 32 bits long, while the G.984 GEM header is 40 bits. For these fields, we simply prepend a suitable number of zeros, in the same way that we did for shortened FEC code words, to bring the total to 64 bits. We generate the HEC field on the padded value and discard the leading zeros before transmission. At the receiving end, we prepend a new set of zeros before decoding, then discard the zeros from the result.

APPENDIX II

PLOAM MESSAGES

The PLOAM channel comprises a small set of relatively simple messages, with the idea that they can be supported in fairly low-level firmware, if not actually in hardware. They are used for low-level negotiations between OLT and ONU, including ONU discovery when no higher level (OMCI) management channel is available.

Many of the PLOAM messages are elaborated in the various sections where they are key actors; this appendix is dedicated to the nuts and bolts. As we have done before, we describe G.987 XG-PON and G.984 G-PON separately: there is some overlap, but not enough to combine the descriptions.

II.1 PLOAM MESSAGES IN G.987 XG-PON

The PLOAM message in G.987 XG-PON is 48 bytes long. With a history of ATM in the original world of B-PON, there was concern that 48-byte messages might be mistaken as a retrograde step, but 48 was the correct size for the messages that needed to be defined, and 48 bytes it is. Variable-length messages were considered but rejected because fixed-length messages are easier to parse and the difference in efficiency is inconsequential.

As illustrated in Figure II.1, all G.987 XG-PON PLOAM messages have the same structure.

Gigabit-capable Passive Optical Networks, First Edition. Dave Hood and Elmar Trojer.
© 2012 John Wiley & Sons, Inc. Published 2012 by John Wiley & Sons, Inc.

Length 2	ONU-ID
1	Message type
1	Sequence number
36	Message contents –depends on message type –may include trailing padding
8	MIC: message integrity check

Figure II.1   G.987 XG-PON generic PLOAM message.

ONU-ID   The ONU-ID consumes 2 bytes of the generic PLOAM message. The first 6 bits are zero padding; the 10-bit ONU-ID occupies the least significant end of the field. The values 0–1022 are reserved for ONUs with those IDs. The value 1023 (0x3FF) is used upstream by an ONU in the process of discovery, an ONU that has not yet been assigned a TC-layer ONU-ID; it is also used as a broadcast address when the OLT transmits PLOAM messages downstream.

Skipping the message type field for the moment, we discuss the sequence number next. As to the MIC, we cover that in the security Section 4.6.1.

Sequence Number   The sequence number assists in correlating transactions and identifying lost messages and may be helpful in troubleshooting. The OLT increments the sequence number with each message sent downstream, rolling over from 255 to 1, skipping 0. The OLT runs a separate sequence number counter for each unicast ONU channel and one for broadcast. If the ONU sends one or more upstream messages in response to an OLT message, it copies the OLT's sequence number into the upstream message; otherwise it uses the value 0—that's why 0 is not valid in the downstream direction. At ONU initialization, the OLT starts with sequence number 1 for each ONU and also for broadcast.

Message Type   As might be expected, the message type indicates the type of the message. Table II.1 is a complete list of the PLOAM message types. As we see, some of the messages are transactional; others are one-way messages only. The order of listing corresponds roughly to the sequence in which they might be used. In preparation for the subsequent section on G.984 G-PON PLOAM messages, which are listed in Table II.2, note that this is a reasonably short list.

The acknowledgment column of Table II.1 indicates whether the ONU answers a downstream message with an acknowledgment upstream PLOAM. U means the message is acknowledged if and only if it is unicast to the specific ONU. The profile and ranging_time messages are acknowledged if the ONU is in operation state O5 and if the message is unicast to the specific ONU; otherwise not.

TABLE II.1 G.987 XG-PON PLOAM Message Types

Downstream	Unicast	Broadcast	ACK	Upstream
Profile	X	X	U	
				Serial_number_ONU[a]
Assign_ONU-ID		X	No	
Ranging_time	X	X	U	
Request_ registration[a]	X			Registration[a]
Assign_alloc-ID	X		Yes	
Key_control	X	X		Key_report
Sleep_allow	X	X	No	
				Sleep_request
				Acknowledgment
Deactivate_ONU-ID[a]	X	X	No	
Disable_serial_number		X	No	

[a]All PLOAM messages contain a MIC, which is normally derived from cryptographic keys known to both OLT and ONU. Before communication is established, this is not possible, nor is it possible for downstream broadcast messages. The default PLOAM integrity key PLOAM_IK is $0x55_{16}$, where the notation indicates 16 bytes of value 0x55. The default key is used on all broadcast messages and on unicast messages identified in Table II.1 with an asterisk. G.987.3 amendment 1 specifies the use of the default PLOAM_IK for the unicast deactivate_ONU-ID message also.

II.1.1 Profile Message: Downstream

An XG-PON bandwidth grant (Section 4.1.2.3) contains a 2-bit profile index. The ONU must not respond to a grant if it does not know the profile with the specified index. The OLT periodically broadcasts profile PLOAM messages to define all of the profiles it intends to use. When it initializes, either from power-up or from a deactivation command, the ONU discards any previous profile information it may have learned and starts over.

In most cases, profiles are broadcast to all ONUs. It is also possible that a PON could contain ONUs of limited capability, or perhaps ONUs at the extreme range of the optical budget, for which the OLT wants to establish special profiles. In this case, a profile message could be unicast to the selected ONU. Of course, the initial discovery step is necessarily done with a broadcast profile, because the ONU has no ONU-ID at the beginning.

A subsequent broadcast profile with the same index would overwrite any unicast profiles that might have been established, so the OLT must use the unicast option judiciously, probably by reserving one index just for unicast. The OLT is also required by the standard to send a profile update or a new profile at least twice— assuming a stable population of ONUs—before using it.

Figure II.2 depicts the profile PLOAM message. Section 4.1.3 discusses the burst header fields in further detail.

VVVV—Version The OLT changes the version field whenever it changes anything else in the profile. Every time it receives a new profile message, the ONU must check the version and profile index to see whether anything has changed from the corresponding profile that it already knows. This allows the ONU to detect changes by checking a single byte, rather than the entire message content. It is not expected that the OLT would change the profile very often; if ever it does, it must ensure that all ONUs have had a chance to learn the new profile before using it in a BWmap.

PP—Profile Index   The 2-bit profile field in the BWmap selects one of up to four possible profiles to be used by the ONU in its response burst. The same value appears here, where the profile is defined.

Version and profile index are completely independent of each other, as are all of the other fields of the profile message.

F—FEC   If this bit is set, the ONU applies FEC to the upstream burst. Upstream FEC is disabled if $F = 0$. The idea behind FEC control through the burst header definition is that upstream FEC is a long-term commitment, at least for a given ONU. If there are two classes of ONU on the PON, one class that benefits from FEC, one class whose performance is okay without FEC, the OLT can specify two profiles, otherwise identical, for the two classes. It has the choice either to unicast the profiles to the appropriate ONUs and use the same index or to set up two broadcast profiles and distinguish ONU classes through the profile index field of the BWmap.

DDDD—Delimiter Length   The number of bytes in the delimiter field, in the range 1-8.

Delimiter   The delimiter itself. Its length is specified by the DDDD field; it is padded with trailing zero bytes if necessary. The delimiter is the final fixed pattern of the burst header, whose purpose is to enable byte synchronization by the OLT receiver.

LLLL—Preamble Length   The number of bytes in the preamble field, in the range 1-8.

RR RRRR—Preamble Repeat Count, range 0–31   The entire preamble is repeated this many times at the beginning of the burst header.

Preamble   Like the delimiter pattern, the preamble template may be as long as 8 bytes. This field is also padded with trailing zero bytes.

PON Tag   The PON tag is an arbitrary octet sequence chosen by the OLT or by the operator and used in the initial derivation of the master session key, as described in Section 4.6.1. With the idea that it could be useful for security purposes, the PON tag should be unique across some reasonable space; as an example, it might be a string derived from the OLT blade serial number and the PON port number. The PON tag should be the same across all profiles and should remain stable indefinitely.

II.1.2   Serial_number_ONU Message: Upstream

When it initializes, the ONU synchronizes to the downstream PON and listens for profile PLOAM messages. When it subsequently sees a serial number grant that

	ONU-ID, unicast or broadcast
	Message type = 0x01
	Sequence number
Length 1	VVVV 00PP
1	0000 000F
1	0000 DDDD
8	Delimiter
1	0000 LLLL
1	00RR RRRR
8	Preamble
8	PON tag
	Padding
	MIC, default or specific key

Figure II.2   Profile message.

specifies a profile whose index it knows, the ONU generates an upstream serial_number_ONU response message, delayed by a random value to reduce the probability of collision with possible responses from other ONUs. Each time it responds to a serial number grant, the ONU uses a new random delay. Section 4.2 explains the discovery message exchange and state model, and Section 4.3.4 describes the timing aspects of the ranging process in detail.

In this message, shown in Figure II.3, the ONU declares its identity using factory-programmed values for vendor ID and serial number. The combination of vendor ID and serial number is expected to be globally unique.

Vendor ID   A four character string, for example, ERSN for Ericsson. The formal reference for the choices is ATIS-0300220.

Vendor-Specific Serial Number   The format and details are unspecified as long as the serial number is unique for the given vendor. The serial number is of course fixed for the lifetime of the ONU.

Random Delay   Expressed with bit time granularity at the 2.5-Gb/s upstream rate. This does not imply that the random delay is either precise or accurate to that level of granularity; it merely indicates the weight of the least significant bit. By measuring the actual delay of the ONU's response and offsetting it with the ONU's reported random delay, the OLT may be able to range the ONU immediately or at least be able to dramatically reduce the width of the ranging window in the subsequent ranging step.

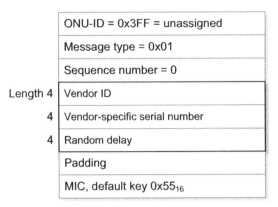

	ONU-ID = 0x3FF = unassigned
	Message type = 0x01
	Sequence number = 0
Length 4	Vendor ID
4	Vendor-specific serial number
4	Random delay
	Padding
	MIC, default key 0x55$_{16}$

Figure II.3 Serial_number_ONU message, G.987.

II.1.3 Assign_ONU-ID Message: Downstream

When the OLT recognizes a serial_number_ONU message, it assigns an ONU-ID to the newly discovered ONU. Figure II.4 illustrates the message.

The assign_ONU-ID message is broadcast to all ONUs. If the vendor ID and vendor-specific serial number fields match an ONU in serial number state O2-3 (Section 4.2.1), that ONU takes on the assigned ONU-ID value as its own TC-layer identifier. Assignable ONU-IDs range from 0 to 1022; henceforth, the ONU is addressable and identifies itself by this value.

At this point in the lifetime of the ONU session, the OLT may not be able to associate the ONU with the correct subscriber, the subscriber whose service records also contain an ONU-ID. The ONU state machine (Section 4.2) also prevents the OLT from changing the ONU-ID later on.

As described in the introductory text to Chapter 4, ONU-ID is an ambiguous term. At the TC-layer PLOAM level, where this message exists, ONU-ID is expressed in the form of real bits in real registers and would be visible, for example, to a snooping device on the PON. At the provisioning and management level, ONU-ID is a name, a reference.

It is the OLT vendor's choice either to maintain a mapping between the management-level ONU-ID and the TC-layer PLOAM ONU-ID or to reinitialize the ONU, rediscover it by serial number, and assign it a different PLOAM ONU-ID on the second pass.

II.1.4 Ranging_time Message: Downstream

When the OLT learns the actual round-trip delay of the ONU during initial discovery, it compensates the delay with the ranging_time message.

From time to time during normal operation, the OLT may also adjust the ranging time to compensate for drift of the ONU's propagation delay.

There is a third application of this message. If a PON is protected at the OLT and/ or trunk fiber, with all ONUs subtended from a common unprotected split point, then

ONU-ID = 0x3FF = broadcast
Message type = 0x03
Sequence number
Assigned ONU-ID value
Vendor ID
Vendor-specific serial number
Padding
MIC, default key 0x55$_{16}$

Length 2 (Assigned ONU-ID value), 4 (Vendor ID), 4 (Vendor-specific serial number)

Figure II.4   Assign_ONU-ID message, G.987.

the equalization delay after a protection switch can be adjusted by broadcasting a single offset value to all ONUs and specifying a relative, rather than an absolute, delay. Section 2.3 describes PON protection.

As we see in Figure II.5, the fields of the ranging_time message include:

P   If this bit is set, the equalization delay is an absolute value, and the S bit is ignored. If $P = 0$, the equalization delay is relative, with its sign determined by S. The first ranging delay during ONU activation must be absolute, since there is no well-defined reference against which to define an offset.

S   If $S = 0$, the ONU's existing equalization delay is to be increased by the value in the equalization delay field; if $S = 1$, the delay is to be decreased.

Equalization Delay   This field specifies the actual value of equalization delay, in 2.5-Gb/s bit times. As with the random delay in the serial_number_ONU message, the 1-bit granularity of this field merely indicates the weight of the least significant bit and implies nothing about either the accuracy or the precision of the value.

The MIC is calculated with the default key 0x55$_{16}$ if and only if the message is broadcast.

ONU-ID, unicast or broadcast
Message type = 0x04
Sequence number
0000 00SP
Equalization delay
Padding
MIC, default or specific key

Length 1 (0000 00SP), 4 (Equalization delay)

Figure II.5   Ranging_time message, G.987.

ONU-ID, unicast
Message type = 0x09
Sequence number
Padding
MIC, default key $0x55_{16}$

Figure II.6 Request_registration message.

II.1.5 Request_registration Message: Downstream

During the initialization sequence or at any time thereafter, the OLT may request the ONU's registration ID. We see in Figure II.6 that there is actually no message content; it is just a request.

II.1.6 Registration Message: Upstream

As discussed in Section 4.4, learning the ONU's registration ID is an integral phase of many operators' ONU identification and registration process. In Section 4.6.1, we see that the registration ID is also built into the XG-PON security mechanism.

When the OLT requests the registration ID, the ONU responds with this message, shown in Figure II.7. As well as a response to a direct request, an ONU in ranging state O4 generates this message in response to a ranging grant (Section 4.2.1). When registration is sent as an autonomous message, its sequence number is zero.

If no one has entered a registration ID into the ONU, its default value is a string of null characters, 0x00 octets. Otherwise, the string entered by the subscriber or the installer should appear in the lower numbered bytes, with unused higher numbered characters padded with nulls. Further semantics of this field are formally unspecified, although it is understood that some operators will concatenate a 24-character identifier and a 12-character password into the registration ID field.

	ONU-ID
	Message type = 0x02
	Sequence number
Length 36	Registration ID
	MIC, default key $0x55_{16}$

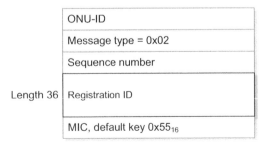

Figure II.7 Registration message.

	ONU-ID, unicast
	Message type = 0x0A
	Sequence number
Length 2	Assigned alloc-ID value
1	Alloc-ID type
	Padding
	MIC, specific key

Figure II.8 Assign_alloc-ID message, G.987.

II.1.7 Assign_alloc-ID Message: Downstream

An alloc-ID used for subscriber traffic must be assigned to its parent ONU with this message and must also be mapped to a T-CONT through OMCI. Refer to Section 6.3.2 for further discussion.

Because the default alloc-ID is assigned automatically and cannot be deassigned, this command would specify a default alloc-ID value only if it were desired to carry subscriber traffic over the same alloc-ID with OMCI. This is legal but not recommended.

Figure II.8 illustrates this message.

The alloc-ID value is a 14-bit field in the least significant bits, preceded by a pair of 0 padding bits.

Two alloc-ID types are defined:

- 1: Assign this alloc-ID to the designated ONU. The formal G.987.3 definition says that the alloc-ID is used for XGEM frames, but there is no other option: *all* alloc-IDs are used for XGEM frames. Conceivably, some other mapping could be defined someday, with some other code point, but it is extremely unlikely.
- 255: Deassign this alloc-ID.

II.1.8 Key_control Message: Downstream

As discussed in detail in Section 4.6.1, G.987 XG-PON has several encryption keys. The purpose of the key_control PLOAM message is to update or verify the key used for unicast payload encryption. This is the key whose index appears in the GEM frame header (Section 4.1.1). The key_control message may be either unicast or broadcast, affecting either one or all of the ONUs on the PON.

Figure II.9 shows the key_control message. There are three fields:

C—Control The value 0 instructs the ONU(s) to generate a new key and return it in one or more key_report PLOAM messages. The value 1 requests that the

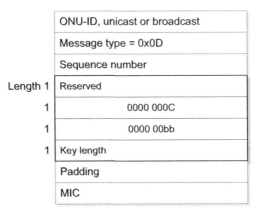

	ONU-ID, unicast or broadcast
	Message type = 0x0D
	Sequence number
Length 1	Reserved
1	0000 000C
1	0000 00bb
1	Key length
	Padding
	MIC

Figure II.9 Key_control message.

ONU confirm the existing key by returning the so-called key name in the key_report PLOAM message.

bb—Key Index The key used in real-time encryption of a unicast XGEM frame is selected by matching this key index, whose values may only be 01 or 10.

Key Length This field specifies the length of the key to be generated by the ONU in bytes. The value 0 specifies a 256-byte key. At the time of writing, all keys were 16 bytes long, 128 bits.

II.1.9 Key_report Message: Upstream

In response to a key_request, the ONU (or all ONUs) generates one or more key_report messages. The key_report message has space for 32 bytes of keying material, so if the requested key length is greater than 32 bytes, several messages must be sent. The PLOAM message sequence number is the same on all such messages. Thirty-two bytes suffice for a 256-bit key, and the current specifications call out only 128-bit keys, so the need for multiple key_report messages is not anticipated in the near term.

Figure II.10 shows the structure of this message.

R—Report Type If R = 0, the message contains an encrypted form of part or all of a new key; if R = 1, the message is a report on an existing key, in which case the key fragment field carries a so-called key name.

bb—Key Index The same value that appeared in the key_control PLOAM message that prompted this response. Only the values 01 and 10 are meaningful. In a GEM frame header, 00 indicates clear text transmission and 11 is reserved.

FFF—Fragment Number The first 32 bytes of keying material are sent as fragment 0. The fragment number increments with each successive 32-byte block of keying material.

Length	Field
	ONU-ID
	Message type = 0x05
	Sequence number
1	0000 000R
1	0000 000bb
1	0000 0FFF
1	Reserved
32	Key fragment
	Padding
	MIC, specific key

Figure II.10 Key_report message.

Key Fragment This field contains 32 bytes of the encrypted key, or the key name. If fewer than 32 bytes are to be sent, they appear in the lower numbered bytes of the message.

Before being fragmented into one or more key fragment fields, a new key is encrypted as

$$\texttt{EncryptedKey} = \texttt{AES_ECB}\,(\texttt{KEK},\ \texttt{CleartextKey}).$$

If the message is a report on an existing key, the single key fragment field carries the key name:

$$\texttt{KeyName} = \texttt{AES_CMAC}\,(\texttt{KEK},\ \texttt{CleartextKey}\,|\,\texttt{0x3331 3431 3539 3236 3533 3538 3937 3933},\ \texttt{128})$$

The security Section 4.6.1 describes the notation conventions, encryption algorithms, and the KEK.

II.1.10 Sleep_allow Message: Downstream

The semantics of the sleep_allow and sleep_request messages are described extensively in Section 4.5 on energy conservation. Here, we limit ourselves to the details of the message format. See Figure II.11.

If A = 1, the ONU(s) is permitted to enter one of its low-power modes at its own discretion. If A = 0, the ONU(s) is required to remain fully active and responsive.

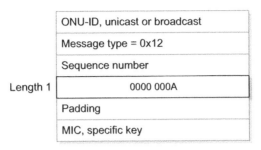

Figure II.11 Sleep_allow message.

II.1.11 Sleep_request Message: Upstream

When the ONU enters or departs a power-saving mode, it signals the OLT with the sleep_request message, as shown in Figure II.12.

The activity level is one of:

0 *Awake.* This value signifies that the ONU is fully responsive and will remain so until it signals a different value.

1 *Doze.* The ONU intends to enter doze mode.

2 *Sleep.* The ONU intends to enter cyclic sleep mode.

II.1.12 Acknowledgment Message: Upstream

Shown in Figure II.13, the acknowledgment message is a multipurpose tool.

Table II.1 indicates which downstream messages routinely receive an acknowledgment response. In the normal case, the completion code would be 0, ok.

If the ONU has nothing to send when the OLT grants an upstream PLOAM opportunity, it transmits an acknowledgment with completion code 1, no message.

If the ONU is busy working on a command, for example, to generate new encryption material, and the OLT requests an upstream PLOAM, the ONU may respond with completion code 2, busy.

In case of error, any unicast PLOAM may elicit a negative acknowledgment, with completion code 3 (unknown message type), 4 (parameter error), or 5 (processing error).

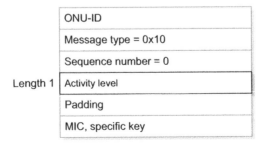

Figure II.12 Sleep_request message.

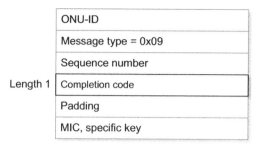

Figure II.13 Acknowledgment message, G.987.

II.1.13 Deactivate_ONU-ID Message: Downstream

Two of the PLOAM messages look similar but have quite different purposes. The deactivate_ONU-ID message of Figure II.14 essentially resets the ONU in terms of its existence on the PON. The ONU stops transmitting, discards all of its PON-specific (TC-layer) provisioning, including its profiles, ONU-ID, equalization delay, and alloc-IDs. It then reenters the activation process. By sending this message as a broadcast, the OLT effectively reinitializes the entire PON.

The difference between this message and the OMCI command to reboot the ONU is that deactivation does not disturb the provisioning of the MIB or the subscriber services.

G.987.3 amendment 1 specifies the use of the default PLOAM key for unicast as well as the broadcast case.

II.1.14 Disable_serial_number Message: Downstream

While the need for a deactivate message suggests some sort of tangle on the PON, the disable_serial_number PLOAM message is used under more dramatic circumstances, namely when there is reason to suspect that an ONU has gone rogue. The message relies on the assumption that the ONU can still recognize downstream PLOAM messages, recognize its own serial number, and shut down its transmitter. Depending on the ONU's failure mode, this will not always be true, of course, but it is still worth having.

ONU-ID, unicast or broadcast
Message type = 0x05
Sequence number
Padding
MIC, default key

Figure II.14 Deactivate_ONU-ID message, G.987.

	ONU-ID = 0x3FF = broadcast
	Message type = 0x06
	Sequence number
Length 1	Disable
4	Vendor ID
4	Vendor-specific serial number
	Padding
	MIC, default key $0x55_{16}$

Figure II.15   Disable_serial_number message, G.987.

As shown in Figure II.15, this is always a broadcast message. The vendor ID and serial number are the same as were reported by the ONU itself during the discovery process.

The disable field has four values:

0xFF   The ONU whose serial number matches is directed to go into emergency stop state O7. The ONU is not permitted to respond to upstream grants. The Emergency stop state is intended to persist across ONU initialization and power cycles.

0x00   The ONU whose serial number matches is directed to restart the activation process, having discarded all of its TC-layer parameters, including its ONU-ID. This effectively performs the same function as a unicast deactivate command, but with the distinction that this message pertains only to an ONU that is in emergency stop state O7.

0x0F   All ONUs are directed into emergency stop state O7. The vendor ID and serial number fields are ignored. Be aware that the semantics of this code point are not the same as those of the same code point in the G.984.3 G-PON message.

0xF0   All ONUs are directed to restart the activation process, having discarded all of their TC-layer information, including their ONU-IDs. This is similar to a broadcast deactivate command, but only for ONUs in state O7.

II.2   PLOAM MESSAGES IN G.984 G-PON

In G.984 G-PON, many PLOAM messages are sent three times as a way to improve the odds of successful delivery. We are skeptical of the value of this, but it does little harm.

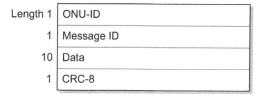

Figure II.16 G.984 G-PON generic PLOAM message.

Actually, that is not quite true. The effect of sending most messages three times is to slow down the PLOAM channel by a factor of 3. G-PON only sends one PLOAM message downstream in each frame anyway, so the combined effect is to make the initialization of a complete PON longer than desirable.

Three-times repetition produces a few bizarrenesses; for example:

- The idea that an ONU ought to send dying gasp three times. What if it cannot?
- Should the OLT ignore DG if it only receives a single copy?
- Should the ONU stop signaling DG after three times, even if it is still dying?
- Some of the triply-repeated OLT messages are even acknowledged (three times) by the ONU, under the theory that an acknowledgment might get lost in transit. Should the ONU send three ACKs, even if it receives only one copy of the downstream message? Should it send nine ACKs?
- What if the OLT receives three ACKs before it gets around to sending the message three times?

Sorry. Sometimes it is hard to resist.

The G.984 G-PON PLOAM message is 13 bytes long. All PLOAM messages have the same structure, which is illustrated in Figure II.16.

In G.984 G-PON, the ONU-ID requires only 1 byte. The value 255 (0xFF) is used upstream by an ONU in the process of discovery, that is, by an ONU that has not yet been assigned a TC-layer ONU-ID. Downstream, the values 254 and 255 are reserved: 255 is the broadcast address, while 254 is precluded as an ONU-ID because it is used as a dummy alloc-ID that invites unregistered ONUs to respond to the corresponding BWmap grant.

Skipping the message ID field for the moment, we just mention that the CRC-8 is computed as the remainder of division modulo 2 of the message (assuming an initial zero CRC-8 byte) by the polynomial $x^8 + x^2 + x + 1$.

The message ID indicates the type of message. Table II.2 is a complete list of G-PON PLOAM message types. As we see, some of the messages are transactional; others are one-way messages only. The order of listing corresponds roughly to the sequence in which they might be used.

TABLE II.2   G.984 G-PON PLOAM Message Types

Downstream	Unicast	Broadcast	ACK	3x	Upstream	3x
Upstream_overhead		X		X		
Extended_burst_length		X		X		
					Serial_number_ONU	
Assign_ONU-ID		X		X		
Ranging_time	X			X		
Request_password	X				Password	X
Assign_alloc-ID	X		X	X	Acknowledge	
Configure_port-ID	X		X	X		
Encrypted_port-ID	X		X	X		
BER interval	X	X	X	X		
					Remote_error_ indication (REI)	
Request_key	X				Encryption key	X
Key_switching_time	X	X	X	X		
No_message		X			No_message	
Popup	X	X		X		
Deactivate_ONU-ID	X	X		X		
Disable_serial_number		X		X		
					Dying_gasp	X
Physical_equipment_ error (PEE)		X			Physical_equipment_ error (PEE)	
PST	X	X			PST	
Change_power_level	X	X				

By comparison with Table II.1, we also see that G.987 XG-PON exploited the experience gained with G.984 G-PON to pare down the message set considerably.

II.2.1   Upstream_overhead Message: Downstream

The G-PON ONU is not permitted to transmit if it has not received an upstream_overhead message (Fig. II.17) that defines its burst header. For detailed discussion of the meaning and use of the preamble fields, refer to Sections 4.1.5 and 4.2.2.

Number of guard bits, g   range 0–255.
Number of type 1 preamble bits, p_1   range 0–255.
Number of type 2 preamble bits, p_2   range 0–255.

The *type 3 preamble pattern* byte is repeated as many times as necessary to total p_3 bits, with any partial repetitions at the trailing end, adjacent to the delimiter.

Delimiter pattern—24 bits that conclude the burst header.

xx—Reserved bits

e—If e = 0, the ONUs are to ignore the preassigned delay. If e = 1, the ONUs use the preassigned delay when they are in serial_number state O3 and ranging state O4.

	ONU-ID = 0xFF = broadcast
	Message ID = 0x01
Length 1	Number of guard bits
1	Number of type 1 preamble bits
1	Number of type 2 preamble bits
1	Type 3 preamble pattern
3	Delimiter
1	xxem sspp
2	Pre-assigned delay
	CRC-8

Figure II.17 Upstream_overhead message.

m—The m bit was once used to enable serial number masking. This feature has been deprecated—so we will not describe it—and the m bit is always set to 0.

ss—The ss field was once used to indicate that an ONU should send additional (more than one) serial_number_ONU upstream PLOAM messages in response to a serial number grant. This feature has been deprecated; the ss field is always set to 00.

pp—This field selects one of three power levels for the ONU transmitters. Although it is still legal in G-PON, it is not used.

00	Normal, full power
01	−3 dB
10	−6 dB
11	Reserved

Preassigned delay—Additional delay that can be used to offset the serial number response window, measured in 32-byte units at the upstream rate of 1.2 Gb/s.

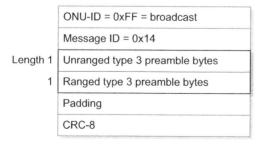

	ONU-ID = 0xFF = broadcast
	Message ID = 0x14
Length 1	Unranged type 3 preamble bytes
1	Ranged type 3 preamble bytes
	Padding
	CRC-8

Figure II.18 Extended_burst_length message.

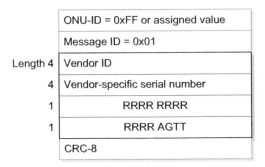

Figure II.19 Serial_number_ONU message, G.984.

II.2.2 Extended_burst_length Message: Downstream

This message allows for a longer burst header during ONU discovery—the unranged type 3 preamble bytes field—and also after the ONU is operational—the ranged type 3 preamble bytes field. For detailed discussion of the meaning and use of the fields, refer to Sections 4.1.5 and II.2.1.

Depicted in Figure II.18, the extended_burst_length message specifies an integer number of bytes, but it is the OLT's responsibility to assure that all of the fields line up with the values specified in the upstream_overhead message such that p_3 is in fact an integer number of bytes.

The maximum total G.984 G-PON burst overhead is 128 bytes.

II.2.3 Serial_number_ONU Message: Upstream

The ONU transmits this message in response to a serial number grant (Section 4.2.2) using the unassigned ONU-ID value 0xFF. In the ranging state, the ONU also responds to a grant with this message, but the ONU now has had a TC-layer ONU-ID assigned to it, which it uses. Figure II.19 illustrates this message.

Vendor_ID—Four characters that identify the ONU vendor.

Vendor-specific serial number—Unspecified but expected to be unique for the given vendor ID.

RRRR RRRR RRRR—The random delay, in 32-byte units, used by the ONU when sending this message.

A—Reserved bit

G—If G = 1, GEM framing is supported by this ONU. This bit is left over from the ATM heritage of B-PON and is always 1 in G-PON.

TT—ONU transmit power level. As noted above, power leveling is rarely or never used. Real-world transmitters just run at their rated power output at all times.

        00   Low power
        01   Medium power

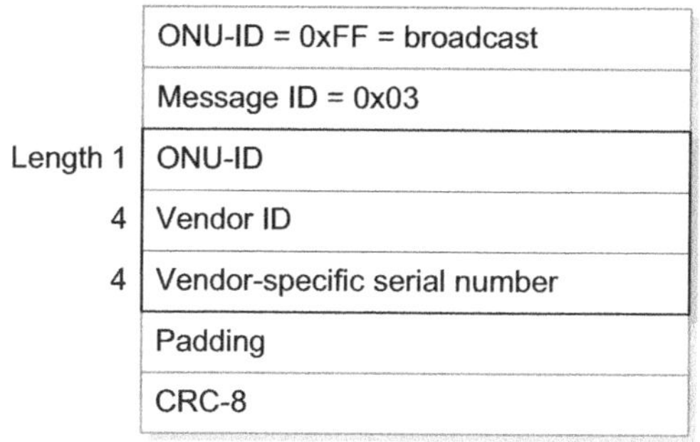

ONU-ID = 0xFF = broadcast
Message ID = 0x03
ONU-ID
Vendor ID
Vendor-specific serial number
Padding
CRC-8

Length 1, 4, 4

Figure II.20   Assign_ONU-ID message, G.984.

10   High (normal) power
11   Reserved.

II.2.4   Assign_ONU-ID Message: Downstream

The act of assigning a TC-layer ONU-ID to a newly discovered ONU causes it to move from serial_number state to ranging state (Section 4.2.2). Once the ONU has departed serial_number state, its ONU-ID cannot be changed short of deactivation. See Figure II.20 for the message layout.

II.2.5   Ranging_time Message: Downstream

With this message, the OLT sets the equalization delay of the ONU, thereby assuring that its upstream transmissions arrive at the OLT at the expected time. This message is used both during initial activation and as necessary during normal operation to correct drift.

Delay is specified with bit-level granularity, bits at the 1.2-Gb/s upstream rate.

Figure II.21 shows this message. If $b = 0$, the associated delay is for the primary or working path. If $b = 1$, the delay is for the protection path.

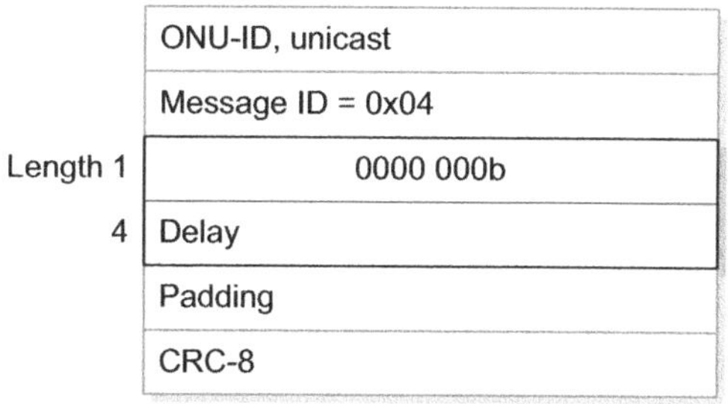

ONU-ID, unicast
Message ID = 0x04
0000 000b
Delay
Padding
CRC-8

Length 1, 4

Figure II.21   Ranging_time message, G.984.

ONU-ID, unicast
Message ID = 0x09
Padding
CRC-8

Figure II.22 Request_password message.

II.2.6 Request_password Message: Downstream

This message (Fig. II.22) causes the ONU to respond with the password message. *Password* is really a misnomer; if these messages are used, they are used to convey registration ID, as described in Section 4.4.

II.2.7 Password Message: Upstream

When the OLT requests the password, the ONU responds with this message. The registration ID contained in the password field is 10 bytes long. It usually comprises an ASCII string, but nothing prevents it from being an arbitrary sequence of octets. Figure II.23 shows the message structure.

II.2.8 Assign_alloc-ID Message: Downstream

The value of the ONU's default alloc-ID—used for OMCI and upstream PLOAM messages—is equal to the TC-layer ONU-ID and can neither be assigned nor deassigned. Alloc-IDs assigned by this message are to be used for subscriber traffic. As described in Sections 4.1.2.3 and 6.3.2.2, alloc-IDs for subscriber traffic must be mapped to a T-CONT through OMCI as well as with this PLOAM message.

The alloc-ID value is 12 bits, packed into 2 bytes with four trailing zeros. See Figure II.24.

Two values are legal in the alloc-ID type field. According to G.984.3, the value 1 indicates that the alloc-ID carries GEM traffic, while the value 255 deallocates this alloc-ID. In the early history of G-PON, an alloc-ID could have been used for other types of traffic, such as ATM or even DBA reports. These code points have been removed.

	ONU-ID
	Message ID = 0x02
Length 10	Password
	CRC-8

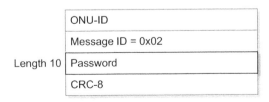

Figure II.23 Password message.

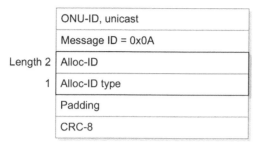

Figure II.24 Assign_alloc-ID message, G.984.

II.2.9 Acknowledge Message: Upstream

As shown in Table II.2, some downstream G-PON PLOAM messages are acknowledged. The message body simply copies as much as possible of the original downstream message. This may assist the OLT in correlating the acknowledgment with the original message.

One of the perceived deficiencies of this message, which was corrected in G.987 XG-PON, is that there is no form of negative acknowledgment to indicate problems that may be known to the ONU, nor any way for the ONU to signal that it is busy and cannot yet respond to a request, for example, to supply a new encryption key.

Figure II.25 shows the message structure.

II.2.10 Configure_port-ID Message: Downstream

Subscriber traffic GEM ports are configured only through OMCI. The purpose of this message (Fig. II.26) is to assign a GEM port to the OMCC using codepoint a = 1. If it ever became necessary to change the OMCC GEM port, the previous value could be deallocated by setting a = 0. But according to G.984.3, this is not really necessary: if a new OMCC GEM port is configured, the ONU is expected to implicitly deconfigure the previous one.

The value of the GEM port-ID is 12 bits, carried in 2 bytes with four trailing zeros.

This message is not needed in G.987 XG-PON because the OMCC GEM port is implicitly assigned to have the same value as the ONU-ID.

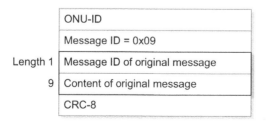

Figure II.25 Acknowledge message, G.984.

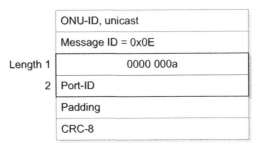

Figure II.26 Configure_port-ID message.

II.2.11 Encrypted_port-ID Message: Downstream

By setting a = 1 (Fig. II.27), the OLT directs that the GEM port specified in the port-ID field be encrypted. Cleartext transport is indicated by a = 0. For historic reasons, the value of bit b is required to be 1.

The value of the GEM port is carried in 2 bytes, 12 bits with four trailing zeros.

This message is not needed in G.987 XG-PON because port encryption is provisioned through OMCI.

II.2.12 BER Interval Message: Downstream

The ONU acknowledges this message (Fig. II.28), which is sent three times. The message may be either unicast or broadcast. Its purpose is to define the interval over which bit errors are accumulated at the ONU(s). The unit of measurement is the frame time, 125 μs, so that a value of 8000 corresponds to 1 s, and 15 min. would be expressed as 7,200,000.

Bit errors are detected with the BIP-8 field of the downstream frame. At intervals specified by the BER interval message, the ONU reports the count of bit errors with an upstream REI message.

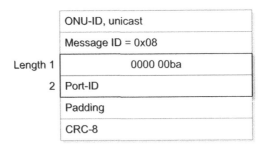

Figure II.27 Encrypted_port-ID message.

ONU-ID, unicast or broadcast
Message ID = 0x12
BER interval
Padding
CRC-8

Length 4 (beside BER interval)

Figure II.28 BER interval message.

II.2.13 REI (Remote Error Indication) Message: Upstream

This message (Fig. II.29) reports the number of errors accumulated in the BIP-8 error check of the downstream frames during the BER interval. The ONU sends the message autonomously once per BER interval.

ssss—The sequence number is a 4-bit field that increments with each REI message. The four most significant bits of the byte are 0.

II.2.14 Request_key Message: Downstream

With this message, shown in Figure II.30, the OLT requests the ONU to generate new keying material and send it upstream in one or more encryption_key messages, each response message being repeated three times.

ONU-ID
Message ID = 0x08
BIP errors counter
0000 ssss
Padding
CRC-8

Length 4 (beside BIP errors counter)
1 (beside 0000 ssss)

Figure II.29 Remote error indication message.

ONU-ID, unicast
Message ID = 0x0D
Padding
CRC-8

Figure II.30 Request_key message.

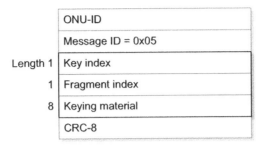

Figure II.31 Encryption_key message.

II.2.15 Encryption_key Message: Upstream

Shown in Figure II.31, the encryption_key PLOAM message carries new keying material to the OLT, new material generated in response to a request_key message. In contrast to the corresponding key_report message of G.987 XG-PON, this message conveys the G-PON key in cleartext form.

Eight bytes of keying material can be carried in each message, so it is expected that several fragments will be needed—128-bit keys are the norm—each indicated by its own fragment index, starting at 0.

The key index is a counter that the ONU increments each time it generates a new key. This is a way for the OLT to assure itself that all fragments are part of the same key.

II.2.16 Key_switching_time Message: Downstream

After having requested a new key from one or all ONUs, the OLT synchronizes key switching with this message (Fig. II.32). Switching occurs at the beginning of the frame whose (super)frame count matches the value carried in the frame counter message field. Coordinated switching between OLT and ONU avoids payload loss during key update.

The G-PON frame counter is 30 bits long; two zero padding bits are inserted at the most significant end of the frame counter field.

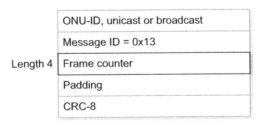

Figure II.32 Key_switching_time message.

Downstream

ONU-ID = 0xFF = broadcast
Message ID = 0x0B
Padding
CRC-8

Upstream

ONU-ID
Message ID = 0x04
Padding
CRC-8

Figure II.33   No_message message.

II.2.17   No_message Message: Both Directions

The G-PON frame structure is defined such that the OLT sends a PLOAM message downstream unconditionally in every frame. If it has nothing else to say, the OLT sends a No_message message, depicted on the left in Figure II.33.

The same is true in the upstream direction, if the OLT grants a PLOAM slot to an ONU that has no message to send.

II.2.18   Popup Message

II.2.18.1   Downstream, Unicast

Section 4.2.2 explains that, if a G.984 G-PON ONU experiences loss of downstream signal or loss of synchronization, it enters popup state O6. If it recovers the downstream signal before a timer expires (recommended value 100 ms), and if the OLT is quick enough to recognize the loss and deal with it, the popup message (Fig. II.34) may bring the ONU back into operation state O5 without taking it through the reranging process.

II.2.18.2   Downstream, Broadcast

The broadcast form of this message causes all ONUs that may be in popup state O6 to enter ranging state O4. ONUs in other states ignore the message.

It is not specified whether the ONU should discard its TC-layer provisioning, but the target state is O4, which implies that the ONU should retain everything.

II.2.19   Deactivate_ONU-ID Message: Downstream

When it receives this message, illustrated in Figure II.35, the individual ONU (or all ONUs) are directed to discard their TC-layer provisioning and move to standby state

ONU-ID, unicast or broadcast
Message ID = 0x0C
Padding
CRC-8

Figure II.34   Popup message.

ONU-ID, unicast or broadcast
Message ID = 0x05
Padding
CRC-8

Figure II.35 Deactivate_ONU-ID message, G.984.

O2. From state O2, the ONU responds to serial number grants, is recognized by the OLT, and is reactivated onto the PON.

II.2.20 Disable_serial_number Message: Downstream

This message (Fig. II.36) is an attempt by the OLT to silence an ONU that may have become a rogue, assuming that the ONU can still recognize the command and can still control its transmitter.

The disable control field has three values:

0xFF The ONU is denied upstream access and placed in emergency stop state O7 (Section 4.2.2).

0x0F All ONUs in emergency stop state O7 are moved to standby state O2, where they can be discovered and activated anew onto the PON. The vendor ID and serial number fields are ignored. Be aware that the semantics of this code point are not the same as those of the same code point in the G.987.3 XG-PON message.

0x00 The ONU with the specified vendor ID and serial number is placed in standby state O2, where it can be rediscovered and reactivated onto the PON. If the ONU is not in emergency stop state O7, it ignores this message.

	ONU-ID = 0xFF = broadcast
	Message ID = 0x06
Length 1	Disable control
4	Vendor ID
4	Vendor-specific serial number
	Padding
	CRC-8

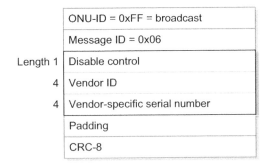

Figure II.36 Disable_serial_number message, G.984.

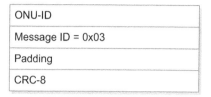

ONU-ID
Message ID = 0x03
Padding
CRC-8

Figure II.37 Dying_gasp message.

Downstream

ONU-ID = 0xFF = broadcast
Message ID = 0x0F
Padding
CRC-8

Upstream

ONU-ID
Message ID = 0x06
Padding
CRC-8

Figure II.38 Physical_equipment_error message.

II.2.21 Dying_gasp Message: Upstream

The DG message of Figure II.37 signals the OLT that the ONU expects to drop off the PON. The purpose of the message is to assist in possible future troubleshooting by assuring the OLT, or the EMS, that the optical distribution network is not defective.

II.2.22 Physical_equipment_error (PEE) Message: Both Directions

G.984.3 states that the OLT sends this PLOAM message once per second (Fig. II.38) when it is unable to send either GEM frames or OMCI messages. It is unclear that such an OLT would still be able to send PLOAM messages, or what value the ONU could derive from this message.

Likewise, if the ONU is truly unable to send either GEM frames or OMCI messages, it probably cannot send this message either. OMCI defines alarm messages for less fundamental equipment faults.

II.2.23 PST (Protection Switch) Message: Both Directions

The PST message of Figure II.39 is intended to be used in somewhat the same way as the K bytes in SDH $1 + 1$ or 1:1 protection. A G-PON is not the same as SDH, however, and if negotiated protection switching were to be supported, these messages would probably require further study. The interested reader is invited to consult G.841 and G.983.5.

Line Number—0 or 1 is the specification, although multi-PON reach extenders are the most likely application of these messages.

K1—As specified in G.841.

K2—As specified in G.841.

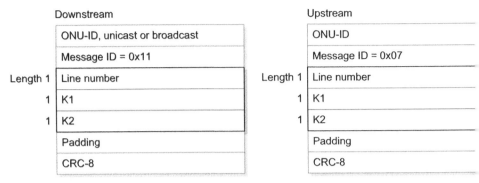

Figure II.39 PST message.

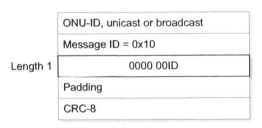

Figure II.40 Change_power_level message.

II.2.24 Change_power_level Message: Downstream, Unicast or Broadcast

Although the message definition remains, as shown in Figure II.40, this function is rarely, if ever, used. In practice, ONUs transmit at their full rated power at all times.

The ID bits have two values:

10—Increase ONU transmitted power.

01—decrease ONU transmitted power.

The other values are reserved.

UPDATE

As we went to press, G.984.3 amendment 3 and G.987.3 amendment 1 were being consented in ITU. Within the scope of this appendix, the changes were:

G.984.3 adds three new optional PLOAM messages, PON-ID, swift_POPUP and ranging_adjustment.

G.987.3 adds two new optional PLOAM messages, PON-ID and discovery_control. The former Acknowledge PLOAM message is now named Acknowledgement. The latter change has been captured in this book.

REFERENCES

INTERNATIONAL TELECOMMUNICATIONS UNION, TELECOMMUNICATION STANDARDIZATION SECTOR

Published versions of ITU-T recommendations are available at no cost at www.itu. int/itu-t/recommendations. Although baseline dates are listed below, amendments exist for many recommendations. It is always advisable to check the ITU website for the current version. Prior to ITU-T consent, work on updated and new recommendations is accessible to ITU members only.

G-PON

G.984.1 (2008), *Gigabit-capable passive optical networks (G-PON): General characteristics.*

G.984.2 (2003), *Gigabit-capable passive optical networks (G-PON): Physical media dependent (PMD) layer specifications.*

G.984.3 (2008), *Gigabit-capable passive optical networks (G-PON): Transmission convergence layer specification.*

G.984.4 (2008), *Gigabit-capable passive optical networks (G-PON): ONT management and control interface specification,* superseded by G.988.

G.984.5 (2007), *Gigabit-capable passive optical networks (G-PON): Enhancement band for gigabit-capable passive optical networks.*

G.984.6 (2008), *Gigabit-capable passive optical networks (G-PON): G-PON optical reach extension (G.984.re).*

G.984.7 (2010), *Gigabit-capable passive optical networks (GPON): Long reach.*

Gigabit-capable Passive Optical Networks, First Edition. Dave Hood and Elmar Trojer.
© 2012 John Wiley & Sons, Inc. Published 2012 by John Wiley & Sons, Inc.

XG-PON

G.987 (2010), *10-Gigabit-capable passive optical network (XG-PON) systems: Definitions, abbreviations, and acronyms.*

G.987.1 (2010), *10 Gigabit-capable passive optical network (XG-PON): General requirements.*

G.987.2 (2010), *10-Gigabit-capable passive optical networks (XG-PON): Physical media dependent (PMD) layer specification.*

G.987.3 (2010), *10-Gigabit-capable passive optical networks (XG-PON): Transmission convergence (TC) specifications.*

G.987.4 (2012), *10 Gigabit-capable passive optical networks (XG-PON): Reach extension.*

G.988 (2010), *ONU management and control interface (OMCI) specification*, also pertains to G.984 G-PON and other access technologies.

Other

E.164 (2010), *The international public telecommunication numbering plan.*

G.652 (2009), *Characteristics of a single-mode optical fibre and cable.*

G.704 (1998), *Synchronous frame structures used at 1544, 6312, 2048, 8448 and 44 736 kbit/s hierarchical levels.*

G.711 (1988), *Pulse code modulation (PCM) of voice frequencies.*

G.841 (1998), *Types and characteristics of SDH network protection architectures.*

G.983 series, B-PON recommendations, various dates.

G.986 (2010), *1 Gbit/s point-to-point Ethernet-based optical access system.*

G.991 series, HDSL and SHDSL, various dates.

G.992.1 (1999), *Asymmetric digital subscriber line (ADSL) transceivers.*

G.992.2 (1999), *Splitterless asymmetric digital subscriber line (ADSL) transceivers.*

G.992.3 (2009), *Asymmetric digital subscriber line transceivers 2 (ADSL2).*

G.992.4 (2002), *Splitterless asymmetric digital subscriber line transceivers 2 (splitterless ADSL2).*

G.992.5 (2009), *Asymmetric digital subscriber line (ADSL) transceivers—Extended bandwidth ADSL2 (ADSL2plus).*

G.993.1 (2004), *Very high speed digital subscriber line transceivers (VDSL).*

G.993.2 (2006), *Very high speed digital subscriber line transceivers 2 (VDSL2).*

G.993.5 (2010), *Self-FEXT cancellation (vectoring) for use with VDSL2 transceivers.*

G.997.1 (2009), *Physical layer management for digital subscriber line (DSL) transceivers.*

G.998 series, *DSL bonding*, various dates.

G.sup.39 (2008), *Optical system design and engineering considerations.*

H.248 series, *Gateway control protocol*, various dates.

M.3100 (2005), *Generic network information model.*

X.731 (1992), *Information technology—Open systems interconnection—Systems management: State management function.*

Y.1413 (2004), *TDM-MPLS network interworking—User plane interworking.*

Y.1415 (2005), *Ethernet-MPLS network interworking—User plane interworking.*

Y.1453 (2006), *TDM-IP interworking—User plane interworking.*

Y.1731 (2008), *OAM functions and mechanisms for Ethernet based networks.*

BROADBAND FORUM

Published BBF technical reports (TRs) are available at no cost at http://www.broadband-forum.org/technical/trlist.php. Work in progress is accessible to BBF members only.

TR-059, *DSL evolution—Architecture requirements for the support of QoS-enabled IP services.*

TR-069, *CPE WAN management protocol.*

TR-101, *Migration to Ethernet based DSL aggregation.*

TR-104, *Provisioning parameters for VoIP CPE.*

TR-124, *Functional requirements for broadband residential gateway devices.*

TR-142, *Framework for TR-069 enabled PON devices.*

TR-156, *Using GPON access in the context of TR-101.*

TR-167, *GPON-fed TR-101 Ethernet access node.*

TR-200, *Using EPON in the context of TR-101.*

TR-247, *ONT conformance test plan.*

TR-255, *GPON interoperability test plan.*

INSTITUTE OF ELECTRICAL AND ELECTRONICS ENGINEERS

Six months after formal publication, IEEE 802 series references are available at no cost at http://standards.ieee.org/getieee802/. Work in progress is accessible to IEEE members only.

The IEEE naming convention is that lowercase letters, for example, in 802.1ad, designate amendments to the parent standard (802.1). When the parent is republished, the amendment material is folded in and disappears as a stand-alone document. An uppercase designation, for example, in 802.1D, indicates a top-level document that stands on its own.

IEEE 802.1ad-2005, *IEEE standards for local and metropolitan area networks—Virtual bridged local area networks—Revision—Amendment 4: Provider bridges.*

IEEE 802.1ag-2007, *IEEE standard for local and metropolitan area networks— Virtual bridged local area networks—Amendment 5: Connectivity fault management.*

IEEE 802.1AX, *IEEE standard for local and metropolitan area networks—Link aggregation.*

IEEE 802.1D-2004, *Standard for local and metropolitan area networks: Media access control (MAC) bridges.*

IEEE 802.1Q-2010, *Standard for local and metropolitan area networks—Virtual bridged local area networks.*

IEEE 802.1X-2010, *IEEE standard for local and metropolitan area networks—Port-based network access control.*

IEEE 802.3-2008, *Carrier sense multiple access with collision detection (CSMA/CD) access method and physical layer specifications,* includes 802.3ah EPON; 10G-EPON (802.3av) will be incorporated when 802.3 is next republished.

IEEE 802.11-2007, *IEEE standard for Information technology—Telecommunications and information exchange between systems—Local and metropolitan area networks—Specific requirements—Part 11: Wireless LAN medium access control (MAC) and physical layer (PHY) specifications.*

IEEE 802.16-2009, *IEEE standard for local and metropolitan area networks Part 16: Air interface for broadband wireless access systems.*

IEEE 1588-2008, *Standard for a precision clock synchronization protocol for networked measurement and control systems.*

IEEE P1904.1, *Service interoperability in Ethernet passive optical networks (SIEPON).*

INTERNET ENGINEERING TASK FORCE

IETF requests for comments (RFCs) are available at no cost from a number of servers. The official source is http://www.ietf.org/rfc.html, but it is less convenient than some of the others, for example, http://www.faqs.org/faqs/.

RFC 791, *Internet protocol.*

RFC 793, *Transmission control protocol.*

RFC 1213, *Management information base for network management of TCP/IP-based internets: MIB-II.*

RFC 2030, *Simple network time protocol (SNTP) version 4 for IPv4, IPv6 and OSI.*

RFC 2236, *Internet group management protocol, version 2 (obsolete).*

RFC 2460, *Internet protocol, version 6 (IPv6) specification.*

RFC 2474, *Definition of the differentiated services field (DS field) in the IPv4 and IPv6 headers.*

RFC 2578, *Structure of management information version 2 (SMIv2).*

RFC 2597, *Assured forwarding PHB group.*

RFC 2698, *A two rate three color marker.*

RFC 2710, *Multicast listener discovery (MLD) for IPv6.*

RFC 2865, *Remote authentication dial-in user service (RADIUS).*

RFC 3032, *MPLS label stack encoding.*

RFC 3261, *SIP: Session initiation protocol.*

RFC 3315, *Dynamic host configuration protocol for IPv6 (DHCPv6).*

RFC 3376, *Internet group management protocol, version 3.*

RFC 3550, *RTP: A transport protocol for real-time applications.*

RFC 3584, *Coexistence between version 1, version 2, and version 3 of the Internet-standard network management framework.*

RFC 3736, *Stateless dynamic host configuration protocol (DHCP) service for IPv6.*

RFC 3748, *Extensible authentication protocol (EAP).*

RFC 3810, *Multicast listener discovery version 2 (MLDv2) for IPv6.*

RFC 3916, *Requirements for pseudo-wire emulation edge-to-edge (PWE3).*

RFC 3985, *Pseudo wire emulation edge-to-edge (PWE3) architecture.*

RFC 4115, *A differentiated service two-rate, three-color marker with efficient handling of in-profile traffic.*

RFC 4291, *IP version 6 addressing architecture.*

RFC 4385, *Pseudowire emulation edge-to-edge (PWE3) control word for use over an MPLS PSN.*

RFC 4448, *Encapsulation methods for transport of Ethernet over MPLS networks.*

RFC 4541, *Considerations for internet group management protocol (IGMP) and multicast listener discovery (MLD) snooping switches.*

RFC 4553, *Structure-agnostic time division multiplexing (TDM) over packet (SAToP).*

RFC 4604, *Using Internet group management protocol version 3 (IGMPv3) and multicast listener discovery protocol version 2 (MLDv2) for source-specific multicast.*

RFC 4605, *Internet group management protocol (IGMP)/Multicast listener discovery (MLD)-based multicast forwarding ("IGMP/MLD proxying").*

RFC 4717, *Encapsulation methods for transport of asynchronous transfer mode (ATM) over MPLS networks.*

RFC 4733, *RTP payload for DTMF digits, telephony tones, and telephony signals.*

RFC 4734, *Definition of events for modem, fax, and text telephony signals.*

RFC 4816, *Pseudowire emulation edge-to-edge (PWE3) asynchronous transfer mode (ATM) transparent cell transport service.*

RFC 4861, *Neighbor discovery for IP version 6 (IPv6).*

RFC 4862, *IPv6 stateless address autoconfiguration.*

RFC 4928, *Avoiding equal cost multipath treatment in MPLS networks.*

RFC 5086, *Structure-aware time division multiplexed (TDM) circuit emulation service over packet switched network (CESoPSN).*

RFC 5087, *Time division multiplexing over IP (TDMoIP),* includes AAL1, AAL2.

RFC 5462, *Multiprotocol label switching (MPLS) label stack entry: "EXP" field renamed to "Traffic class" field.*

RFC 5605, *Managed objects for ATM over packet switched networks (PSNs).*

RFC 5905, *Network time protocol (version 4): Protocol and algorithms specification.*

RFC 6080, *A framework for session initiation protocol user agent profile delivery.*

RFC 6106, *IPv6 router advertisement options for DNS configuration.*

OTHER

Alliance for Telecommunications Industry Solutions, ATIS 0300220 (2011), *Representation of the communications industry manufacturers, suppliers, and related service companies for information exchange.*

Alliance for Telecommunications Industry Solutions, ATIS 0900101 (2006), *Synchronization interface standard,* formerly ANSI T1.101.

Edmon, E. et al., "Today's broadband fiber access technologies and deployment considerations at SBC", in *Broadband Optical Access Networks and Fiber-to-the-Home*, C. Lin, Ed, Wiley, Hoboken, NJ, 2006.

Federal Information Processing Standard, FIPS 197 (2001), *Federal information processing standard 197, Advanced encryption standard.*

International Electrotechnical Commission, IEC 61754-4-1 (2003), *Fibre optic connector interfaces—Part 4-1: Type SC connector family—Simplified receptacle SC-PC connector interface.*

Keiser, G. *Optical Fiber Communications*, 3rd ed. McGraw-Hill, New York.

Metro Ethernet Forum, MEF 8 (2004), *Implementation agreement for the emulation of PDH circuits over metro Ethernet networks.*

National Institute of Science and Technology, NIST special publication 800-38A (2001), *Recommendation for block cipher modes of operation—Methods and techniques.*

National Institute of Science and Technology, NIST special publication 800-38B (2005), *Recommendation for block cipher modes of operation: The CMAC mode for authentication.*

Payne, R. et al., "Optical networks for the broadband future," in *Broadband Optical Access Networks and Fiber-to-the-Home*, C. Lin, Ed. Wiley, Hoboken, NJ, 2006.

Telcordia, GR-326-CORE (2010), *Generic requirements for singlemode optical connectors and jumper assemblies.*

Telcordia, GR-499-CORE (2009), *Transport systems generic requirements (TSGR): Common requirements.*

Telcordia, GR-909-CORE (2004), *Generic criteria for fiber in the loop systems.*

ACRONYMS

10GE	10 Gb/s Ethernet
10G-EPON	IEEE 802.3 Ethernet passive optical network, 10 Gb/s downstream, 1 or 10 Gb/s upstream
1G-EPON	IEEE 802.3 Ethernet passive optical network, 1 Gb/s in both directions
3GPP	Third Generation Partnership Project, responsible for IMS and LTE
AAA	Authentication, authorization, accounting
AAL	ATM adaptation layer
AAL1	ATM adaptation layer 1, intended for constant bit rate clients, ITU-T I.363.1
AAL2	ATM adaptation layer 2, TDM clients with multiplexed or intermittent transmission, ITU-T I.363.2
AAL5	ATM adaptation layer 5, asynchronous data clients, ITU-T I.363.5
ABF	Air-blown fiber
AC	Alternating current
ACH	Associated channel header, MPLS, pseudowires, IETF RFC 5586
ACK	Acknowledge
ACL	Access control list, multicast, sometimes also called a whitelist
ACS	Access control server, BBF TR-69
ADSL	Asymmetric digital subscriber line, defined in ITU-T G.992 family
ADSS	All-dielectric self-supporting (aerial fiber cable)
AES	Advanced encryption system, (U.S.) federal information processing standards publication 197
AES_CMAC	AES in cipher-based message authentication mode
AES_ECB	AES in electronic codebook mode
AES-CTR	AES in counter mode

Gigabit-capable Passive Optical Networks, First Edition. Dave Hood and Elmar Trojer.
© 2012 John Wiley & Sons, Inc. Published 2012 by John Wiley & Sons, Inc.

AF	Assured forwarding, IP traffic priority class, IETF RFCs 2597, 3260
AGC	Automatic gain control
AIS	Alarm incoming signal. Indicates a failure that need not be reported as an alarm because the upstream equipment has already notified management about the problem.
AM	Amplitude modulation
AMI	Alternate mark inversion, DS1 line coding format
ANCP	Access network control protocol
ANSI	American National Standards Institute
ANI	Access network interface (PON interface at the ONU), G.988
AOR	Address of record, SIP
APC	Angled physical contact, automatic power control
APD	Avalanche photodiode
A-PON	ATM passive optical network
AR	Acknowledgment request
ARC	Alarm reporting control, ITU-T M.3100 and G.988
ARP	Address resolution protocol, IETF RFC 826
ARQ	Automatic repeat request (protocol)
ASCII	American standard code for information interchange
ASE	Amplified spontaneous emission
ASM	Any-source multicast; see also SSM
ATA	Analog telephony adaptor
ATC	Automatic threshold control
ATIS	Alliance for Telecommunications Industry Solutions
ATM	Asynchronous transfer mode
ATU-C	ADSL termination unit, central office end; see also xTU-C
ATU-R	ADSL termination unit, remote end; see also xTU-R
AVC	Attribute value change notification, ITU-T G.988
AWG	American wire gauge (conversion table in Section 2.5), arrayed waveguide
B3ZS	Bipolar with suppression of strings of three or more consecutive zeros, DS3 line coding format
B8ZS	Bipolar with suppression of strings of eight or more consecutive zeros, DS1 line coding format
BBF	Broadband Forum
BCH	Bose–Chaudhuri–Hocquenghem, family of error detecting and correcting codes
BE	Best efforts, default IP traffic priority class, IETF RFC 2474. Also used with same meaning in other traffic forwarding contexts

BER	Bit error rate, ratio
BIP	Bit interleaved parity
BITS	Building integrated timing supply
BM	Burst mode
BMR	Burst-mode receiver
BNG	Broadband network gateway
BOL	Beginning of life
BOSA	Bidirectional optical subassembly
B-PON	Broadband passive optical network, ITU-T G.983
BRAS	Broadband remote access server
BW	Bandwidth
BWmap	Bandwidth map, ITU-T G.984.3, G.987.3
c	Speed of light in vacuum, 2.99793×10^8 m/s
CAC	Connection admission control
Capex	Capital expenditure; see also opex
CAS	Channel associated signaling, DS1 and E1
CBU	Cellular backhaul unit
CCM	Continuity check message, IEEE 802.1ag
CCSA	China Communications Standards Association
CDR	Clock and data recovery
CE	Customer edge
CES	Circuit emulation service, one form of pseudowire
CESoEth	Circuit emulation service over Ethernet
CESoPSN	Circuit emulation service over packet switched network
CEV	Controlled environment vault
CFI	Canonical format indicator, IEEE 802.1
CFM	Connectivity fault management, IEEE 802.1ag, ITU-T Y.1731
CID	Consecutive identical digit
CM	Common mode, continuous mode
CMAC	Cipher-based message authentication, AES mode
CO	Central office, also called local exchange
CPA	Clock phase aligner
CPE	Customer premises equipment
CPRI	Common public radio interface
CPU	Central processing unit
CRC	Cyclic redundancy check, ITU-T I.363.5
CTP	Connection termination point
CTR	Counter, a mode of AES operation

C-VLAN	Customer virtual local area network (tag or value); see also S-VLAN
CWDM	Coarse wavelength division multiplexing; see also DWDM
CWMP	CPE WAN management protocol, BBF TR-69
DA	Destination address, Ethernet and IP
DAD	Duplicate address detection, IPv6
dB	decibel
DBA	Dynamic bandwidth assignment, ITU-T G.984.3, G.987.3; dynamic bandwidth allocation
dBm	decibel, normalized to 1 mW
DBR	Dynamic bandwidth queue occupancy report
DC	Direct current
DEI	Drop eligibility indicator, bit in VLAN S-tag, IEEE 802.1ad
DELT	Double-ended loop test, xDSL; see also SELT
DFB	Distributed feedback laser
DG	Dying gasp, ITU-T G.984.2, G.984.4, G.987.2, G.988.
DGD	Differential group delay
DHCP	Dynamic host configuration protocol, IETF RFC 2131. By default associated with IPv4 but generically applicable also to IPv6
DLC	Digital loop carrier
DM	Differential mode
DMT	Discrete multitone, xDSL modulation
DNS	Domain name server
DOCSIS	Data over cable service interface specification
DoS	Denial of service, form of security attack
DoW	Drift of equalization delay window, ITU-T G.984.3, G.987.3
DS0	Digital signal, level 0, 64 kb/s
DS1	Digital signal, level 1, North American plesiochronous digital hierarchy, 1.544 Mb/s
DS2	Digital signal, level 2, North American plesiochronous digital hierarchy
DS3	Digital signal, level 3, North American plesiochronous digital hierarchy
DSCP	Differentiated services code point, IP QoS
DSL	Digital subscriber line
DSLAM	Digital subscriber line access multiplexer
DSP	Digital signal processor, processing
DSX-1	Digital cross-connect level 1, DS1
DTMF	Dual-tone multifrequency, POTS signaling
DWDM	Dense wavelength division multiplexing; see also CWDM

e	Electron charge, 1.60218×10^{-19} C
E1	TDM protocol at 2.048 Mb/s
EAP	Extended authentication protocol, used in IEEE 802.1X
ECB	Electronic code book, a mode of AES operation
ECC	Error-correcting code
ECID	Emulated circuit identifier, pseudowire
ECMP	Equal cost multipath, MPLS forwarding
EDFA	Erbium-doped fiber amplifier
EF	Expedited forwarding, IP traffic priority class, IETF RFC 3246
EMI	Electromagnetic interference
EMS	Element management system
EOL	End of life
EONU	Embedded ONU for reach extender management, sometimes EONT
EPON	Ethernet PON, IEEE 802.3, sometimes used generically to include both 1 Gb/s and 10 Gb/s versions
EqD	Equalization delay, ITU-T G.984.3, G.987.3
ER	Extinction ratio
ESF	Extended superframe, DS1
ETSI	European Telecommunications Standards Institute
EXP	Experimental, field in the header of an MPLS frame, now called TC, traffic class, IETF RFC 5462
FAT	Fiber access terminal
FCAPS	Fault, configuration, accounting, performance, security—the range of OAM functions
FCS	Frame check sequence, IEEE 802.1
FDH	Fiber distribution hub
FDL	Facility data link, DS1
FEC	Forward error correction
FET	Field effect transistor
FFT	Fast Fourier transform
FIPS	Federal Information Processing Standard
FITL	Fiber in the (access) loop
FMC	Fixed-mobile convergence
FP	Fabry–Pérot resonator structure, laser
FPGA	Field-programmable gate array
FSAN	Full Service Access Network
FTB	Fiber termination box
FTTB	Fiber to the building, to the business

FTTC	Fiber to the curb, to the cabinet
FTTH	Fiber to the home
FWI	Forced wakeup indication, ITU-T G.987.3
Gb/s	Gigabits per second
GbE	Gigabit Ethernet
GBW	Gain–bandwidth (product)
GEM	G-PON encapsulation method, ITU-T G.984.3. Term also used generically for G.987.3 XGEM
GE-PON	Gigabit Ethernet passive optical network, IEEE 802.3, usually used for 1 Gb/s version (10 Gb/s version called 10G-EPON)
GHz	Gigahertz, 10^9 cycles per second
GMII	Gigabit media-independent interface, Ethernet chip interconnection
G-PON	Gigabit passive optical network, ITU-T G.984. Term also used generically to include ITU-T G.987 XG-PON
GPS	Global positioning system
GTC	G-PON transmission convergence layer, ITU-T G.984.3. Term sometimes used generically to include G.987 XGTC
GVD	Group velocity dispersion, units ps/nm · km
h	Planck's constant, 6.6256×10^{-34} J · s
HDB3	High-density bipolar, order 3, E1 and E3 line coding format
HDLC	High-level data link control
HDSL	High-speed digital subscriber line
HDTV	High-definition television
HEC	Header error control (ATM), hybrid error control/correction
HLend	Header length, downstream, ITU-T G.987.3
HSPA	High-speed (wireless) packet access
HTTP(S)	Hypertext transfer protocol secure
IANA	Internet Assigned Numbers Authority, http://www.iana.org
IC	Integrated circuit
ICMP	Internet control message protocol
ID	Identifier
IEEE	Institute of Electrical and Electronics Engineers
IETF	Internet Engineering Task Force
IGMP	Internet group management protocol. Multicast control, currently at version 3, RFC 3376. Term also used generically to include MLD
IL	Insertion loss
ILOS	Intermittent loss of signal, ITU-T G.987.3
IMS	IP multimedia subsystem, 3GPP architectural framework

InGaAs	Indium–gallium–arsenide, optical semiconductor material
IP	Internet protocol
IPTV	Internet protocol television, usually implying multicast
IPv4	Internet protocol version 4
IPv4oE	IP version 4 over Ethernet
IPv6	Internet protocol version 6
ISI	Intersymbol interference
ISO	International Standards Organization, best known in datacoms for its seven-layer model
IT	Information technology
ITU-T	International Telecommunications Union, Telecommunication standardization sector
IWTP	Interworking termination point, OMCI, ITU-T G.988
k	Boltzmann's constant 1.38054×10^{-23} J/K
kb/s	kilobits per second
KEK	Key encryption key, ITU-T G.987.3
kHz	kilohertz
KI	Key index, ITU-T G.987.3
km	kilometer
L2TP	Layer 2 tunneling protocol
LA	Limiting amplifier
LAN	Local area network
LASER	Light amplification by stimulated emission of radiation
LBM	Looopback message, IEEE 802.1ag, ITU-T Y.1731
LBR	Looopback reply, IEEE 802.1ag, ITU-T Y.1731
LDI	Local doze indicator, ITU-T G.987.3
LDP	Label distribution protocol, MPLS, RFC 5036
LED	Light-emitting diode
LF	Last fragment, XGEM framing
LFACS	Loop facilities assignment and control system, Telcordia product
LLID	Logical link identifier, IEEE 802.3 EPON
LOB	Loss of bursts, G.987.3
LODS	Loss of downstream synchronization, G.987 framing
LOF	Loss of frame
LOS	Loss of signal
LSI	Local sleep indication, ITU-T G.987.3
LSP	Label switched path, MPLS
LSR	Label switching router, MPLS

LTE	Line terminating equipment, long-term evolution (mobile)
LTE-A	Long-term evolution—advanced (mobile)
LTM	Link trace message, IEEE 802.1ag, ITU-T Y.1731
LTR	Link trace reply, IEEE 802.1ag, ITU-T Y.1731
LWI	Local wakeup indication, ITU-T G.987.3
M2M	Machine-to-machine communications
mA	Milliampere, 10^{-3} Amperes
MAC	Media access control, medium access control
MAN	Metro Ethernet network
Mb/s	Megabits per second
MC	Multicast
MDU	Multiple dwelling unit (ONU)
ME	Managed entity, OMCI, ITU-T G.988
MEF	Metro Ethernet Forum
MEP	Maintenance endpoint, IEEE 802.1ag, ITU-T Y.1731
MF	Multifrequency (signaling)
MGC	Media gateway control
MGCP	Media gateway control protocol (see ITU-T H.248)
MHz	Megahertz
MIB	Management information base, ITU-T G.988
MIC	Message integrity check, PLOAM and OMCI
MIP	Maintenance intermediate point, IEEE 802.1ag, ITU-T Y.1731
MLD	Multicast listener discovery, IPv6. Version 2 (MLDv2) is current
MoCA	Multimedia over Co-ax Alliance
MPD	Monitoring photodiode
MPLS	Multiprotocol label switching
MSAP	Multiple service access platform
MSK	Master session key, ITU-T G.987.3
MSTP	Multiple spanning tree protocol
MTU	Maximum transfer unit (Ethernet), multitenant unit (ONU type)
NA	Not applicable, numerical aperture, nonassured
NAT	Network address translation
NE	Network element
NG-1	Next-generation 1, same as XG-PON 1, ITU-T G.987
NIST	National Institute of Science and Technology (USA)
NOC	Network operations center
NRZ	Non-return-to-zero line code
NS	Not specified

NTP	Network time protocol
OA	Optical amplifier, optical amplification
OAM	Operations, administration, maintenance
ODF	Optical distributing frame
ODN	Optical distribution network, ITU-T G.987
OEO	Optical–electrical–optical (regeneration)
OISG	OMCI implementors' study group
OLT	Optical line termination
OMCC	ONU management and control channel, ITU-T G.988
OMCI	ONU management and control interface, ITU-T G.988
ONT	Single-family ONU. This term is sometimes used generically to refer to all ONUs, but not in this book
ONU	Optical network unit
Opex	Operational expense; see also capex
OSNR	Optical signal-to-noise ratio
OSP	Outside plant
OTDR	Optical time domain reflectometry
OTL	Optical trunk line, OLT to splitter or reach extender
OTN	Optical transport network
PA	Postamplifier
PABX	Private automatic branch exchange
PC	Personal computer
PCBd	Physical control block, downstream, ITU-T G.984.3
PCM	Pulse code modulation
PCP	Priority code point, Ethernet
PDH	Plesiochronous digital hierarchy, DS1 and E1 families
PDL	Polarization-dependent loss
PDU	Protocol data unit
PEE	Physical equipment error, G.984.3
PHY	Physical layer, bits on the fiber or their electrical equivalents
PIN	Positive–intrinsic–negative (doping of layers in photodetectors)
PLC	Planar lightwave circuit
Plend	Payload length, downstream, G.984, a misnomer
PLI	Payload length indicator, (X)GEM framing
PLL	Phase-locked loop
PLO	Physical layer overhead, ITU-T G.984.3
PLOAM	Physical layer operations, administration, maintenance, ITU-T G.984.2, G.987.2. PLOAMd: downstream, PLOAMu: upstream

PLSu	Power leveling upstream, ITU-T G.984.3
PM	Performance monitoring, ITU-T G.987.3 and G.988
PMD	Physical medium-dependent function
PON	Passive optical network
POTS	Plain old telephone service
PPP	Point-to-point protocal
PPPoE	Point-to-point protocol over Ethernet
PPS	Pulses per second
PPTP	Physical path termination point, ITU-T G.988
PQS	PLOAM queue status, ITU-T G.987.3
PSB	Physical layer synchronization block, ITU-T G.987.3. PSBd: downstream, PSBu: upstream
PST	Protection switch
PSTN	Public switched telephone network (POTS)
PSync	Physical layer synchronization
PT	Payload type
PTI	Payload-type indicator
PTM	Packet transfer mode (xDSL)
PW	Pseudowire
PWE3	Pseudowire end to end
QAM	Quadrature amplitude modulation
QED	Quod erat demonstrandum (which was to be demonstrated)
QoS	Quality of sevice
QPSK	Quadrature (quaternary, quadriphase) phase-shift keying
RA	Router advertisement, IPv6
RADIUS	Remote authentication dial in user service, IETF RFC 2138
RDI	Remote defect indication
RE	Reach extender
REI	Remote error indication
RF	Radio frequency (video overlay)
RFC	Request for comments, conventional name for IETF standards
RFID	Radio frequency identification
RG	Residential gateway
RMS	Root mean square
ROADM	Reconfigurable optical add–drop multiplex
ROSA	Receiver optical subassembly
RPT	Remote protocol terminator, form of reach extender
RS	Reed–Solomon

RSSI	Received signal strength indicator
RSTP	Rapid spanning tree protocol
RTCP	Real-time control protocol
RTP	Real-time protocol
SA	Sleep allowed, ITU-T G.987.3; source address, Ethernet and IP
SBC	Set by Create, OMCI
SBS	Simulated Brillouin scattering
SBU	Small business unit (ONU)
SC	Standard optical connector
SC/APC	Standard connector with angled physical contact
SC/UPC	Standard connector with ultrafine polished end-face
SDH	Synchronous digital hierarchy; the North American version is known as synchronous optical network, sonet
SDO	Standards development organization
SDU	Service data unit
SELT	Single-ended loop test (xDSL); see also DELT
SER	Symbol error rate
SES	Severely errored second (PM)
SF	Superframe (DS1), signal fail
SFC	Superframe counter
SFF	Small-form-factor fixed optical module
SFP	Small-form-factor pluggable optical module
SFU	Single-family unit (ONU)
SHDSL	Symmetric high-speed digital subscriber line, used for some DS1 services
SIEPON	Service interoperability for EPON, IEEE working group P.1904.1
SIM	Subscriber identity module
SIP	Session initiation protocol
SK	Session key, ITU-T G.987.3
SLA	Service-level agreement
SLAC	Subscriber line audio processing circuit (POTS)
SLIC	Subscriber line interface circuit (POTS)
SMF	Single-mode fiber
SNMP	Simple network management protocol
SNR	Signal-to-noise ratio
SNTP	Simple network time protocol
SOA	Semiconductor optical amplifier
SON	Self-organizing network, self-optimizing network

SP	Strict priority
SPR	Snooping with proxy reporting, IGMP/MLD function
SR-DBA	Status reporting dynamic bandwidth allocation, ITU-T G.984.3, G.987.3
SRS	Stimulated Raman scattering
SSM	Source-specific multicast; see also ASM. (Also Synchronization status message, not used in this book.)
SSRC	Synchronization source, RTP for pseudowires
STB	Set-top box, G-PON's ultimate multicast client
STP	Spanning tree protocal
S-VLAN	Service provider virtual local area network (tag or value); see also C-VLAN
TC	Transmission convergence layer, ITU-T G.984.3, G.987.3; traffic class
TCA	Threshold crossing alert, PM
TCI	Tag control information, Ethernet
T-CONT	Transmission container, OMCI equivalent of TC layer alloc-ID, ITU-T G.988
TCP	Transmission control protocol, reliable delivery of data over IP, IETF RFC 793
TDM	Time division multiplex services, typically DS1 or E1
TDMA	Time division multiple access
TDMoIP	Time division multiplex services over Internet protocol
TFF	Thin-film filter
THz	Terahertz, 10^{12} cycles per second
TIA	Transimpedance amplifier
TISPAN	Telecommunications and Internet converged services and protocols for advanced networking, a technical committee of ETSI
TIW	Transmission interference warning, ITU-T G.987.3
TLS	Transparent LAN service
TM-DBA	Traffic monitoring dynamic bandwidth assignment
TO-can	Transistor outline metal can package
ToD	Time of day
TOS	Type of service
TOSA	Transmitter optical subassembly
TP	Termination point
TPID	Tag protocol identifier, Ethernet
trTCM	Two-rate, three-color marker
TTL	Time to live, IP or MPLS. Decremented at each hop. If TTL reaches zero, the packet is discarded

TV	Television
UDP	User datagram protocol, IETF RFC 768
UE	(Wireless) user equipment
UNI	User network interface, typically Ethernet (RJ-45 connector) or POTS (RJ-11)
UPC	Ultrapolished optical connector
UPS	Uninterruptible power supply
URI	Uniform resource identifier
URL	Uniform resource locator
US	United States (of America)
USB	Universal serial bus
UV	Ultraviolet
VA	Volt–ampere; in DC circuits, a watt
VC	Virtual circuit (ATM)
VCC	Virtual circuit (channel) connection
VCI	Virtual circuit identifier
VDSL	Very high speed digital subscriber line. VDSL2 (ITU-T G.993.2) is current; G.993.1 is obsolete
VEIP	Virtual Ethernet interface point (OMCI, ITU-T G.988)
VF	Voice frequency, often used to refer to POTS or special voiceband services
VID	VLAN identifier
VLAN	Virtual local area network
VoIP	Voice over Internet protocol
VP	Virtual path (ATM)
VPC	Virtual path connection
VPI	Virtual path identifier
VTU-R	VDSL termination unit, remote
WAN	Wide area network, the DSL side of a DSL modem or the PON side of an ONU
WBF	Wavelength-blocking filter
WDM	Wavelength division multiplexing
WiMax	Worldwide interoperability for microwave access
WRR	Weighted round robin queuing, ITU-T G.988
xDSL	ADSL or VDSL, advanced digital subscriber line or very high speed DSL
XG-1	10 Gb/s G-PON 1, with 2.5 Gb/s upstream rate, ITU-T G.987
XG-2	10 Gb/s G-PON 2, with 10 Gb/s upstream rate
XGEM	G-PON encapsulation method, as defined in G.987 XG-PON. Often just GEM

XG-PON   10 Gb/s gigabit passive optical network, ITU-T G.987
XGTC   XG-PON transmission convergence layer, ITU-T G.987.3
xTU-C   ATU-C or VTU-C, ADSL/VDSL termination unit, central office end
xTU-R   ATU-R or VTU-R, ADSL/VDSL termination unit, remote end

INDEX

Gigabit-capable Passive Optical Networks, First Edition. Dave Hood and Elmar Trojer.
© 2012 John Wiley & Sons, Inc. Published 2012 by John Wiley & Sons, Inc.

Printed and bound by CPI Group (UK) Ltd, Croydon, CR0 4YY

16/04/2025